Microbial Metabolomics

David J. Beale · Konstantinos A. Kouremenos
Enzo A. Palombo
Editors

Microbial Metabolomics

Applications in Clinical, Environmental, and Industrial Microbiology

Springer

Editors
David J. Beale
Land and Water
CSIRO
Dutton Park, QLD
Australia

Enzo A. Palombo
Swinburne University of Technology
Hawthorn, VIC
Australia

Konstantinos A. Kouremenos
Metabolomics Australia, Bio21 Molecular Science & Biotechnology Institute
The University of Melbourne
Parkville, VIC
Australia

ISBN 978-3-319-46324-7 ISBN 978-3-319-46326-1 (eBook)
DOI 10.1007/978-3-319-46326-1

Library of Congress Control Number: 2016950891

Printed on acid-free paper

This Springer imprint is published by Springer Nature
The registered company is Springer International Publishing AG
The registered company address is: Gewerbestrasse 11, 6330 Cham, Switzerland

Preface

Systems biology has greatly increased our understanding of many cellular functions and cell–cell interactions. Our awareness of the enormous range of activities performed by microorganisms has been greatly enhanced by major developments in the fields of genomics, proteomics and transcriptomics. However, understanding these functions does not fully describe the global metabolic features of single cells or cell populations (mono-specific or multi-species). Characterization of the metabolome brings us closer to the phenotype of the cell and more closely reflects the activities of microbes in response to environmental stimuli.

Microbial metabolomics is an emerging field that has developed rapidly in recent years. This development has been paralleled and supported by important advances in analytical instrumentation and technologies, in particular chromatographic and mass spectrometric methods, coupled with new and more powerful computational tools.

This book brings together contributions from global experts from diverse areas that have facilitated the exciting advances in microbial metabolomics, with special attention given to the development of relevant hardware and software platforms. Thus, the principles of these technologies will be a major focus of the book.

The main application of metabolomics is likely to be in the field of clinical and veterinary microbiology with a focus on disease-causing microorganisms. However, there is a great potential to apply metabolomics to help better understand complex biological systems that are dominated by multi-species microbial populations exposed to changing growth and nutritional conditions. In particular, environmental (e.g. water and soil), food (e.g. microbial spoilage and food pathogens), agriculture and industrial applications are seen as developing. As such, this book looks at the application metabolomics from clinical, environmental and industrial perspectives.

Dutton Park, QLD, Australia — David J. Beale
Parkville, VIC, Australia — Konstantinos A. Kouremenos
Hawthorn, VIC, Australia — Enzo A. Palombo

Contents

1 **Introduction to Microbial Metabolomics** 1
Silas G. Villas-Boas

2 **Microbes, Metabolites and Health** 13
Trevor J. Lockett, Anthony R. Bird, Claus Christophersen, Julie M. Clarke, Michael A. Conlon and David L. Topping

3 **Exploring the Bioactive Landscape of the Gut Microbiota to Identify Metabolites Underpinning Human Health** 49
Páraic Ó Cuív, Sriti Burman, Sian Pottenger and Mark Morrison

4 **Using Metabolomic Approaches to Characterize the Human Pathogen *Leishmania* in Macrophages** 83
Joachim Kloehn, Eleanor C. Saunders and Malcolm J. McConville

5 **Exometabolomics for Linking Soil Carbon Dynamics to Microbial Communities** 119
Andrea Lubbe and Trent Northen

6 **Soil Microbial Metabolomics** 147
Michael W. Heaven and Devin Benheim

7 **Community Metabolomics in Environmental Microbiology** 199
Oliver A.H. Jones, Gavin Lear, Aalim M. Welji, Gavin Collins and Christopher Quince

8 **Metabolomics: Applications to Food Safety and Quality Research** 225
Farhana R. Pinu

9 **Microbial Metabolomics in Biomass Waste Management** 261
Avinash V. Karpe, David J. Beale, Ian H. Harding and Enzo A. Palombo

10 **Beyond Metabolomics: A Review of Multi-Omics-Based Approaches** 289
David J. Beale, Avinash V. Karpe and Warish Ahmed

Index 313

Chapter 1
Introduction to Microbial Metabolomics

Silas G. Villas-Boas

1 Introduction

The ultimate goal of metabolomics is the unbiased and nontargeted analysis of cell metabolites combined with the detection, identification, and sometimes also the absolute quantification of a multitude of compounds in biological samples. Despite this being a very ambitious goal due to all the analytical challenges associated with the analysis of metabolomes (and our inability to design completely unbiased analytical methods), there has been an enormous development in the field of metabolomics in the past 16 years that has placed metabolomics in the spotlight of cutting-edge technologies for life sciences in the twenty-first century (Reeves and Rabinowtiz 2011; Nguyen et al. 2012; Russell and Duncan 2013; Zhang et al. 2014).

Microbiology greatly benefited from developments in metabolomics since its conception (Allen et al. 2003; Castrillo et al. 2003; Villas-Bôas et al. 2005a, b, 2006; Wang et al. 2006; van der Werf et al. 2007). Microbial systems are usually simpler and highly controllable compared to higher (multicellular) organisms, which make them the ideal platform for developing, applying, and validating metabolomics tools. Moreover, the development of genome-scale metabolic models for a wide range of microbial species—based on genome sequence data—has offered the best scaffold for interpretation of metabolomics data and, more importantly, for integration of metabolomics data with other omics (Lee et al. 2005).

However, microbial metabolomics is not free of its challenges. The microbial metabolome is sparsely distributed into three very distinct matrices: (i) inside the cell (intracellular); (ii) outside the cell, in the medium (extracellular); and (iii) in the

S.G. Villas-Boas (✉)
School of Biological Sciences, University of Auckland, 3A Symonds
Street, Auckland 1142, New Zealand
e-mail: s.villas-boas@auckland.ac.nz

D.J. Beale et al. (eds.), *Microbial Metabolomics*,
DOI 10.1007/978-3-319-46326-1_1

culture headspace (volatile compounds). Although this feature is not unique to microbial systems, for microorganisms the metabolite profile of each matrix can be equally important and it is not always easy to assess them separately; and most importantly, we know very little about the metabolic mechanism driving their distribution (Granucci et al. 2015). These challenges combined with the great diversity of microbial cell types provide a glimpse to the main technical issues that microbial metabolomics still faces today.

In this chapter, the state-of-the-art of microbial metabolomics field will be briefly reviewed, including the main applications of metabolomics in microbiology, current challenges, major achievements, and a future perspective for this field.

2 Microbial Metabolomics: From Early Metabolite Profiles to Metabolome Databases

A summarized timeline of main developments in microbial metabolomics is provided in Table 1. The first peer-reviewed articles that one finds when searching scientific literature databases using the keyword "metabolome" is focused on the two best known microorganisms: *Escherichia coli* (Tweeddale et al. 1998) and *Saccharomyces cerevisiae* (Oliver 1997; Oliver et al. 1998). Therefore, we could say that the concept of metabolomics started with microorganisms. Nonetheless, microbial metabolite profiling precedes the concept of metabolomics (Collar 1991; Montel et al. 1991; Andersen et al. 1995; Smedsgaard and Frisvad 1996). Since the early 1990s, metabolite profiling has been applied to microbiology mainly for phenotypic characterization of closely related microbial strains or species (Collar 1991; Montel et al. 1991; Andersen et al. 1995; Smedsgaard and Frisvad 1996) as well as a tool for microbial identification (Montel et al. 1991).

The first two publications which appear in metabolomics originate from the group of Professor Thomas Ferenci at the University of Sydney in Australia, describing the metabolic response of *E. coli* to different environmental perturbations (Tweeddale et al. 1998, 1999). The authors assessed the metabolic response of *E. coli* to different growth rates when grown on glucose and minimal medium (Tweeddale et al. 1998) and to oxidative stress (Tweeddale et al. 1999). The global metabolite pool (metabolome) of *E. coli* was analyzed through two-dimensional thin-layer chromatography of all ^{14}C-glucose labeled compounds extracted from bacterial cells. Since then, the microbial metabolomics field has progressed significantly, mostly taking advantage of developments in mass spectrometry technology and major advances in plant metabolomics.

Although the application of metabolomics apparently started around microbial systems, plant metabolomics was the main area that initially launched metabolomics to the world and drove the main developments around mass spectrometry-based metabolomics. The pioneering works on mass spectrometry-based plant metabolomics came from the Max Planck Institute in Golm, Germany (Fiehn et al. 2000;

Table 1 Timeline of important developments in microbial metabolomics field

Year	Development	References
Era pre-metabolome		
1991	Metabolite profiling applied to metabolism assessment and systematics of microorganisms	Collar (1991) Montel et al. (1991)
1995	Metabolite profiling to phenotype microbial cultures	Andersen et al. (1995)
1996	Direct electrospray mass spectrometry for fungal taxonomy and secondary metabolite profiling	Smedsgaard and Frisvad (1996)
Metabolome era		
1997	The word metabolome first proposed	Oliver 1997
1998	Metabolome as a functional genomic tool and first paper traceable using the keyword "metabolome"	Oliver et al. (1998) Tweeddale et al. (1998)
1999		Soga and Ross (1999) Tweeddale et al. (1999)
2000		Fiehn et al. (2000) Roessner et al. (2000) Soga and Heiger (2000)
2001		Fiehn (2001) Soga and Imaizumi (2001)
2002		Fiehn (2002) Roessner et al. (2001) Soga et al. (2002a, b
2003		Allen et al. (2003) Castrillo et al. (2003) Maharjan and Ferenci (2003) Villas-Bôas et al. (2003)
2004		Mashego et al. (2003) Wittmann et al. (2004)
2005		Villas-Bôas et al. (2005a, b)
2012	Publication of the yeast metabolome database	Jewison et al. (2012)
2013	Publication of the *E. coli* metabolome database	Guo et al. (2013)

Fiehn 2001, 2002; Roessner et al. 2000, 2001) and paved the way for modern metabolomics. However, the microbial metabolomics field was also particularly influenced by proposed approaches and methodologies developed by the group of Professor Stephen Oliver at that time at the University of Manchester, UK (Allen et al. 2003; Castrillo et al. 2003). Oliver's group was not only the first group to propose the concept of metabolic fingerprinting using direct infusion mass spectrometry and *S. cerevisiae* as the model organism (Castrillo et al. 2003); they also demonstrated the importance of extracellular metabolites found in the culture media for phenotyping purposes, referring to it as the metabolic footprint (Allen et al. 2003). Other important contributions to early developments in microbial metabolomics include the global metabolite profiling of intra- and extracellular metabolites of microbial cells and cultures by methyl chloroformate (MCF) derivatization and gas chromatography–mass spectrometry (GC-MS) analysis (Villas-Bôas et al. 2003, 2005a, b); assessment of sampling and sample preparation for microbial metabolomics (Maharjan and Ferenci 2003; Wittmann et al. 2004; Villas-Bôas et al. 2005a, b); quantitative microbial metabolomics using capillary electrophoresis–mass spectrometry (CE-MS) (Soga and Ross 1999; Soga and Heiger 2000; Soga and Imaizumi 2001; Soga et al. 2002a, b); and the application of stable isotope dilution theory in metabolome characterization of *S. cerevisiae* cultures, where each metabolite concentration is quantified relative to the concentration of its U-^{13}C-labeled equivalent (Mashego et al. 2003).

In the past 14 years, metabolomics has been successfully applied to a wide range of microbiological applications, from microbial functional genomics (Allen et al. 2003; Castrillo et al. 2003; Moxley et al. 2009), phenotyping (Bundy et al. 2005; Villas-Bôas et al. 2008), to the detection of microbial contaminations in foods (Jahangir et al. 2008; Cevallos-Cevallos et al. 2011), fermentation broth (Sue et al. 2011); and to characterize microbial symbiotic associations with different hosts (Martin et al. 2007; Russell and Duncan 2013). In addition, the microbial metabolomics field includes two important microbial metabolome databases developed at the University of Alberta (Canada): The Yeast Metabolome Database (YMDB) (http://www.ymdb.ca/) and the *E. coli* Metabolome Database (ECMDB) (http://ecmdb.ca/) (Jewison et al. 2012 and Guo et al. 2013). These are databases of metabolites found in cultures or cells of *S. cerevisiae* and *E. coli*, respectively. The YMDB contains 2027 small molecules with 980 associated enzymes and 138 associated transporters and the ECMDB contains 3755 small molecules with 1402 associated enzymes and 387 associated transporters. For each metabolite there is a description of the compound, its chemical characteristics, synonyms and links to spectral and chemical databases. Therefore, these two databases represent a landmark for microbial metabolomics, offering powerful Web-based tools to aid in microbial metabolomics data analysis and interpretation.

3 Challenges

As mentioned above, microbial metabolomics still faces some important challenges, with the efficient quenching of microbial cell metabolism during sampling combined with separation of intra- and extracellular metabolites being the most important. Microbial cells are usually diffused and diluted in a large volume of culture medium containing high levels of unutilized substrates and metabolic waste products. Adding to that is the fact that the ratio between microbial cell biomass and extracellular medium is often extremely low, making it often difficult to obtain a reasonable amount of intact cells for intracellular metabolite analysis per sample unit. Therefore, sampling and sample preparation in microbial metabolomics have to be carefully considered and tested before any study begins. To date, there are no standard protocols that can be widely applied to different microbial cell types and experimental conditions.

Ideally, microbial metabolomics needs robust sampling methods that quench the metabolic activity of microbial cells simultaneously with sampling and allow the separation of microbial biomass from the culture medium without affecting the metabolic state and integrity of the cells. Unfortunately, to a large degree, this has not been satisfactorily achieved as yet. There have been numerous studies comparing different sampling (quenching) methods for different microbial cells (Villas-Bôas et al. 2005b; Bolten et al. 2007) and proposing new methods (Villas-Bôas and Bruheim 2007; Canelas et al. 2008). However, many of those studies show controversial results (mainly regarding the popular cold methanol water solution for sampling microbial cells) and most protocols seem to be suitable only to specific cell types (Faijes et al. 2007; Tredwel et al. 2011).

Another challenge that metabolomics still faces specific to microbial metabolomics is the metabolome coverage by different analytical methods. It is common knowledge in metabolomics today that there is no single analytical method capable of detecting and identifying the whole metabolome of a cell. Therefore, metabolomics is achieved through the combination of different analytical tools in order to get as much information as possible about the metabolite profile of a population of cells or organism. Considering that microbial samples are usually short in biomass concentration, obtaining enough samples from a single experiment to be analyzed by different instruments or methods can be an issue. Moreover, identification of microbial secondary metabolites (and characterization of novel ones) is still a relatively low throughput process, similar to plant secondary metabolites.

Finally, microbial metabolomics suffers from a huge knowledge gap regarding the metabolic relationship between intra- and extracellular metabolites (Granucci et al. 2015). The transport of substrates to the intracellular medium is an essential function to any live cell and it has been extensively studied (Aguilar-Barajas et al. 2011; Jojima et al. 2010; Hofman-Bang 1999). However, the transport of chemical molecules through the plasmatic membrane is not exclusively to provide nutrients and other substrates to the cells, but to also remove metabolic waste products in order to maintain intracellular homeostasis. Therefore, the secretion of metabolites is also an essential biochemical function to all living cells, but it is very little studied if compared to the process of nutrient uptake.

The secretion of macromolecules such as proteins (enzymes) and nucleic acids have been widely studied in microorganisms (Alvarez-Martinez and Christie 2009; Chapon-Hervé et al. 1997; Genin and Boucher 1994). The best known mechanisms of macromolecule secretion are the xcp secretion pathway (Chapon-Hervé et al. 1997), the transmembrane protein family PulD (Genin and Boucher 1994), and the type IV secretion systems (T4SS) (Alvarez-Martinez and Christie 2009). On the other hand, recent studies have shown that ABC transmembrane proteins (ATP-binding cassette) and MFS (Major Facilitator Superfamily) play a very important role in the secretion of secondary/toxic metabolites in microbial systems such as endogenous and exogenous antibiotics (Stergiopoulos et al. 2002; Martín et al. 2005). It is believed that small metabolites, mainly those end products of microbial fermentation (e.g., acetate, lactate, butyrate, ethanol, butanol, acetone, etc.) are excreted passively through the plasma membrane, or secreted by specialized mechanisms such as response reaction to hypo-osmotic stress (Krämer 1994) and uniport, synport, and antiport transport systems (Krämer 1994). Although there is little experimental data to validate these mechanisms as being responsible for the metabolic efflux, all of them are based on the concept of metabolic overflow, which says that under specific metabolic conditions, a massive excretion of some metabolic intermediates is observed due to intracellular accumulation of one or more intermediates from one or more metabolic pathways (Krämer 1994). Although this concept seems to be appropriate to explain the secretion of some metabolic intermediates, it does not apply to many cases studied during continuous culture (chemostats) (Krämer 1994). Microbial metabolomics data obtained during continuous culture in our laboratory challenge the concept of metabolic overflow (Granucci et al. 2015). For instance, in several studies in which microbial samples were collected before, during and after an environmental perturbation (during continuous culture), we observed that some intracellular metabolites are actively secreted to the extracellular medium in response to an environmental stimulus, demonstrating that the secretion of a given metabolite to the extracellular medium does not take place exclusively as a result of its accumulation in the intracellular medium. Our results suggest that microbial cells very often remove some intracellular metabolites, even though these metabolites could be key intermediates of central carbon metabolism such as pyruvate, phosphoenolpyruvate, and several amino acids. Therefore, a better understanding about the relationship between intra- and extracellular metabolites is essential if we want to make better use of extracellular metabolomics data.

4 Cutting-Edge Applications

4.1 *Assessing the Metabolic State of a Microbial Community*

It is evident today that the metabolite profile of a pure microbial culture is very distinct from that of a mixed culture and the metabolite profile of a mixed culture is not a simple sum up of the metabolite profiles of the different species composing the community (Sue et al. 2011; Kimes et al. 2013; Lv 2013; Ponnusamy et al. 2013). Microbial cells sense the presence of a different microbial species in their environment and immediately respond to the foreign presence by changing their metabolic state. This change in metabolic state can be easily detected through metabolomics (e.g., metabolic footprint analysis) (Sue et al. 2011). Therefore, metabolomics has become an important tool for assessing the metabolic state of microbial communities both in vitro (Sue et al. 2011; Ponnusamy et al. 2013) as well as in nature (Badri et al. 2013; Kimes et al. 2013). Nevertheless, it is essential to any metabolomic study involving microbial communities to keep track of the microbial community dynamics in terms of its composition both qualitative and quantitatively.

4.2 *Host–Microbe Interactions*

The relationship of metabolite with the microbial population profiles derived from culture-independent molecular techniques can help to unravel the inherent and intimate host–microbe relationships. This approach has been increasingly applied to study the molecular mechanisms behind host–microbe interactions, with demonstrated potential to elucidate the causes of various pathologies resulting from host–microbial associations (Ming et al. 2012; Xie et al. 2012; Marcobal et al. 2013; Storr et al. 2013), including human enteric infections (Han et al. 2010). Gut microorganisms and their metabolites affect the luminal environment and alter development, motility, and homeostasis of the gut (Han et al. 2010). Therefore, insights regarding the metabolic principles coordinating gut microbiota will guide future manipulation of the gut microbiota to promote human health.

4.3 *New Diagnosis of Microbial Infection Through Metabolomics*

The identification of the cause of microbial infection is not always simple and usually requires days of culturing and sometimes evasive sampling procedures. The discovery of metabolic biomarkers specific to different infective agents through metabolomics can pave the way for development of fast, sensitive, and accurate methods of diagnosis through metabolite profiles. Metabolomics has been successfully applied to distinguish different causative agents of bacterial pneumonia in animal models (Slupsky 2010; Hoerr et al. 2012) as well as for early diagnosis of pediatric septic shock (Mickiewicz et al. 2013). In addition, the opportunity to use noninvasive sampling procedures such as the use of exhaled breath for diagnostic of infectious diseases through metabolite profiling of volatile metabolites promises to be the future of clinical diagnosis (Boots et al. 2012).

5 Future Perspective

The future of microbial metabolomics certainly lies in further advances in analytical techniques showing ultrasensitivity and specificity to allow the detection and identification of intracellular metabolites using very small biomass samples. Mass spectrometry seems to be the technology with best potential to achieve this goal. Important recent steps have been taken towards this goal, such as the development of an ionization method called laser ablation electrospray ionization (LAESI) capable of in situ mass spectrometric analysis of individual cells at atmospheric pressure (Zenobi 2013). Mass spectrometry imaging is another novel technique which makes use of matrix-assisted laser desorption/ionization (MALDI) and secondary ion mass spectrometry (SIMS) imaging to visualize metabolites within a spatial resolution of <1 μm, at ambient pressures, and with very high mass accuracy and mass resolution (Zenobi 2013).

Consequently, single-cell metabolomics should also soon become a reality for microbial systems. Impressive advances in single-cell metabolomics have been achieved recently, but they have so far failed to generate meaningful insights into the phenotypic diversity of single cells within a population (Zenobi 2013). The main reason for that is probably the lack of sensitivity of the analytical methods available to measure low-concentrated metabolites within a single cell. Therefore, more extensive coverage of microbial metabolomes is the main requirement for future advances in microbial metabolomics, which is intrinsically linked to the development of ultrasensitive analytical methods. The size of the model yeast *S. cerevisiae* metabolome is around 1168 metabolites and the yeast metabolome database (http://ymdb.ca) lists over 2000 molecules. Still, most current metabolomics methods are capable of detecting and identifying less than 10 % of the full yeast metabolome (Zenobi 2013). Therefore, ultrasensitive analytical methods, combined with faster identification of metabolites from high-throughput measurements using small sample sizes (e.g., single cell metabolomics) will pave way for the future of microbial metabolomics.

References

Aguilar-Barajas E, Diaz-Perez C, Ramirez-Diaz MI, Riveros-Rosas H, Cervantes C (2011) Bacterial transport of sulfate, molybdate, and related oxyanions. Biometals 24:687–707

Allen J, Davey HM, Broadhurst D, Heald JK, Rowland JJ, Oliver SG, Kell DB (2003) High-throughput classification of yeast mutants for functional genomics using metabolic footprinting. Nat Biotech 21:692–696

Alvarez-Martinez CE, Christie PJ (2009) Biological diversity of prokaryotic type IV secretion systems. Microbiol Mol Biol Rev 73:775–808

Andersen B, Solfrizzo M, Visconti A (1995) Metabolite profiles of common *Staphylium* species. Mychological Res 99:672–676

Badri DV, Zolla G, Bakker MG, Manter DK, Vivanco JM (2013) Potential impact of soil microbiomes on the leaf metabolome and on herbivore feeding behaviour. N Phytol 198:264–273

Bolten CJ et al (2007) Sampling for metabolome analysis of microorganisms. Anal Chem 79:3843–3849

Boots AW, van Berkel JJBN, Dallinga JW, Smolinska A, Wouters EF, van Schooten FJ (2012) The versatile use of exhaled volatile organic compounds in human health and disease. J Breath Res 6:027108

Bundy JG, Willey TL, Castell RS, Ellar DJ, Brindle KM (2005) Discrimination of pathogenic clinical isolates and laboratory strains of Bacillus cereus by NMR-based metabolomic profiling. FEMS Microbiol Lett 242:127–136

Canelas AB et al (2008) Leakage-free rapid quenching technique for yeast metabolomics. Metabolomics 4:226–239

Castrillo JI, Hayes A, Mohammed S, Gaskell SJ, Oliver SG (2003) An optimized protocol for metabolome analysis in yeast using direct infusion electrospray mass spectrometry. Phytochemistry 62:929–937

Cevallos-Cevallos JM, Danyluk MD, Reyes-De-Corcuera JI (2011) GC-MS based metabolomics for rapid simultaneous detection of *Escherichia coli* O157:H7, *Salmonella Typhimurium*, *Salmonella Muenchen*, and *Salmonella Hartford* in ground beef and chicken. J Food Sci 76: M238–M246

Chapon-Hervé V, Akrim M, Latifi A, Williams A, Bally M (1997) Regulation of the xcp secretion pathway by multiple quorum-sensing modulons in *Pseudomonas aeruginosa* Mol Microbiol 24:1169–1178

Collar C (1991) Biochemical and technological assessment of the metabolism of pure and mixed cultures of yeast and lactic acid bacteria in breadmaking applications. Food Sci Technol Int 2:349–367

Faijes M, Mars AE, Smid EJ (2007) Comparison of quenching and extraction methodologies for metabolome analysis of *Lactobacillus plantarum*. Microb Cell Fact 6(1):1

Fiehn O (2001) Combining genomics, metabolome analysis, and biochemical modeling to understand metabolic networks. Com Funct Genomics 2:155–168

Fiehn O (2002) Metabolomics—the link between genotypes and phenotypes. Plant Mol Biol 48:155–171

Fiehn O, Kopka J, Dormann P, Altmann T, Trethewey RN, Willmitzer L (2000) Metabolite profiling for plant functional genomics. Nat Biotechnol 18:1157–1161

Genin S, Boucher CA (1994) A superfamily of proteins involved in different secretion pathways in gram-negative bacteria: modular structure and specificity of the N-terminal domain. Mol Genomics Genet 243:112–118

Granucci N, Pinu FR, Han T-L, Villas-Bôas SG (2015) Can we predict the intracellular metabolic state of a cell based on extracellular metabolite data? Mol Biosyst 11:3297–3304

Guo AC, Jewison T, Wilson M, Liu YF, Knox C, Djoumbou Y, Lo P, Mandal R, Krishnamurthy R, Wishart DS (2013) ECMDB: the E-coli metabolome database. Nucleic Acid Res 41:D625–D630

Han J, Antunes LCM, Finlay BB, Borcherst CH (2010) Metabolomics: towards understanding host-microbe interactions. Future Microbiol 5:153–161

Hoerr V, Zbytnuik L, Leger C, Tam PPC, Kubes P, Vogel HJ (2012) Gram-negative and Gram-positive bacterial infections give rise to a different metabolic response in a mouse model. J Proteome Res 11:3231–3245

Hofman-Bang J (1999) Nitrogen catabolite repression in *Saccharomyces cerevisiae*. Mol Biotechnol 12:35–73

Jahangir M, Kim HK, Choi YH, Verpoorte R (2008) Metabolomic response of *Brassica rapa* submitted to pre-harvest bacterial contamination. Food Chem 107:362–368

Jewison T, Knox C, Neveu V, Djoumbou Y, Guo AC, Lee J, Liu P, Mandal R, Krishnamurthy R, Sinelnikov I, Wilson M, Wishart DS (2012) YMDB: the yeast metabolome database. Nucleic Acid Res 40:D815–D820

Jojima T, Omumasaba CA, Inui M, Yukawa H (2010) Sugar transporters in efficient utilization of mixed sugar substrates: current knowledge and outlook. Appl Microbiol Biotechnol 85:471–480

Kimes NE, Callaghan AV, Aktas DF, Smith WL, Sunner J, Golding BT, Drozdowska M, Hazen TC, Suflita JM, Morris PJ (2013) Metagenomic analysis and metabolite profiling of deep-sea sediments from the Gulf of Mexico following the Deepwater Horizon oil spill. Front Microbiol 4:50

Krämer R (1994) Secretion of amino acids by bacteria: physiology and mechanism. FEMS Microbiol Rev 13:75–94

Lee SY, Lee D, Kim TY (2005) Systems biotechnology for strain improvement. Trends Biotechnol 23:349–358

Lv HT (2013) Mass spectrometry-based metabolomics towards understanding of gene functions with a diversity of biological contexts. Mass Spectrom Rev 32:118–128

Maharjan RP, Ferenci T (2003) Global metabolite analysis: the influence of extraction methodology on metabolome profiles of *Escherichia coli*. Anal Biochem 313:145–154

Marcobal A, Kashyap PC, Nelson TA, Aronov PA, Donia MS, Spormann A, Fischbach MA, Sonnenburg JL (2013) A metabolomic view of how the human gut microbiota impacts the host metabolome using humanized and gnotobiotic mice. ISME J 7:1933–1943

Martín JF, Casqueiro J, Liras P (2005) Secretion systems for secondary metabolites: how producer cells send out messages of intracellular communication. Curr Opin Microbiol 8:282–293

Martin FPJ, Dumas ME, Wang YL, Legido-Quigley C, Yap IKS, Tang HR, Zirah S, Murphy GM, Cloarec O, Lindon JC, Sprenger N, Fay LB, Kochhar S, van Bladeren P, Holmes E, Nicholson JK (2007) A top-down systems biology view of microbiome-mammalian metabolic interactions in a mouse model. Mol Syst Biol 112:112

Mashego MR, van Gulik WM, Vinke JL, Heijnen JJ (2003) Critical evaluation of sampling techniques for residual glucose determination in carbon limited chemostat culture of *Saccharomyces cerevisiae*. Biotech Bioeng 83:395–399

Mickiewicz B, Vogel HJ, Wong HR, Winston BW (2013) Metabolomics as a novel approach for early diagnosis of pediatric septic shock and its mortality. Am J Respir Crit Care Med 187:967–976

Ming X, Stein TTP, Barnes V, Rhodes N, Guo LN (2012) Metabolic perturbance in autism spectrum disorders: a metabolomics study. J Proteomic Res 11:5856–5862

Montel E, Bridge PD, Sutton BC (1991) An integrated approach to *Phoma* systematics. Mychopathologia 115:89–103

Moxley JF et al (2009) Linking high-resolution metabolic flux phenotypes and transcriptional regulation in yeast modulated by the global regulator Gcn4p. Proc Natl Acad Sci USA 106:6477–6482

Nguyen QT, Merlo ME, Medema MH, Jankevics A, Breitling R, Takano E (2012) Metabolomics methods for the synthetic biology of secondary metabolism. FEBS Lett 586:2177–2183

Oliver SG (1997) Yeast as a navigational aid in genome analysis. Microbiology 7:405–409

Oliver SG, Winson MK, Kell DB, Baganz F (1998) Systematic functional analysis of the yeast genome. Trends Biotechnol 16:373–378

Ponnusamy K, Lee S, Lee CH (2013) Time-dependent correlation of the microbial community and the metabolomics of traditional Barley Nuruk starter fermentation. Biosci Biotechnol Biochem 77:683–690

Reeves ML, Rabinowtiz JD (2011) Metabolomics in systems microbiology. Curr Opin Biotechnol 22:17–25

Roessner U, Wagner C, Kopka J, Trethewey RN, Willmitzer L (2000) Simultaneous analysis of metabolites in potato tuber by gas chromatography-mass spectrometry. Plant J 23:131–142

Roessner U, Luedemann A, Brust D, Fiehn O, Linke T, Willmitzer L, Fernie AR (2001) Metabolic profiling allows comprehensive phenotyping of genetically or environmentally modified plant systems. Plant Cell 13:11–29

Russell WR, Duncan SH (2013) Advanced analytical methodologies to study the microbial metabolome of the human gut. Trac-Trends Anal Chem 52:54–60

Slupsky CM (2010) NMR-based analysis of metabolites in urine provides rapid diagnosis and etiology of pneumonia. Biomark Med 4:195–197

Smedsgaard J, Frisvad JC (1996) Using direct electrospray mass spectrometry in taxonomy and secondary metabolite profiling of crude fungal extracts. J Microbiol Meth 25:5–17

Soga T, Ross GA (1999) Simultaneous determination of inorganic anions, organic acids, amino acids and carbohydrates by capillary electrophoresis. J Chromatogr A 837:231–239

Soga T, Heiger DN (2000) Amino acid analysis by capillary electrophoresis electrospray ionization mass spectrometry. Anal Chem 72:1236–1241

Soga T, Imaizumi M (2001) Capillary electrophoresis method for the analysis of inorganic anions, organic acids, amino acids, nucleotides, carbohydrates and other anionic compounds. Electrophoresis 22:3418–3425

Soga T, Ueno Y, Naraoka H, Ohashi Y, Tomita M, Nishioka T (2002a) Simultaneous determination of anionic intermediates for *Bacillus subtilis* metabolic pathways by capillary electrophoresis electrospray ionization mass spectrometry. Anal Chem 74:2233–2239

Soga T, Ueno Y, Naraoka H, Matsuda K, Tomita M, Nishioka T (2002b) Pressure-assisted capillary electrophoresis electrospray ionization mass spectrometry for analysis of multivalent anions. Anal Chem 74:6224–6229

Stergiopoulos I, Zwiers L-H, Waard, MAD (2002) Secretion of natural and synthetic toxic compounds from filamentous fungi by membrane transporters of the ATP-binding cassette and major facilitator superfamily. Eur J Plant Pathol 108:719–734

Storr M, Vogel HJ, Schicho R (2013) Metabolomics: is it useful for inflammatory bowel diseases? Curr Opin Gastroenterol 29:378–383

Sue T, Obolonkin V, Griffiths H, Villas-Bôas SG (2011) An exometabolomics approach to monitoring microbial contamination in microalgal fermentation processes by using metabolic footprint analysis. Appl Environ Microbiol 77:7605–7610

Tredwell GD et al (2011) The development of metabolomic sampling procedures for *Pichia pastoris* and baseline metabolome data. Plos One 6(1):e16286

Tweeddale H, Notley-McRobb L, Ferenci T (1998) Effect of slow growth on metabolism of *Escherichia coli*, as revealed by global metabolite pool (Metabolome) analysis. J Bacteriol 180:5109–5116

Tweeddale H, Notley-McRobb L, Ferenci T (1999) Assessing the effect of reactive oxygen species on *Escherichia coli* using a metabolome approach. Redox Rep 4:237–241

van der Werf MJ, Overkamp KM, Muilwijk B, Coulier L, Hankemeier T (2007) Microbial metabolomics: toward a platform with full metabolome coverage. Anal Biochem 370:17–25

Villas-Bôas SG, Delicado DG, Åkesson M, Nielsen J (2003) Simultaneous analysis of amino and nonamino organic acids as methyl chloroformate derivatives using gas chromatography-mass spectrometry. Anal Biochem 322:134–138

Villas-Bôas SG, Moxley JF, Åkesson M, Stephanopoulos G, Nielsen J (2005a) High-throughput metabolic state analysis: the missing link in integrated functional genomics of yeasts. Biochem J 388:669–677

Villas-Bôas SG, Højer-Pedersen J, Åkesson M, Smedsgaard J, Nielsen J (2005b) Global metabolite analysis of yeasts: evaluation of sample preparation methods. Yeast 22:1155–1169

Villas-Bôas SG, Noel S, Lane GA, Attwood G, Cookson A (2006) Extracellular metabolomics: a metabolic footprinting approach to assess fiber degradation in complex media. Anal Biochem 349:297–305

Villas-Bôas SG, Bruheim P (2007) Cold glycerol-saline: the promising quenching solution for accurate intracellular metabolite analysis of microbial cells. Anal Biochem 370:87–97

Villas-Bôas SG et al (2008) Phenotypic characterization of transposon-inserted mutants of *Clostridium proteoclasticum* B316T using extracellular metabolomics. J Biotech 134:55–63

Wang Q, Wu C, Chen T, Chen X, Zhao X (2006) Integrating metabolomics into systems biology framework to exploit metabolic complexity: strategies and applications in microorganisms. Appl Microbiol Biotech 70:151–161

Wittmann C, Krömer JO, Kiefer P, Binz T, Heinzle E (2004) Impact of the cold shock phenomenon on quantification of intracellular metabolites in bacteria. Anal Biochem 327:135–139

Xie BG, Waters MJ, Schirra HJ (2012) Investigating potential mechanisms of obesity by metabolomics. J Biomed Biotech #805683.

Zenobi R (2013) Single-cell metabolomics: analytical and biological perspectives. Science 342:1243259

Zhang AH, Sun H, Yan GL, Wang P, Han Y, Wang XJ (2014) Metabolomics in diagnosis and biomarker discovery of colorectal cancer. Cancer Lett 345:17–20

Chapter 2
Microbes, Metabolites and Health

Trevor J. Lockett, Anthony R. Bird, Claus Christophersen, Julie M. Clarke, Michael A. Conlon and David L. Topping

1 Introduction

High throughput DNA sequencing has revolutionised the analysis of complex communities of microorganisms (microbiota). In humans, this paradigm shift, allowing analysis of the membership of microbial communities without the need to culture individual organisms, has spawned an avalanche of research and a number of important collaborative programs characterising not only the catalogue of species comprising the microbiota associated with different ecological niches in the body in states of both health and disease (Group et al. 2009) but also their genetic potential (Qin et al. 2010). Nowhere in the human body are the microbes more numerous or the microbiota more complex than in the colon of the gastrointestinal tract (Savage 1977). With its trillions of bacteria, comprising over one thousand individual microbial species present in varying abundances, the microbiota of the human gut and the genes they carry (collectively the microbiome) provides the host with a vast metabolic potential over and above that afforded by the 20,000–25,000 protein coding genes of the human genome.

The composition of the human colonic microbiota is remarkably diverse between individuals with each person's microbiota being unique (Qin et al. 2010; Yatsunenko et al. 2012). Despite this diversity at the taxonomic level, when analysed for functional potential by whole genome analysis at the gene level, conservation of bacterial genes between individuals was high suggesting that

T.J. Lockett (✉)
CSIRO Health and Biosecurity, North Ryde, NSW, Australia
e-mail: Trevor.Lockett@csiro.au

A.R. Bird · J.M. Clarke · M.A. Conlon · D.L. Topping
CSIRO Health and Biosecurity, Kintore Ave, Adelaide, SA, Australia

C. Christophersen
Edith Cowan University, Joondalup, WA, Australia

D.J. Beale et al. (eds.), *Microbial Metabolomics*,
DOI 10.1007/978-3-319-46326-1_2

diverse microbial populations could deliver very similar functional outcomes for both the microbial community and the host (Huttenhower et al. 2012).

There is growing evidence that the gut microbiota impacts significantly on the health of the host. Animal studies have shown that the gut microbiota plays a key role in the post-natal maturation of both the gut itself and the gut immune system (Smith et al. 2007; Chung et al. 2012). In humans, changes in gut microbiota structure have been associated with gut-associated disorders including: inflammatory bowel diseases (Gevers et al. 2014; Frank et al. 2007; De Cruz et al. 2015) and colorectal cancer (Garrett 2015); metabolism-related disorders such as obesity (Ley et al. 2006), metabolic syndrome (Le Chatelier et al. 2013; Haro et al. 2016) and diabetes (Qin et al. 2012) but also with systemic disorders including cardiovascular disease, neurodegenerative diseases including Alzheimer's (Alam et al. 2014; Naseer et al. 2014) and Parkinson's (Goldman et al. 2014; Scheperjans et al. 2015) diseases and neuro-developmental and psychiatric conditions including autism spectrum disorder (Louis 2012; Mulle et al. 2013) and schizophrenia (Nemani et al. 2015).

Of these disorders, the role of the microbiota has probably been studied most extensively in obesity which is also a well-recognized risk factor for many of the other chronic diseases mentioned above. Obese humans, when compared to normal weight individuals, displayed a reduced ratio of microbes from the *Bacteroidetes* phylum relative to those of the *Firmicutes* phylum in their faeces and this ratio increased in obese individuals who lost weight (Ley et al. 2006). Important insights into the functional significance of these associations have come from the use germ free, gnotobiotic and gene mutant animal models. This same relationship of *Bacteroidetes* to *Firmicutes* was also observed in obese mice carrying two copies of the *ob* mutation in the leptin gene (*ob/ob* mice) relative to their *ob*/+ and +/+ litter mates. In this animal model, biochemical analyses revealed that these different microbiota varied in the efficiency of energy harvest from the diet and that both the obese-prone and lean phenotypes was transferrable to germ free mice by colonization with the microbiota from obese and lean donors respectively (Turnbaugh et al. 2006). This same result was observed when human faecal microbiota from female twins discordant for obesity were used to colonize germ free mice (Ridaura et al. 2013). Mice practise coprophagy. Co-housing of mice, conventionalised with the obese twin's microbiota, with other mice conventionalised with the lean twin's microbiota prevented acquisition of the obese phenotype. Microbial community analysis revealed that this prevention of obesity was associated with an invasion of *Bacteroidetes* from the lean microbiota into the microbiota of the obese-conventionalised animals (Ridaura et al. 2013). Interestingly this effect was diet-dependent and was most marked when the animals were fed a diet high in fruit and vegetables. When the animals were fed a diet high in saturated fat and low in fruit and vegetables, levels of *Bacteroidetes* invasion from the lean conventionalised animals was very low (Ridaura et al. 2013) highlighting the importance of diet as a selective force in gut microbiota structure/function determination. While microbial population changes associated with diseases have been extensively studied, the mechanisms by which these changes elicit their physiological effects on the host are just beginning to be unraveled.

Metabolites and other products made by the gut microbiota play a key role in the dialogue between the gut microbiota and the host. Contributions of the gut microbiota to the nutritional wellbeing of the host are well known and include the provision of vitamin K, B group vitamins, B12, biotin, riboflavin, thiamine and folate (Koenig et al. 2011; Said 2011) and amino acids (Yatsunenko et al. 2012). However, gut microbes also make a vast array of proteins, peptides and low molecular weight metabolites and chemically transform others of host or bacterial origin, all of which impact on the structure and functionality of the microbiota and the health of the host (Nicholson et al. 2012). Some bacteria help shape their local community by producing bacteriophage or bacteriocins to kill off specific competitors. The production of bacteriocins by commensal *Escherichia coli* can, for example, inhibit the invasion of their niche by pathogenic *E. coli* O157:H7 (Schamberger and Diez-Gonzalez 2002). Others form feeding chains where one species performs an initial break down of a starting nutrient source to release intermediate breakdown products that can themselves provide substrates for other types of bacteria as exemplified using in vitro fermentation studies to examine the degradation of dietary glycans (Leitch et al. 2007a, b) and host-derived mucins (Png et al. 2010). Others can induce the host to create local environments that provide a colonisation and proliferation advantage for themselves and potentially other bacteria (Collier et al. 2008), while others can have direct effects on host functions e.g. the production of gamma amino butyric acid (GABA) (Barrett et al. 2012), the major inhibitory neuro transmitter in the mammalian central nervous system (CNS) with a major role in the regulation of muscle tone in humans. Although GABA produced peripherally does not cross the blood brain barrier and hence is unlikely to impact on CNS functions, it may modify gut motility and blood pressure (Inoue et al. 2003).

Likewise the host helps shape the gut microbiota through the secretion of bioactive molecules into the lumen of the gut. Defensins, (Gallo and Hooper 2012) secreted from the Paneth cells and most enterocytes of the gut mucosa, and IgA (Cerutti 2008), secreted from plasma cells in the lamina propria and through the gut epithelium, provide protection against overgrowth of many pathogenic microbial species. Further, using a germ free/gnotobiotic maternal separation mouse model of stress it has been demonstrated recently that stress can alter the colonic environment leading to an altered colonic microbiota. While this altered microbiota was insufficient to recapitulate the stress-associated changes in host physiology upon microbial transplantation to non-stressed germ free animals, transplantation was associated with a transfer of anxiety-like behavior and behavioral despair but the molecular effectors are yet to be defined.

So what are the key influencers of gut microbiota structure and function and how might they elicit their impacts on health? While the gut of the near term foetus may have some microbes present, the numbers are very low (Jimenez et al. 2008). The mode of delivery (vaginal vs. caesarean section) impacts significantly on the structure of the gut microbiota of the neonate (Dominguez-Bello et al. 2010) with vaginally delivered infants developing bacterial communities resembling those of their mother's vaginal microbiota, while the gut microbiota of caesarean-delivered infants' reflected more closely those of their mother's skin. Prior to weaning the gut microbiota is relatively simple, dominated by Bifidobacteria and lactic acid bacteria. This changes significantly with the introduction of solid food, acquiring something close to its mature composition by the time of their first birthday (Backhed et al. 2015; Favier et al. 2002; Koenig et al. 2011). The base microbial population structure of a human's colon is relatively stable through childhood and much of adult life but begins undergoing change and simplification with aging and associated changes in health status, diet and lifestyle (Claesson et al. 2012). Despite this stability of an individual's core microbiota, components still show a dynamic response, both in composition and biological functionality, to changing dietary and environmental conditions (Clemente et al. 2012). Diet composition, in particular the amounts and nature of non-digestible components of the diet, can modify the of gut microbiota markedly (Gibson and Roberfroid 1995; Flint et al. 2012). Dietary fibre is the best understood of these. Epidemiological studies have associated diets high in dietary fibre with a reduced risk of a number of disorders including colorectal cancer and inflammatory bowel diseases (Hou et al. 2011). Here we discuss what is known about dietary fibre, the early research implicating it in promotion of gastrointestinal health, its interaction with the gut microbiota to produce a range of bioactive metabolites and our growing understanding of how these metabolites elicit their associated health benefits in the host.

2 Dietary Fibre and Gut Health: A Brief History

The high community awareness of the health potential of dietary fibre is relatively new. This interest was triggered by early work of a number of British physicians in South and East Africa. They saw that native Africans ate a diet high in whole grain foods and were essentially free of the non-infectious diseases which affected Europeans living in the same geographical area who ate highly refined foods (Burkitt 1973). Initially, they focused on large bowel disorders including constipation, diverticular disease, appendicitis and related problems. They linked these differences to their dietary fibre intakes and suggested that the disease profile of the Europeans reflected a simple dietary fibre deficiency. These early studies relied on observation (not measurement) and were quite limited in scope. They were limited also by the early definition of dietary fibre as plant cell wall material resistant to human small intestinal enzymes i.e. indigestible and largely insoluble plant material. The total indigestibility of these non-starch polysaccharides (NSP) explained their faecal bulking and laxating effects and protection against large bowel diseases very neatly (Cummings et al. 1992). However, fibre (as NSP) has proved very disappointing for protecting against other diseases of affluence, especially colorectal cancer (CRC), where it was expected to lower risk substantially. While some studies (e.g. the European Prospective Investigation into Cancer and Nutrition, EPIC) have shown a dose-dependent reduction in CRC risk (Murphy et al. 2012), others have not (Park et al. 2005; Lanza et al. 2007). The Australian paradox also fits with these inconsistencies. Australia seems to be one of the few countries where fibre consumption has increased substantially over time with a reported population average of 28 g/person/day (Baghurst et al. 1996) but CRC morbidity and mortality rates remain high (AIHW 2012). This paradox is leading to a revision of the relative importance of the different components which contribute to fibre intake in low risk populations and also the way in which dietary fibre actually works in the large bowel.

Much of the credit for the popularization of the dietary fibre hypothesis is due to Dennis Burkitt (Trowell and Burkitt 1987) but it seems that the focus on fibre as an indigestible bulking agent may be misplaced. It is now impossible to quantify actual food consumption rates in the original African population. Changes due to urbanization and higher incomes have altered the traditional rural lifestyle patterns of native South Africans so that their activity levels and total dietary fibre (TDF) consumption have fallen. TDF intakes have declined from an estimated 25–35 g/person/day (2 generations ago) to 15–20 g/person/day (now) (Segal et al. 2000). One would have expected this to lead to an increase in CRC rates but this does not seem to have happened. Comparison of the diet of indigenous Africans in South Africa and African Americans in the United States shows that they are quite similar in their low TDF content (O'Keefe et al. 2007) but the latter have one of the highest global rates of CRC of any ethnic group. One suggestion is that the high content of animal products in the African American diet is responsible. It is true that CRC risk has been linked to greater consumption of red and processed meat (Norat et al. 2005) but the work of Walker and others suggest very strongly that the type of fibre consumed by Africans may be equally important (Walker et al. 1986). The original concept underpinning the differential in disease risks was the high whole grain consumption of the Africans. Put simply, the notion of "fibre" came later, essentially as roughage (Topping and Illman 1986). However, it has emerged that the culinary habits of the Africans are an important factor. Many native South Africans consume reheated or cold maize porridge on a daily basis, particularly in rural areas. Cooking gelatinises the starch, increasing its small intestinal digestibility but cooling leads to self-association of the chains. This process (retrogradation) leads to resistance to amylolysis i.e., the formation of resistant starch (RS). Resistance to amylase is a key feature of TDF meaning that RS is also a fibre component. It has been shown that hot traditionally cooked maize meal contains 18 g of RS/100 g while cooked and cooled maize contains a higher RS content than that of a hot maize meal (Heneker et al. 1998). Cooking is not the only factor. The amylose content of South African maize is 37–40 % of total starch which is relatively high (37.1–39.9 %) (van der Merwe et al. 2001). Amylose is a relatively compact starch polymer which is slower to gelatinize than amylopectin and also quicker to retrograde meaning that it is digested more slowly than low amylose starches and also yields more RS in food products.

One of the most important observations driving the reappraisal of the relative importance of fibre (i.e., NSP and RS) was the lack of relationship between TDF and bowel habit in native South African children. This has led to the current focus on the role of the microbiota in human health, specifically through the production of short chain fatty acids (SCFA). Knowledge of the existence of RS is not new nor is the presence of SCFA in the large bowel of omnivores (Elsden et al. 1946). It was known that TDF polysaccharides are lost during whole of gut transit (Schneeman 1986). For NSP, the extent of loss varies by source and ranges from ~0 (for cellulose) to >90 % for pectin. In contrast very little starch appears in normal human faeces which, coupled with its susceptibility to amylolysis, gave the obvious conclusion that small intestinal starch digestion was complete. The existence of potentially significant levels of indigestible starch (i.e., amylase resistant starch—RS) in foods was shown by Englyst and colleagues (Englyst et al. 1982). There was earlier (indirect) evidence from Levitt's laboratory (Anderson et al. 1981) in intact humans showing that, following consumption of a low-fibre convenience cereal product, breath H_2 evolution rose significantly. Breath H_2 is a marker for large bowel fermentation and their observations can now be interpreted as showing the presence of RS and also that fermentation was occurring. Both are now established as significant factors in human gut physiology (Topping and Clifton 2001). RS is defined in terms of starch which escapes into the large bowel and the products of its fermentation. RS is today recognized as a component of fibre and not a separate entity.

There are striking similarities between large bowel fermentation by the microbiota of humans and other omnivores and that of obligate herbivores (Topping and Clifton 2001). The end products are similar i.e., SCFA (principally acetate, propionate and butyrate), gases (CH_4, H_2 and CO_2) and the energy to fuel bacterial growth. In humans and relevant model species (most notably pigs), SCFA levels are highest in the proximal large bowel and decline with passage of the digesta stream. This is consistent with the high production rates in the caecum and proximal colon i.e., where substrate availability is greatest. The fall in SCFA levels along the colon reflects substrate depletion but also their uptake by the large bowel so that <10 % of total production is excreted in faeces. This distributional profile has very important implications for gut health and is believed to relate to the higher prevalence of non-infectious disease in the distal colon and rectum (Topping and Clifton 2001; Cats et al. 1996).

SCFA have a number of general effects which contribute to large bowel health (Topping and Clifton 2001; Brouns et al. 2002; Wong et al. 2006). Their production acidifies the digesta which inhibits the overgrowth of potential pathogens and limits the absorption of potentially toxic compounds e.g. NH_3. The latter cannot enter colonocytes as NH_4^+ so lower pH values limit their exposure to this and other cytotoxic and genotoxic agents (Topping 2007). A study in normal humans consuming their usual diets with supplements of NSP or NSP + RS showed an increase in faecal NH_3 excretion which correlated closely with greater butyrate excretion (McOrist et al. 2011). This is consistent with diminished absorption by (and less exposure to) colonocytes. SCFA enhance visceral blood flow through dilation of resistance vessels in the colon which facilitates O_2 supply (Topping and Clifton 2001). The history of research in dietary fibre is littered with misconceptions, revisions and reappraisals. The current focus on the microbiome is another case in point. The emergence of modern technologies has given us an opportunity to identify the species which are present and how they are altered by diet. Previously, this was almost impossible using the techniques of traditional microbiology. However, it should be noted that it is the products of bacterial action in the large bowel which influence human health. While they have these general effects, the individual acids have specific actions which benefit visceral function, health and wellbeing.

3 Dietary Fibre: a Complex Mixture

Dietary fibre is not a single entity but an umbrella term for a large, diverse group of substances, predominately but not exclusively carbohydrate polymers of plant origin, that pass through the small intestine and undergo partial or complete fermentation in the colon.

The global consensus definition of dietary fibre (Codex 2010), has broadened from a botanical and structural focus (plant cell wall polymers) and now captures other salient properties of fibre, notably compositional diversity and physiological functionality.

The major classes of dietary fibre are NSP, RS and non-digestible oligosaccharides although inclusion of the latter can vary with jurisdiction (Codex 2010). Fibre also includes analogous compounds, non-digestible carbohydrate polymers, of animal origin, and plant secondary metabolites that are intimately associated with plant cell wall polysaccharides, notably lignin.

Fibres, both natural and synthetic, differ markedly in their physiological properties and capacity to promote human health (Dahl and Stewart 2015; Bird 2007; Cummings and Stephen 2007; Buttriss and Stokes 2008; Brownlee 2014; Edwards et al. 2015). While physiological functionality is included in the definition of fibre, it is applied only to foods containing carbohydrate polymers sourced from either raw food materials or chemically synthesized. A demonstrable beneficial physiological effect in humans is a requirement for foods to be advertised as containing added/functional fibre.

A prerequisite property and defining feature of fibre is that it is potentially fermentable by colonic microbiota. In recent times there has been growing scientific and commercial interest in highly fermentable fibres, notably non-digestible oligosaccharides and resistant starches, owing to their favourable effects on gut microbiota composition and activity, and host health (Bird 2007; Conlon and Bird 2015; Bird and Topping 2001; Broekaert et al. 2011, de Menezes et al. 2013).

NSP, mainly cellulose and hemicelluloses, and lignin, are the fundamental structural elements of plant cell walls and have been integral to all descriptions of dietary fibre (Miller Jones 2014). Cellulose and hemicelluloses are widely distributed in the plant kingdom and are ubiquitous in the plant foods that humans eat. NSP typically comprise long chains of monosaccharides, mostly glucose moieties, joined by β-1,4 linkages. Cellulose is a high molecular weight homopolymer consisting of linear β-1,4 linked-chain of glucose units. The chains are closely packed and the hydrogen bonds that form between adjacent chains ensure that this structure is highly resistant to digestion by human hydrolases, which specifically attack monosaccharide units linked by α-1,4 and α-1,6 bonds (Tungland and Meyer 2002; Englyst et al. 2013). Hemicelluloses, in contrast, are smaller, branched, heteropolymers comprising a variety of monomeric units, commonly xylose and arabinose, and mannose, glucose, rhamnose and glucuronic acid, joined by various types of chemical bonds. Hemicelluloses comprise a large variety of glycans, both water insoluble and soluble (Tungland and Meyer 2002; Englyst et al. 2013).

NSP generally conform to the notion of "insoluble fibre" and are concentrated in highly fibrous foods such as cereal brans (Bird 2007). Fibres such as β-glucans and arabinoxylans tend to be water soluble (water extractable). Other examples of soluble non-starch polysaccharides include gum arabic (acacia gum), guar gum, psyllium, pectins and alginates (Tungland and Meyer 2002; Englyst et al. 2013). However, solubility is not strictly a fixed property. Mixed-linkage glucans, such as those found in oats and barley, exhibit varying degrees of solubility depending on the source, analytical methodology and grain processing (Tosh et al. 2008; Agbenorhevi et al. 2011).

RSs are starches which escape small intestinal digestion. Like highly digestible starches, RS is a polymer of glucose monomers joined α-1,4 and α-1,6 glycosidic linkages at branch points and hence are not intrinsically resistant to digestion by enteric enzymes of humans. Rather they are rendered less accessible to the amylases of the upper gastrointestinal tract by some feature of the starch itself, e.g. its physicochemical state and/or the physical complexity of the food matrix in which the starch is presented e.g. whole grains or some feature of the host e.g. compromised dentition or a rapid food transit through the upper GI tract (mouth, stomach, small intestine) (Bird et al. 2000). In particular, the duration of exposure of ingested starch to hydrolytic processes in the upper gut, which is governed by gut motility and digesta transit rate, is a major determinant of the amount of dietary starch reaching the colonic microbiota. RS therefore represents the sum of the undigested starch and intermediates of starch digestion that are not absorbed in the small intestine and hence enter the large bowel of healthy humans (Bird et al. 2010, 2012). While most fibres are reasonably stable to culinary practices, RS can be formed or easily depleted by cooking and other processes involved in food production and preparation (Bird et al. 2012).

Non-digestible oligosaccharides are important contributors to the total fibre content of many plant foods. They are low molecular weight fibres that have a degree of polymerization (chain length; DP) between 3 and 9 (de Menezes et al. 2013; Englyst et al. 2013). Fructans are polyfructoses consisting of 3–20 units with a terminal glucose unit (Gibson et al. 2004; Roberfroid 2005). Food sources rich in fructans include Jerusalem artichoke, onions, garlic and bananas. Legumes, such as soybeans, contain galactooligosaccharides (raffinose, stachyose and verbascose) (Muir et al. 2009). Cereal fructans consist of straight and branched chain fructose polymers (Jenkins et al. 2011). Commercial sources of dietary fructans, commonly derived from chicory, inulin and fructooligosaccharides (FOS), are extensively used as food ingredients (Roberfroid 2007). Inulin is a large molecule with a DP of between 2 and 60 fructose residues whereas FOS comprises smaller molecules with a DP between 2 and 8. 'Oligofructose' is a partial enzymatic hydrolysate of inulin.

Fibre also encompasses a range of other plant substances, including lignin, cutin, waxes, suberin, tannins, saponins and phytosterols, that are naturally associated with botanical cellular structures (Bird 2007; Englyst et al. 2007; Miller Jones 2014). Lignin is a heterogenous polyphenolic ether, not a carbohydrate, that is covalently bonded to cellulose (Tungland and Meyer 2002). Most of these substances are quantitatively minor constituents of human diets and are not metabolized extensively by the gut microbiota (Topping and Clifton 2001). However, they may impair fermentation of plant structural and storage polysaccharides (Stephen 1994; Edwards et al. 2012).

In certain disease settings, dietary carbohydrates that are normally digestible in the upper gut ('available carbohydrate') can also make their way to the colon and essentially function as dietary fibre. For instance, lactose can be an important substrate for the microbiota in those individuals with lactase deficiency (Deng et al. 2015; Lukito et al. 2015; Wahlqvist 2015; Windey et al. 2015). The contribution of this carbohydrate is dependent on the degree of lactase deficiency. Other bowel diseases and conditions that compromise the functional capacity of the upper gut, such as celiac and Crohn's diseases, will also increase the amount of malabsorbed carbohydrate, and other dietary constituents, reaching the colonic microbiota (Conlon and Bird 2015).

Physicochemical and functional properties of fibre, including solubility in water, viscosity and gel formation, and water holding capacity, have formed the basis of various classification schemes. Soluble/insoluble fibre classification has proven popular, however it is inconsistent with the accepted definitions for fibre and is an unreliable predictor of physiological function. 'Solubility' relates to an in vitro environment, not the gut lumen. The proportion of fibre that is solubilised is often a function of the assay conditions used and in many instances the results are method specific (Cummings and Stephen 2007; Englyst et al. 2007). The proportion of soluble fibre can also change during food processing and passage of the fibre through the upper gut (Arrigoni 2001; Comino et al. 2016).

Dietary fibres can be classified on their fermentability (Dahl and Stewart 2015; Slavin 2001). But, gauging the rate and extent of fibre fermentation in vivo is challenging, which is why knowledge of fermentation of fibre-rich substrates is based largely on studies deploying in vitro techniques. These commonly use human faecal inocula in static and flow-through incubation systems to simulate conditions and events in the complex, anaerobic and highly dynamic micro-ecosystem of the large bowel. As such, they provide limited indicative data on fibre degradation by the colonic microbiota.

Generally, about 70 % of the fibre that is consumed by people on mixed diets is considered to be fermented to completion in the colon (Slavin 2001; Topping and Clifton 2001). However, individual fibres vary markedly in their fermentability although most are partially fermentable. Water soluble fibres, such as those from vegetables & fruits, are usually more easily fermented compared to insoluble fibres, such as cellulose, which is quite resistant to fermentation. For instance, only about

30 % of the fibre contained in wheat bran is fermented completely in humans (Topping and Clifton 2001). Not surprisingly, degradation of the smaller molecular weight carbohydrates and soluble NSPs occurs in the more proximal regions, whereas the breakdown of fibre that is more difficult to ferment occurs further along the large bowel. In general, the colonic microbiota metabolises individual fibres in the following order: mono- and oligosaccharides are preferentially fermented followed by starches, soluble NSPs (soluble β-glucans, arabinoxylans) and then insoluble NSPs (Bach Knudsen 2015). However, there are many exceptions to the rule.

There are myriad factors that influence the rate and extent of fermentability of carbohydrates aside from the colonic microbiota per se. Physicochemical attributes are the primary intrinsic determinants, notably chemical structure (monosaccharide composition, linkage type) and molecular size (chain length) but diet, food form (matrix and particle size) and physiological factors that alter the gut microbial habitat, such as digestive and absorptive capacity and digesta transit time, are also important, as is the amount (dose/concentration) and composition of fibre-containing foods in the diet.

The fibre content of most western diets is exceedingly low and consequently, so too are fibre intakes. Many populations consume inadequate amounts of fibre, falling well short of intake targets recommended by health authorities (i.e., ~14 g total fibre/1000 kcal or ~30 g dietary fibre/day for an adult) (Dahl and Stewart 2015; Baghurst et al. 1996). Furthermore, there is a limited diversity of fibre components in most modern diets. Insoluble NSP predominate (Baghurst et al. 1996; Conlon and Bird 2015). Oligosaccharides are consumed in much smaller amounts, whereas resistant starch is essentially nonexistent in most modern diets simply because modern processed and staple foods contain only small amounts (Bird et al. 2012).

Ancestral hunter-gatherer diets provided five times more fibre than do contemporary diets (Eaton 2006, Leach 2007). The range of fibre sources would also have been considerably greater than that consumed by modern populations, which suggests that our early forebears harboured a much larger, richer and more diverse gut microbiota.

Cereals and cereal products are the major source of fibre for most human populations, but the variety of cereal sources is limited (Baghurst et al. 1996). Wheat predominates and its increasing worldwide popularity explains why insoluble NSP such as cellulose and arabinoxylans, are the major type of dietary fibre for most people. Fruits and vegetables are the next most important source of fibre but they too tend to supply mainly NSP (Baghurst et al. 1996). There is also an increasing range of synthetic fibres (e.g. chemically modified resistant starches such as cross-bonded starches) in processed foods. Most are derivatives of plant polymers.

NSP account for the major portion of dry matter (20–45 %; equivalent to 10–25 g/day) supplied to the colon of individuals consuming western diets (Conlon and Bird 2015). Monosaccharides and oligosaccharides account for a further 10 %, and RS, which includes partial hydrolysis products, is unlikely to be more than 10 % (Conlon and Bird 2015); Small quantities of sugar alcohols also serve as fermentable substrates but the amount and type is very much diet dependent (Grembecka 2015; Halmos et al. 2015; Muir et al. 2009; Tennant 2014).

Interactions between different fibre components in a food as well as its physical form (matrix), in particular the structural integrity of plant cell walls, will influence the rate and extent of fermentation of the component fibres. Milling, processing, preparation and storage conditions of foods can alter their fibre content and functionality and hence their impact on the gut microbiota (Tuohy et al. 2012). For instance, soluble fibre may be lost through leaching during the cooking process. Heat treatment may solubilise certain fibres, thereby increasing the ratio of soluble to insoluble fibre. Short chain fructans can be degraded by yeast enzymes and also exposure to acidic conditions. A sufficiently large particle size is needed to ensure an intact cell wall structure resists extensive fermentation. Coarse bran loses both its water holding capacity and some of its laxative effects when it is finely ground simply because it is more extensively fermented in the large bowel (Topping and Bird 1999).

Another important consideration is functional capacity of the upper gut. Small intestinal digesta transit rate and nutrient assimilation can differ markedly among people. For instance, some individuals are especially efficient in digesting and fermenting starch (Thornton et al. 1987). Furthermore, some fibres may undergo degradation in the upper digestive tract owing to the action of bacterial glycanases in that region of the gut (Bach Knudsen 2015). However, the loss of fermentable substrate is probably small.

Poorly fermented carbohydrates may alter the metabolic capacity of the gut microbiota through various mechanisms. For instance, by acting as a physical scaffold for colonic microbes, resident and transitory, it may facilitate the fermentation process. Fibres in general also influence fermentation events by augmenting intestinal motility and flow of the faecal stream (Topping and Bird 1999; James et al. 2015).

4 Dietary Fibre and Its Interaction with the Gut Microbiota

The dietary fibres described in the previous section have distinct effects on the gut microbiome. This can be because of their solubility, structure or where in the gut the fibre is fermented. This differential effect will lead to selection of certain groups of bacteria, specific fermentation outputs, like SCFA and gas production, and likely inhibition of pathogenic bacteria.

Some of the most widely studied prebiotic fibres are inulin, fructooligosaccharies (FOS) and galactooligosaccharides (GOS) with the latter two encompassing numerous different types of fibres. All of these fibres are soluble and highly fermentable which means they are fermented in the terminal ileum or the proximal colon (Eswaran et al. 2013). FOS and GOS were both found to have a bifidogenic effect and increased levels of *Lactobacillus* in both animals and humans (Davis et al. 2011; Djouzi and Andiueux 1997; Kleessen et al. 2007a). These effects were accompanied with higher levels of SCFA and a lower pH in the large bowel, however FOS were found to favour an increase in butyrate whereas GOS advocated for higher propionate levels (Djouzi and Andiueux 1997). FOS and GOS also had differential effects on gas production in rats, with FOS feeding increasing both hydrogen and methane excretion, but GOS only induced higher methane excretions (Djouzi and Andiueux 1997). Inulin seems to have differential effects in animals and humans. This difference was found albeit using a human microbiota-associated rat model. In the rat model, no difference in *Bifidobacterium* spp. was observed after inulin intake, but higher levels of fibre degrading bacteria *Roseburia* spp. and *Eubacterium rectale* were observed and a reduction in potentially pathogenic bacteria (*Clostridium* spp.) (Kleessen et al. 2007a; Van den Abbeele et al. 2011). SCFA concentrations increased and pH levels reduced with inulin intake in animals. Human intervention trials have not been able to replicate these changes. In these studies, higher levels of *Bifidobacterium*, *Faecalibacterium prasnitzii*, reduced numbers of *Bacteroides-Prevotella* and potentially pathogenic bacteria were observed (Kleessen et al. 2007b; Ramirez-Farias et al. 2009; Costabile et al. 2010). In a later study, Ramirez-Farias et al. (2009) showed that it was certain species of *Bifidobacterium*; *B. adolescentis*, *B. longum* and *B. bifidum* that changed with inulin intake. Another significant difference between the animals and human studies was the absences of changes observed in faecal SCFA levels in humans given inulin (Kleessen et al. 2007b; Costabile et al. 2010). Another way to evaluate the effect of certain fibres is to remove them from the diet. This was done by Halmos et al. (2015) in a randomised control crossover trial of a low Fermentable Oligo- Di- and Monosaccharides And Polyols (FODMAP) diet (no inulin) and a diet containing a typical level of FODMAP's found in an Australian diet. This study showed that by avoiding FODMAP the overall bacterial abundance, and the relative abundance of *Clostridium* cluster XIVa (butyrate-producing bacteria) and the mucus-associated bacterium *Akkermansia muciniphila* was reduced. However, these changes did not result in differences in faecal SCFA levels, although pH was higher on the low FODMAP diet.

A possible reason for not noticing any differences in faecal SCFA levels may well be because >95 % of the SCFAs produced in the human colon are taken up by the gut epithelia (Kleessen et al. 2007b; Costabile et al. 2010). This does not exclude that there might be a difference in SCFA levels in the proximal colon however, when SCFAs are measured in faeces, levels are diminished due to uptake along the colon.

RS, can also change the gut microbiome and the metabolic components produced, however the RS type plays an important role. All of the RS types are generally digested in the proximal and the transverse colon, but some are digested throughout the colon like RS type 1 and 4 (Clarke et al. 2011b; Eswaran et al. 2013).

Animal studies have revealed great benefit and potential for RS as a prebiotic fibre. A limited number of studies have been conducted using RS type 1; in animals (rats), this type of RS did promote higher abundances of *Bifidobacterium* spp. and increased SCFA levels at month five compared to control (Kleessen et al. 1997). The RS type 2 fibre is often used in animal studies and seems to have a broader effect on the gut microbiota. It was shown in rat and pig studies that RS type 2 can increase the abundances of *Bifidobacterium* spp., *Lactobacillus* spp., *Enterobacteriaceae* and *Ruminococcus bromii* compared to control animals (Conlon et al. 2012; Abell et al. 2011; Kleessen et al. 1997; Le Blay et al. 1999; Martin et al. 1998). These microbial changes were accompanied by higher SCFA levels due to RS type 2 feeding with certain studies observing specific increases in propionate (Conlon et al. 2012; Abell et al. 2011) and butyrate (Conlon et al. 2012; Abell et al. 2011; Le Blay et al. 1999; Martin et al. 1998) excretion. RS type 3 have also been shown in rat and pig studies to select for *Bifidobacterium* spp., *Lactobacillus* spp. and *Faecalibacterium prausnitzii*, but lower the abundances of gamma-Proteobacteria (Conlon et al. 2012; Haenen et al. 2013). Again, these microbial changes were in conjuction with increased SCFA levels, especially butyrate and propionate. The gut microbiota can also be modified with RS type 4 with a rat study conducted with chemically modified RS showing increases in *Lactobacillus* spp. and *Parabacteroides distasonis* compared to control (Abell et al. 2011). This study also recorded increases in SCFA levels.

Human studies have confirmed many of the changes observed in the animal gut, however with a high inter-individual variation for both microbial and biochemical composition. Studies in humans are generally conducted using RS type 2, 3 and 4. RS type 2 and 3 have very similar effects in the human gut promoting higher levels of *R. bromii*, *F. prausnitzii* and *E. rectale* compared to a Non-Starch Polysaccharide (NSP) control (Abell et al. 2008; Martinez et al. 2010; Walker et al. 2011). For two of these studies, SCFA were also measured and the RS consumption was associated with higher levels of faecal SCFA levels, especially butyrate (Abell et al. 2008; Walker et al. 2011; McOrist et al. 2011). This increase in faecal butyrate levels successfully correlated with the increasing faecal abundance of *F. prausnitzii* and *E. rectale* (Abell et al. 2008; Walker et al. 2011). Human research studies have been conducted with RS type 4 showing beneficial effects on the gut microbiota with higher abundances of Actinobacteria, Bacteroidetes and a reduction in Firmicutes (Martinez et al. 2010), more specifically *Clostridium* cluster XIVa and IV, *Lactobacillus*, *B. adolescentis*, *P. distasonis* and *R. bromii* (Clarke et al. 2011b, Martinez et al. 2010; West et al. 2013; Le Leu et al. 2015). Reductions were also observed in bacteria, *Ruminococcus gnavis*, *R. torques* and *E. coli*, previously associated with gut diseases (Le Leu et al. 2015). RS type 4 consumption also resulted in higher SCFA level in the faeces of volunteers consuming these fibres. These substantial shifts in bacterial composition imply that the different RS types have the ability to select for certain groups of bacteria, despite a relative high inter-individual variation. Combined, these studies strongly suggest a key role for *R. bromii* in the degradation of RS and these bacteria have also been found to be the predominant ones to colonise RS (Leitch et al. 2007a; Ze et al. 2012). It is also worthwhile keeping in mind that these prebiotic effects are not universal, as some volunteers are non-responders. The response in any volunteer will always depend on the initial composition of the gut microbiota which helps to explain the often large inter-individual variation seen in human intervention studies examining the gut microbiome.

Fibres, like NSP, with a very limited prebiotic effect because up to 75 % of the ingested fibre is not fermented in the gut (Backman 2009; Roberts et al. 2010) and no selective promotion of specific bacterial populations (Abell et al. 2008; Walker et al. 2011) can still affect the gut microbiota. It is suggested that NSP can interact with the intestinal bacteria via contrabiotic effects, whereby they prevent potentially harmful interactions between bacteria and the gut epithelium that occur upon dysbiosis (Backman 2009; Parsons et al. 2014; Roberts et al. 2010, 2013). The contrabiotic effect has been tested for a range of soluble NSP's and they are not all equally effective. NSP from plantain bananas and broccoli seems to have greater effect than NSP from apple and leek (Roberts et al. 2010). These tests have mainly been conducted in vitro testing the ability of the fibres to block the attachment of adherent, invasive *E. coli* (AIEC). Soluble plantain NSP's have not only been able to block AIEC but also adherent enteric pathogens, like *Shigella* spp., *Clostridium difficile*, enterotoxigenic *E. coli* and *Salmonella* spp. (Parsons et al. 2014; Roberts et al. 2013). Parsons et al. (2014) also demonstrated that NSP from plantain bananas can block Salmonellosis in chicken when added to their feed and translocation of *Salmonella* Typhimurium across isolated follicle-associated epithelium from human ileum. It is believed that microfold (M)-cells overlying Peyer's patches in the human ileum and lymphoid follicles in the colon are potential points of entry for AIEC (Roberts et al. 2010). It is suggested that the contrabiotic effect is mediated via an interaction between the soluble NSP and the epithelial cell causing an electrogenic chloride secretion preventing the adhesion of the pathogens (Parsons et al. 2014). Another and more simple explanation is the direct interaction between plantain NSP and *E. coli*, *C. difficile* and *Salmonellae* as they can all utilise plantain NSP as an energy source (Roberts et al. 2010, 2013).

5 Dietary Fibre, Short Chain Fatty Acids and Their Impact on Gut Health

Collectively, the microbes present within the GI tract can produce a vast array of bioactive compounds, many of which can influence the functions of host tissues. A comprehensive review of the many known gut microbe-derived metabolites and their functions can be found elsewhere (Nicholson et al. 2012). We will focus our discussion on the impacts of some of the primary fermentation end-products, particularly the SCFA, which are considered beneficial and largely a result of the breakdown of dietary polysaccharides, and some toxic products such as ammonia, phenols and hydrogen sulphide which are largely derived from fermentation of dietary protein.

The wide degree of variation in microbiota composition between individuals would be expected to be reflected in different gut microbial metabolite profiles and susceptibility to diseases. For example, levels of SCFA in the large bowel of humans vary by ten-fold, and although dietary and environmental factors contribute to the variation, microbial differences appear to play a major part (McOrist et al. 2011). Our understanding of what represents a normal or healthy complement of gut microbial taxonomic units is poor, as is our understanding of the factors which can disturb the commensal microbial populations and potentially impact health. Studies examining the cause(s) and potential treatments of inflammatory bowel disease (IBD) have provided our greatest insights into this area.

Gut microbes play a central role in the causation of IBD. For example, it is not possible to induce ulcerative colitis (UC) in germ-free animals (Sellon et al. 1998). Also, antibiotics can improve symptoms of IBD (Khan et al. 2011), as can diversion of faeces away from the large bowel of individuals with UC by creation of a temporary ileostomy. Conversely, it is also possible to induce diversion colitis in individuals with an ileostomy (Harig et al. 1989), which suggests removal of healthy microbes from the colon also has a detrimental effect. It is also now well documented that the gut microbiota populations of individuals with IBD are significantly altered compared to community controls (Manichanh et al. 2006; De Cruz et al. 2012; Rajilic-Stojanovic et al. 2013), adding further support for a role in causation. A key change is a reduced diversity, largely a result of a decrease in bacteria of the gram positive *Firmicutes* phylum, especially *Clostridium* clusters IV and XIV, whereas some bacteria belonging to the *Proteobacteriaceae*, particularly *E. coli* and *Enterobacteriaceae*, are increased in relative abundance. Other pathogens such as *Fusobacterium* sp., *Peptostreptococcus* sp., *Helicobacter* sp., *Campylobacter* sp. and *C. difficile* also tend to increase in number (Rajilic-Stojanovic et al. 2013).

Much of the diversity of the human gut microbiota seems to be a consequence of the need to use their enzymatic and metabolic machinery to breakdown and utilise the wide range of polysaccharides available to us as food. In the large bowel, the cells of the gut mucosa are reliant on fermentation by the resident bacteria to generate SCFA in order to maintain their integrity and function, and polysaccharides derived from the diet are the principal substrates for this. Of particular interest then, are the observations suggesting an altered microbial capacity to generate SCFA occurs in IBD (Galecka et al. 2013; Machiels et al. 2014). For example, significant reductions can be seen for the key butyrate producers *E. rectale*, *Roseburia* spp. (Rajilic-Stojanovic et al. 2013) and *F. prausnitzii*, the latter of which also has some other anti-inflammatory properties (Sokol et al. 2008). *R. bromii* plays an important role in the initial breakdown of dietary resistant starch (Abell et al. 2008), a substrate that is readily converted to SCFA, and reduced abundance of this species is also associated with UC (Rajilic-Stojanovic et al. 2013). Added to this, there are now a number of studies which have shown that treatments which increase butyrate in the large bowel can improve colitis symptoms using experimental animal models (Komiyama et al. 2011; Vieira et al. 2012; Le Leu et al. 2013). In humans, rectal administration of butyrate has been shown to improve diversion colitis symptoms (Harig et al. 1989), helping to prevent atrophy of colonic tissues (Luceri et al. 2016).

Enteropathogenic microbes such as certain strains of *E. coli* and *Enterococcus* may contribute significantly to IBD and other gut conditions. In addition to the evidence of raised numbers in affected individuals there is experimental evidence to support such activities. To add weight to an important role for SCFA production, and the microbes which mediate this, in helping to protect against effects of these bacteria, the toxic effects of enterohaemorrhagic *E. coli* O157:H7 can be inhibited in vivo through increased production of acetate (Fukuda et al. 2011b). The acetate derived from bacteria of the genus *Bifidobacterium* was able to mediate this protection, suggesting potential for the use of appropriate strains as probiotics to treat conditions where *E. coli* or other similar pathogens contribute to disease. It is important to note that some strains of *E. coli* may benefit health e.g. by maintaining insulin-like growth factor signalling to muscle and preventing wasting associated with intestinal infections by certain pathogenic microbes (Schieber et al. 2015). Nevertheless, increasing numbers of gram negative bacteria such as *E. coli* can increase the risk of inflammation as components of their cell well, lipopolysaccharide, promote inflammatory responses in tissues.

Despite numerous microbial species having been linked to IBD, it has not been possible to consistently pinpoint any particular microbes as a cause. It is quite likely that many gut microbes have the potential to trigger IBD under the right conditions, especially those where host defences, particular the mucus gut barrier, are weakened (Swidsinski et al. 2009). A large proportion of host immune defences reside within the gut to deal with the threat of luminal microbes invading the tissues. One of the first lines of defence is the mucus barrier which lines the gastrointestinal tract. Microbes are present within the outer layer of this mucus, and many use it as a nutritional substrate, but they are not normally present within the innermost layer. A breakdown of mucus barrier integrity is implicated in IBD, in part because the numbers of bacteria that contribute to large bowel mucus turnover have been shown to be altered in those with the disease, i.e. *A. muciniphila*, *Ruminococcus torques* and *Ruminococcus gnavus* (Png et al. 2010). Interestingly, some of these bacteria have also been found to be altered in the stool of children with autism spectrum disorder, who also often report a significant level of GI discomfort (Wang et al. 2011).

The maintenance of tight junctions between cells of the GI tract is also a critical line of defence by helping to prevent passage of unwanted microbes or molecules into host tissues. SCFA, but particularly butyrate, appear to play an important role in the maintenance of tight junctions (Fukuda et al. 2011b; Suzuki et al. 2008). A breakdown of this defence, a so-called 'leaky gut', may be a significant contributor to inflammation in IBD and is also linked to a growing number of other conditions (Michielan and D'Inca 2015). The recognition that a 'dysbiosis' of gut microbe populations is a likely contributor to IBD has recently led to the testing of a seemingly radical new treatment, faecal microbial transplantation (FMT), in which an individual receives stool donated by healthy individuals. It is hypothesised that the healthy complement of microbes will become established in the large bowel, promote re-establishment of tight junctions, a protective mucus lining, repopulate the mucus layer with commensal bacteria and thereby curtail microbe-mediated inflammation. Although relatively unproven for treatment of IBD and other conditions where microbes contribute to the etiology, FMT is now firmly established as a successful means of treating *C. difficile* infections (which often resist all other known treatments) through rectal administration and through ingestion of capsules containing stool (Borody and Khoruts 2012; Mattila et al. 2012).

Gut microbes and their products also appear to play a role in the aetiology of colorectal cancer (CRC). Individuals with IBD have an elevated risk of CRC, suggesting that microbe-associated inflammation can contribute to oncogenesis. In otherwise healthy individuals there is a significantly greater risk of CRC when diets high in red or processed meat and low in fibre, which are typical of western style diets, are consumed (Norat et al. 2005; Murphy et al. 2012). A range of mechanisms has been proposed to explain this. Dietary protein can escape digestion and undergo fermentation in the large bowel, generating toxic by-products which can include ammonia, phenols, cresols, amines, and hydrogen sulphide (Windey et al. 2012). N-nitroso compounds (NOC) are alkylating agents which can also be produced from meat proteins by the action of gut microbes and have also been suggested as contributing to CRC through their ability to form DNA adducts in human colonic cells (Bingham et al. 1996). There is growing experimental evidence in animals and humans that consuming high levels of dietary protein, particularly as red meat, increases the level of DNA damage in the colon (which is regarded as a required step in oncogenesis) (Toden et al. 2007; Winter et al. 2011; Conlon et al. 2012; O'Callaghan et al. 2012; Le Leu et al. 2015). These studies also show that addition of fermentable dietary fibre to the diet in the form of resistant starch protects against this damage through mechanisms that appear to be at least partly mediated by the actions of SCFA produced by the microbiota. Gut microbes, including members of *Lactobacillus*, *Bifidobacterium* and *Bacteroides* genera, mediate secondary bile acid formation and turnover (Nicholson et al. 2012). These acids, particularly deoxycholic acid, are potentially carcinogenic and there is evidence that their levels are higher when SCFA levels are lower (Ou et al. 2012). Numerous species of gut microbes have been implicated in CRC, and these may act through a variety of mechanisms, such as the production of compounds like toxins or forms of reactive oxygen which damage tissues, or through exclusion of beneficial bacteria (Yu and Fang 2015). Some of the bacteria implicated include *Fusobacterium nucleatum*, invasive bacteria found to be strongly associated with the neoplastic tissues of CRC patients (Castellarin et al. 2012), *Bacteroides fragilis*, *Enterococcus faecalis*, *Helicobacter pylori* and *Strepococcus bovis* (Yu and Fang 2015).

6 Short Chain Fatty Acids, Inflammation and Allergy

The SCFA products of dietary fibre fermentation by the gut microbiota are pleiotropic metabolites that elicit their biological effects via both intra- and extra-cellular mechanisms. While un-ionized SCFA can cross cellular membranes by simple diffusion, with pKa values of around 4.8, the majority of these acids exist in the ionised state at normal colonic pH. Therefore, most uptake of SCFA by the tissues of the colonic mucosa is facilitated by membrane bound transporter proteins. These include the low affinity/ high capacity transporters SLC16A1 (proton-coupled monocarboxylate transporter-1, MCT-1), SLC16A3 (MCT-3) and SLC16A7 (MCT-16) which are important when SCFA concentrations are relatively high and high affinity/ low capacity sodium-coupled monocarboxylate transporter-1 (SMCT-1; or SLC5A8) which are important when digesta and tissue SCFA concentrations are low. Low affinity receptors such as MCT-1 are distributed throughout the gut with expression highest in the large intestine of most mammals and reside on the apical (luminal) surface of the cell. In contrast, the high affinity transporter SMCT-1, which also promotes water absorption, is expressed on the apical epithelial membrane of the distal ileum and large intestine of mice, with expression highest in the colo-rectum (Iwanaga and Kishimoto 2015). SMCT-1 is also found on cells within the lamina propria (the tissue immediately below the mucosal layers) where SCFA, particularly butyrate, influences immune cells and enteric neurons (Vadivel et al. 2014). The relative affinity of SMCT-1 for its different SCFA substrates is: butyrate > propionate > lactate > acetate, which may contribute to its proposed tumour suppressor role (Ganapathy et al. 2005).

Past research has focused on the ability of SCFA to modulate biological responses by directly inhibiting histone deacetylases (HDACs) which results in increased histone acetylation and altered gene expression. The potent HDAC inhibitory effects of butyrate on the expression of genes that regulate proteins involved in cellular apoptosis, cell cycle regulation and DNA repair that may protect from colorectal oncogenesis (Fung et al. 2012) has received considerable attention, although HDAC inhibition may be SCFA and tissue specific. Acetate has recently been found to increase brain histone acetylation-state in rodents (Soliman and Rosenberger 2011). The effects on the immune response of SCFA on HDAC inhibition are largely anti-inflammatory (Tan et al. 2014) affecting not only cells of the innate immune system (Lin et al. 2015) but also increasing *Foxp3* gene expression and the numbers and immunosuppressive function of T regulatory cells in mice (Tao et al. 2007).

SCFA also modulate biological responses by activating G-protein-coupled receptors (GPCRs), mainly GPR41 (FFAR3), GPR43 (FFAR2) and GPR109A, which are expressed on the apical membrane of colonic epithelial and enteroendocrine cells (see reviews by Vadivel et al. 2014; Tan et al. 2014). GPR41 and GPR43 are activated by all three SCFA with propionate the most potent agonist for both GPR41 and GPR43, and with acetate more selective for GPR43 (Le Poul et al. 2003a). GPR109A is only activated by butyrate (Thangaraju et al. 2009) and is expressed in adipocytes, intestinal epithelial cells and some immune cells (see review by Singh et al. 2014). GPR43 was first described in neutrophils (Le Poul et al. 2003a) and has since been described in other immune cells, the gastrointestinal tract including endocrine L-cells of the ileum and colon producing intestinal peptide YY (PYY) and glucagon-like peptide 1 (GLP-1) (Tolhurst et al. 2012). GPR41 is more broadly expressed in a number of tissues (Le Poul et al. 2003b).

The gut intestinal barrier consists of a layer of epithelial cells and products including mucus, immunoglobulins and other antimicrobial agents and microbes that interact to provide a selective barrier to protect the host. The barrier enables the uptake of water, electrolytes, nutrients and other molecules while preventing the entry of antigens and microorganisms (Camilleri et al. 2012). The epithelium forms the main physical barrier between the gut lumen and the mucosal tissues with absorption occurring either transcellularly by active transport or passive diffusion, or through the paracellular spaces between the cells by passive movement. The selective paracellular barrier is maintained by three adhesive protein complexes: desmosomes, adherens junctions, and tight junctions (Groschwitz and Hogan 2009). The tight junctions seal the intercellular space and involve claudins and occludins that form the paracellular pore, and scaffold and other cytoplasmic proteins including zonula occludens (ZO-1, 2 and 3).

Intestinal barrier dysfunction enabling gut translocation of bacteria causing inflammation and autoimmune disease has also been suggested as a causative factor in a range of diseases including irritable bowel syndrome (Piche 2014; Vivinus-Nebot et al. 2014), inflammatory bowel disease (Michielan and D'Inca 2015; Ploger et al. 2012) (IBD), food allergies (Perrier and Corthesy 2011), celiac disease, asthma (Hijazi et al. 2004) and types 1 (T1D) and 2 (T2D) diabetes (de Kort et al. 2011). Butyrate, but not acetate, decreased bacterial translocation in cell models (Lewis et al. 2010) and enhanced intestinal barrier function in vitro by modifying the expression of the tight junction protein Claudin-1 (Wang et al. 2012). Elsewhere, acetate has been shown to enhance gut barrier function and protect mice treated with *E. coli* O157:H7 from translocation of Shiga toxin from the gut lumen (Fukuda et al. 2011a).

Impaired tight junction function has been found in patients with Crohn's disease (Zeissig et al. 2007) (CD) and ulcerative colitis (Heller et al. 2005) (UC) and first degree relatives of patients with Crohn's disease have increased intestinal permeability (Peeters et al. 1997). Paracellular permeability is increased in patients with quiescent IBD and IBS-like symptoms associated with persistent subclinical intestinal inflammation (Vivinus-Nebot et al. 2014). Butyrate-producing bacteria are reduced in the faeces of CD (Takahashi et al. 2016) and UC (Machiels et al. 2013) patients compared to healthy controls. The physical and chemical protective mucus layer is thinner and more variable in thickness in UC disease patients (Pullan et al. 1994); in rodents, diets containing resistant starch delivering high concentrations of butyrate to the large bowel increased mucus thickness (Toden et al. 2014). However, the use of butyrate enemas to control or maintain relapse in IBD patients has had mixed results (Breuer et al. 1997; Scheppach 1996; Steinhart et al. 1996; Hamer et al. 2010) which may be due to difficulty maintaining prolonged mucosal contact. Butyrate metabolism is impaired in the inflamed mucosa of IBD patients, which may be due to a reduction in butyrate uptake due to a downregulation of the MCT-1 butyrate transporter in IBD (Thibault et al. 2007), leading to the suggestion that butyrate deficiency is a causative factor for the inflammation (Thibault et al. 2010). Butyrate delivered to the colon by butyrylated starch ameliorated colitis in an adoptive transfer mouse model whereas starches delivering acetate and propionate had no protective effect (Furusawa et al. 2013).

Increased gastric and small intestinal permeability precedes the development of diabetes in the BB rat (Meddings et al. 1999) and increased permeability resulting in autoimmune destruction of the pancreatic beta cells may play a role in the development of diabetes. Furthermore, glucagon-like peptide-2 (GLP-2), an enteroendocrine peptide involved with satiety control released in response to the ingestion of a prebiotic, improves barrier function and reduces LPS concentrations in mice (Cani et al. 2009). Dietary resistant starch (RS) provides substrate for colonic fermentation and the production of SCFA. These diets are associated with improved bowel health, reduced abdominal fat and improved insulin sensitivity and increased serum GLP-1 is likely to play a role in mediating these effects (Keenan et al. 2015). Diets providing high levels of colonic SCFA may have potential to improve metabolic control in type 2 diabetes (Puddu et al. 2014).

SCFA have been shown to affect the immune system through multiple mechanisms that contribute to colonic health and homeostasis (see reviews by (Tan et al. 2014; Fung et al. 2014; Vinolo et al. 2011b). These mechanisms include modulating chemotaxis and the activity of macrophages and neutrophils (Maslowski et al. 2009; Vinolo et al. 2011a); the production of reactive oxygen species by neutrophils (Maslowski et al. 2009); suppressing pro-inflammatory and inhibiting anti-inflammatory cytokine production by macrophages and neutrophils (TNF-α, IL-2, IL-6 and IL-10), the interaction of neutrophils with endothelial epithelium (Vinolo et al. 2009); and modulating NF-KB activity by inhibition of HDAC activity (Inan et al. 2000).

SCFA also induce T_{reg} cells from naive CD4+ T-cell precursors in the colonic lamina propria by inhibiting histone deacetylases and facilitating *Foxp3* expression (Furusawa et al. 2013, Arpaia et al. 2013). SCFA have been shown to expand the existing T_{reg} cell pool via a GPR43-dependent mechanism (Smith et al. 2013). T_{reg} cells have an important role moderating allergic and inflammatory responses (Sakaguchi et al. 2008) and may also be induced through the production of TGFβ1 by *Clostridia* species from clusters IV, XIVa, and XVIII (Atarashi et al. 2013). In addition, butyrate promotes anti-inflammatory properties in macrophages and dendritic cells in the colonic lamina propria and enables them to induce differentiation of IL-10-producing T_{reg} cells via GPR109A signaling (Singh et al. 2014).

The route of administration may determine the effects of SCFA. When individual SCFA are administered by drinking water they are largely absorbed in the small intestine and acetate and propionate delivered by this route are more effective at accumulating colonic T_{reg}s than butyrate, suggesting acid absorbed in the small intestine may be important for migration of T_{reg} cells into the colon from lymphoid tissues (Smith et al. 2013). Butyrate delivered in the drinking water increased peripheral but not thymic or colonic Treg cells, whereas butyrate delivered by enema increased Treg numbers in the colonic lamina propria (Arpaia et al. 2013). Likewise, butyrate delivered luminally to the colon via butyrylated starch induced colonic Treg differentiation but propionate delivered from propionylated starch was less effective and acetate from acetylated starch had no effect (Furusawa et al. 2013).

The anti-inflammatory effects of acetate administered in drinking water have been demonstrated in animal models of colitis, arthritis and asthma. Acetate ameliorated disease symptoms implying a major role for the acetate-GPR43 axis (Maslowski et al. 2009). Rodent studies suggest a maternal diet high in acetate may reduce the incidence of asthma in offspring by influencing the expression of certain genes in the fetal lung (Thorburn et al. 2015).

Early studies showed that SCFA are rapidly taken up in the large bowel with the absorption of sodium and chloride (Rajendran and Binder 1994) and with the secretion of bicarbonate (Fleming et al. 1991). SCFA have since been shown to induce the activity of sodium-hydrogen exchange transporters (NHE) in intestinal cells that promote the absorption of fluid and Na-H, SCFA-HCO_3^- and Cl-SCFA exchanges (see review Binder 2010). Multiple isoforms of NHEs have been identified in intestinal epithelial cells but the activity of NHE3 (Musch et al. 2001) is increased by SCFA and with NHE2 is found in the small and large intestine of humans (Hoogerwerf et al. 1996). Butyrate also stimulates intestinal NHE8 expression enhancing sodium uptake in vitro (Xu et al. 2015). The transport protein (s) responsible for SCFA-HCO_3^- exchange have not been described (Binder 2010) although a low-affinity, high capacity butyrate-HCO_3^- process and a high affinity, low-capacity proton-monocarboxylate co-transporter have been described (Lecona et al. 2008).

Unlike several other intestinal transport processes, SCFA-stimulated absorption of fluid and electrolytes is not inhibited by mucosal cAMP (Ramakrishna et al. 1990). Mucosal cAMP is produced in response to enterotoxins secreted by many bacterial pathogens responsible for acute infectious diarrhoea, including *Vibrio cholera* (Kimberg et al. 1971), *S. typhimurium, Shigella dysenteriae* type 1 toxin, *Campylobacter jejuni* (see review by Binder 2009). Butyrate has been found to restore the activities of NHE2 and 3 isoforms inhibited by cAMP in vitro (Subramanya et al. 2007) and also decrease by 40 % the cholera-toxin induced increase in cAMP (Subramanya et al. 2007). Hence, delivering SCFA to the colon has potential to reduce fluid and electrolyte loss and improve the clinical response to oral rehydration solution (ORS) in patients with acute infectious diarrhoea.

SCFA can be produced in the colon by the fermentation of ingested RS, and RS added to ORS promotes fluid and electrolyte uptake in a perfused-gut rat model of cholera-toxin (CT) diarrhoea (Subramanya et al. 2006). It also reduces fluid loss and recovery time in human adults (Ramakrishna et al. 2000) and duration of diarrhoea in children (Raghupathy et al. 2006). Acylated starches are largely resistant to small intestinal digestion and deliver specific SCFA rapidly to the large bowel (Clarke et al. 2008, 2011c). Acetylated starch was more effective at promoting fluid and electrolyte uptake than RS or acylated starch delivering propionate or butyrate in the perfused-gut rat CT model (Clarke et al. 2011a). This may be the result of more rapid and complete release of bound acetate resulting from the more disrupted starch structure compared to starches acylated with propionate and butyrate. The acetylated starch shortened the duration of diarrhoea in patients with acute infectious diarrhoea (Pal et al. 2013) but not faecal volume.

7 Future Prospects

Modern techniques in metagenomics analysis continue to dramatically enhance our understanding of how microbial taxonomy and the genetic potential of the human microbiota change in health and disease and how they respond to changing environmental conditions such as diet. However, this is only part of the equation. Understanding how the function of the microbiota changes, how this impacts on the health and wellbeing of the host and what levers can be pulled to efficiently shape the microbiome and its function for improved health will be important areas for future research.

Dietary fibre provides a glimpse into the multifaceted impacts that food and xenobiotics can have on the microbiota and the host, and highlights the importance of small molecule metabolites as mediators of change. As discussed above, provision of a fermentable substrate alone provides a growth advantage to bacteria in the gut ecosystem capable of metabolizing these substrates. Selective pressure within the gut microbial population is further modified by changes in the chemical environment caused by the accumulation of the metabolic end products of fermentation. Dropping digesta pH resulting from the accumulation of SCFA further advantages some microbes over others while the SCFA themselves, pleiotropic bioactive agents in their own right, provide vehicles to the host for energy salvage, immunological signaling, promotion of gut homeostasis and repair, motility modification, cellular signaling (both via GPCRs and histone deacetylase inhibition) and satiety promotion. But fibre is just one dietary substrate capable of modifying the gut microbial structure and function. Other macronutrients entering the colon can also promote the growth of other bacterial populations leading to the production of other metabolites with different outcomes for other microbes and the host. Yet other dietary molecules or phytonutrients may alternatively act by modifying the efficiency of other microbial processes e.g. the inhibition of RS fermentation efficiency by cutins and waxes (Edwards et al. 2012).

The human metabolome is very complex. Lipids alone have been estimated to have the potential to generate 9000–10,000 different molecular species (Yetukuri et al. 2008). With the gut microbiome carrying around 40 times the number of protein coding genes as the human genome, it is likely that the metabolome produced by the gut microflora will be substantially more complex. While SCFA are undoubtedly interesting and important metabolites, there are many other gut metabolites of microbe, host or dietary origin, both known and yet to be identified, that are capable of eliciting potent physiological effects. These include quorum sensing and virulence factors that regulate the composition of the surrounding microbiota as well as metabolites modulating host functions including the gut immune system (Nicholson et al. 2005, 2012).

The challenge will be to bring together our growing understanding of factors affecting the gut microbiome and the extent and variation in the gut metabolome with knowledge of the associated impacts on host physiology to develop predictive personalized approaches to health management and chronic disease prevention. But there remains much to be done. Expanding our knowledge of the gut microbiome and its metabolic potential will be important. Only some 10 % of the currently sequenced open reading frames in the microbiome are recognized as encoding proteins with an assignable or predicted structural or enzymatic function. Further our understanding of the metabolome is in its infancy while attribution of the physiological impacts of metabolites continues to be assessed on a candidate by candidate basis (Martin et al. 2007). While there is much important work to be done in the expansion of these individual catalogues (genes, their function; metabolites and their impact on physiology) there is now a real need to bring these areas closer together. With the increasing use of high resolution spectroscopy methods to analyse metabolite profiles of microbial origin, opportunities now exist to statistically integrate large and complex metabolic and metagenomic data sets, using informatics to reveal pathways and putative functional relationships previously inaccessible. However, validation of these proposed relationships will be crucial. In this regard, the development of improved in vitro fermentation systems that more accurately represent the processes occurring in the gut digesta and the availability of gut organoid culture systems that accurately recapitulate biological responses of the gut mucosa to bioactive metabolites are providing an exciting prospect of a medium throughput screening of these in silico predictions en route to more classical animal and human substantiation of the strongest candidates. While we are only in the early stages of this new, data-driven revolution, if it lives up to its promise then the path to a more detailed understanding of the gut, its metabolites and their impact on human health becomes clearer and the promise of translating this knowledge to help inform the personalization of health care, a tangible and exciting prospect.

References

Abell GC, Cooke CM, Bennett CN, Conlon MA, Mcorist AL (2008) Phylotypes related to Ruminococcus bromii are abundant in the large bowel of humans and increase in response to a diet high in resistant starch. FEMS Microbiol Ecol 66:505–515

Abell GCJ, Christophersen CT, McOrist AL, Clarke JM (2011) Dietary resistant and butyrylated starches have different effects on the faecal bacterial flora of azoxymethane-treated rats. Br J Nutr 105:1480–1485

Agbenorhevi JK, Kontogiorgos V, Kirby AR, Morris VJ, Tosh SM (2011) Rheological and microstructural investigation of oat beta-glucan isolates varying in molecular weight. Int J Biol Macromol 49:369–377

Australian Institute of Health and Welfare & Australasian Association of Cancer Registries (AIHW) (2012) Cancer in Australia: an overview, 2012 Cancer Series No. 74. Cat. No. CAN 70

Alam MZ, Alam Q, Kamal MA, Abuzenadah AM, Haque A (2014) A possible link of gut microbiota alteration in type 2 diabetes and alzheimer's disease pathogenicity: an update. CNS Neurol Disord-Drug Targets 13:383–390

Anderson IH, Levine AS, Levitt MD (1981) Incomplete absorption of the carbohydrate in all-purpose wheat-flour. N Engl J Med 304:891–892

Arpaia N, Campbell C, Fan X, Dikiy S, van der Veeken J, Deroos P, Liu H, Cross JR, Pfeffer K, Coffer PJ, Rudensky AY (2013) Metabolites produced by commensal bacteria promote peripheral regulatory T-cell generation. Nature 504:451–455

Arrigoni E (2001) Enzymatic modification of dietary fiber sources. Marcel Dekker Inc, New York

Atarashi K, Tanoue T, Oshima K, Suda W, Nagano Y, Nishikawa H, Fukuda S, Saito T, Narushima S, Hase K, Kim S, Fritz JV, Wilmes P, Ueha S, Matsushima K, Ohno H, Olle B, Sakaguchi S, Taniguchi T, Morita H, Hattori M, Honda K (2013) Treg induction by a rationally selected mixture of Clostridia strains from the human microbiota. Nature 500:232–236

Bach Knudsen KE (2015) Microbial degradation of whole-grain complex carbohydrates and impact on short-chain fatty acids and health. Adv Nutr (Bethesda, Md.) 6:206–213

Backhed F, Roswall J, Peng YQ, Feng Q, Jia HJ, Kovatcheva-Datchary P, Li Y, Xia Y, Xie HL, Zhong HZ, Khan MT, Zhang JF, Li JH, Xiao L, Al-Aama J, Zhang DY, Lee YS, Kotowska D, Colding C, Tremaroli V, Yin Y, Bergman S, Xu X, Madsen L, Kristiansen K, Dahlgren J, Jun W (2015) Dynamics and stabilization of the human gut microbiome during the first year of life. Cell Host Microbe 17:690–703

Backman RV (2009) The effects of plantain non-starch polysaccharide upon the gut bacteria. Ph. D., University of Aberdeen, UK

Baghurst PA, Baghurst KI, Record SJ (1996) Dietary fibre, non-starch polysaccharides and resistant starch—a review. Food Aust 48:S3–S35

Barrett E, Ross RP, O'Toole PW, Fitzgerald GF, Stanton C (2012) γ-Aminobutyric acid production by culturable bacteria from the human intestine. J Appl Microbiol 113:411–417

Binder HJ (2009) Mechanisms of diarrhea in inflammatory bowel diseases. Ann NY Acad Sci 1165:285–293

Binder HJ (2010) Role of colonic short-chain fatty acid transport in diarrhea. Annu Rev Physiol 72:297–313

Bingham SA, Pignatelli B, Pollock JRA, Ellul A, Malaveille C, Gross G, Runswick S, Cummings JH, O'Neill IK (1996) Does increased endogenous formation of N-nitroso compounds in the human colon explain the association between red meat and colon cancer? Carcinogenesis 17:515–523

Bird AR (2007) Dietary fibre, health and chronic disease. Diab Manage J 12

Bird AR, Brown IL, Topping DL (2000) Starches, resistant starches, the gut microflora and human health. Curr Issues Intest Microbiol 1:25–37

Bird AR, Conlon MA, Christophersen CT, Topping DL (2010) Resistant starch, large bowel fermentation and a broader perspective of prebiotics and probiotics. Benef Microbes 1:423–431

Bird AR, Topping DL (2001) Resistant starches, fermentation, and large bowel health. Marcel Dekker Inc, New York

Bird AR, Usher S, May B, Topping DL, Morell MK (2012) Resistant starch measurement, intakes, and dietary targets. CRC Press, Taylor & Francis Group, Boca Raton

Borody TJ, Khoruts A (2012) Fecal microbiota transplantation and emerging applications. Nat Rev Gastroenterol Hepatol 9:88–96

Breuer RI, Soergel KH, Lashner BA, Christ ML, Hanauer SB, Vanagunas A, Harig JM, Keshavarzian A, Robinson M, Sellin JH, Weinberg D, Vidican DE, Flemal KL, Rademaker AW (1997) Short chain fatty acid rectal irrigation for left-sided ulcerative colitis: a randomised, placebo controlled trial. Gut 40:485–491

Broekaert WF, Courtin CM, Verbeke K, van de Wiele T, Verstraete W, Delcour JA (2011) Prebiotic and other health-related effects of cereal-derived arabinoxylans, arabinoxylan-oligosaccharides, and xylooligosaccharides. Crit Rev Food Sci Nutr 51:178–194

Brouns F, Kettlitz B, Arrigoni E (2002) Resistant starch and "the butyrate revolution". Trends Food Sci Technol 13:251–261

Brownlee I (2014) The impact of dietary fibre intake on the physiology and health of the stomach and upper gastrointestinal tract. Bioact Carbohydr Diet Fibre 4:155–169

Burkitt DP (1973) Some diseases characteristic of modern western civilization. Br Med J 1:274–278

Buttriss JL, Stokes CS (2008) Dietary fibre and health: an overview. Nutr Bull 33:186–200

Camilleri M, Madsen K, Spiller R, Greenwood-van Meerveld B, Verne GN (2012) Intestinal barrier function in health and gastrointestinal disease. Neurogastroenterol Motil 24:503–512

Cani PD, Possemiers S, van de Wiele T, Guiot Y, Everard A, Rottier O, Geurts L, Naslain D, Neyrinck A, Lambert DM, Muccioli GG, Delzenne NM (2009) Changes in gut microbiota control inflammation in obese mice through a mechanism involving GLP-2-driven improvement of gut permeability. Gut 58:1091–1103

Castellarin M, Warren RL, Freeman JD, Dreolini L, Krzywinski M, Strauss J, Barnes R, Watson P, Allen-Vercoe E, Moore RA, Holt RA (2012) Fusobacterium nucleatum infection is prevalent in human colorectal carcinoma. Genome Res 22:299–306

Cats A, Devries EGE, Mulder NH, Kleibeuker JH (1996) Regional differences of physiological functions and cancer susceptibility in the human large intestine. Int J Oncol 9:1055–1069

Cerutti A (2008) The regulation of IgA class switching. Nat Rev Immunol 8:421–434

Chung HC, Pamp SJ, Hill JA, Surana NK, Edelman SM, Troy EB, Reading NC, Villablanca EJ, Wang S, Mora JR, Umesaki Y, Mathis D, Benoist C, Relman DA, Kasper DL (2012) Gut immune maturation depends on colonization with a host-specific microbiota. Cell 149:1578–1593

Claesson MJ, Jeffery IB, Conde S, Power SE, O'Connor EM, Cusack S, Harris HMB, Coakley M, Lakshminarayanan B, O'Sullivan O, Fitzgerald GF, Deane J, O'Connor M, Harnedy N, O'Connor K, O'Mahony D, van Sinderen D, Wallace M, Brennan L, Stanton C, Marchesi JR, Fitzgerald AP, Shanahan F, Hill C, Ross RP, O'Toole PW (2012) Gut microbiota composition correlates with diet and health in the elderly. Nature 488:178

Clarke JM, Dharmalingam T, Srinivasan P, Young GP, Lockett TJ, Ramakrishna BS (2011a) Acetylated high amylose maize starch improves the efficacy of oral rehydration solution in a rat model of cholera. Gastroenterology 140:S134

Clarke JM, Topping DL, Bird AR, Young GP, Cobiac L (2008) Effects of high amylose maize starch and butyrylated high amylose maize starch on azoxymethane-induced intestinal cancer in rats. Carcinogenesis 29:2190–2194

Clarke JM, Topping DL, Christophersen CT, Bird AR, Lange K, Saunders I, Cobiac L (2011b) Butyrate esterified to starch is released in the human gastrointestinal tract. Am J Clin Nutr 94:1276–1283

Clarke JM, Topping DL, Christophersen CT, Bird AR, Lange K, Saunders I, Cobiac L (2011c) Butyrate esterified to starch is released in the human gastrointestinal tract. Am J Clin Nutr 94:1276–1283

Clemente JC, Ursell LK, Parfrey LW, Knight R (2012) The impact of the gut microbiota on human health: an integrative view. Cell 148:1258–1270

CODEX (2010) Joint FAO/WHO food standards programme. In: Alimentarius (ed) Secretariat of the Codex Alimentarius Commission: CODEX guidelines on nutrition labeling CAC/GL 2–1985 as last amended 2010

Collier CT, Hofacre CL, Payne AM, Anderson DB, Kaiser P, Mackie RI, Gaskins HR (2008) Coccidia-induced mucogenesis promotes the onset of necrotic enteritis by supporting Clostridium perfringens growth. Vet Immunol Immunopathol 122:104–115

Comino P, Collins H, Lahnstein J, Gidley MJ (2016) Effects of diverse food processing conditions on the structure and solubility of wheat, barley and rye endosperm dietary fibre. J Food Eng 169:228–237

Conlon MA, Bird AR (2015) The impact of diet and lifestyle on gut microbiota and human health. Nutrients 7:17–44

Conlon MA, Kerr CA, McSweeney CS, Dunne RA, Shaw JM, Kang S, Bird AR, Morell MK, Lockett TJ, Molloy PL, Regina A, Toden S, Clarke JM, Topping DL (2012) Resistant starches protect against colonic DNA damage and alter microbiota and gene expression in rats fed a Western diet. J Nutr 142:832–840

Costabile A, Kolida S, Klinder A, Gietl E, Bauerlein M, Frohberg C, Landschutze V, Gibson GR (2010) A double-blind, placebo-controlled, cross-over study to establish the bifidogenic effect of a very-long-chain inulin extracted from globe artichoke (*Cynara scolymus*) in healthy human subjects. Br J Nutr 104:1007–1017

Cummings JH, Bingham SA, Heaton KW, Eastwood MA (1992) Fecal weight, colon cancer risk, and dietary-intake of nonstarch polysaccharides (dietary fiber). Gastroenterology 103:1783–1789
Cummings JH, Stephen AM (2007) Carbohydrate terminology and classification. Eur J Clin Nutr 61:S5–S18
Dahl WJ, Stewart ML (2015) Position of the academy of nutrition and dietetics: health implications of dietary fiber. J Acad Nutr Diet 115:1861–1870
Davis LM, Martinez I, Walter J, Goin C, Hutkins RW (2011) Barcoded pyrosequencing reveals that consumption of galactooligosaccharides results in a highly specific bifidogenic response in humans. PLoS One 6:e25200
De Cruz P, Kang S, Wagner J, Buckley M, Sim WH, Prideaux L, Lockett T, McSweeney C, Morrison M, Kirkwood CD, Kamm MA (2015) Association between specific mucosa-associated microbiota in Crohn's disease at the time of resection and subsequent disease recurrence: a pilot study. J Gastroenterol Hepatol 30:268–278
de Cruz P, Prideaux L, Wagner J, Ng SC, McSweeney C, Kirkwood C, Morrison M, Kamm MA (2012) Characterization of the gastrointestinal microbiota in health and inflammatory bowel disease. Inflamm Bowel Dis 18:372–390
de Kort S, Keszthelyi D, Masclee AA (2011) Leaky gut and diabetes mellitus: what is the link? Obes Rev 12:449–458
de Menezes EW, Giuntini EB, Dan MCT, Sarda FAH, Lajolo FM (2013) Codex dietary fibre definition—justification for inclusion of carbohydrates from 3 to 9 degrees of polymerisation. Food Chem 140:581–585
Deng YY, Misselwitz B, Dai N, Fox M (2015) Lactose intolerance in adults: biological mechanism and dietary management. Nutrients 7:8020–8035
Djouzi A, Andiueux C (1997) Compared effects of three oligosaccharides on metabolism of intestinal microflora in rats inoculated with a human faecal flora. Br J Nutr 78:313–324
Dominguez-Bello MG, Costello EK, Contreras M, Magris M, Hidalgo G, Fierer N, Knight R (2010) Delivery mode shapes the acquisition and structure of the initial microbiota across multiple body habitats in newborns. Proc Natl Acad Sci USA 107:11971–11975
Eaton SB (2006) The ancestral human diet: what was it and should it be a paradigm for contemporary nutrition? Proc Nutr Soc 65:1–6
Edwards CA, Garcia AL, Combet E (2012) Colonic metabolism of bioactive molecules potential impact of dietary fiber. CRC Press, Taylor & Francis Group, Boca Raton
Edwards CA, Xie CR, Garcia AL (2015) Dietary fibre and health in children and adolescents. Proc Nutr Soc 74:292–302
Elsden SR, Hitchcock MWS, Marshall RA, Phillipson AT (1946) Volatile acid in the digesta of ruminants and other animals. J Exp Biol 22:191–202
Englyst H, Wiggins HS, Cummings JH (1982) Determination of the non-starch polysaccharides in plant foods by gas-liquid-chromatography of constituent sugars as alditol acetates. Analyst 107:307–318
Englyst K, Quigley M, Englyst H, Parmar B, Damant A, Elahi S, Lawrance P (2013) Evaluation of methods of analysis for dietary fibre using real foods and model foods. Food Chem 140:568–573
Englyst KN, Liu S, Englyst HN (2007) Nutritional characterization and measurement of dietary carbohydrates. Eur J Clin Nutr 61:S19–S39
Eswaran S, Muir J, Chey WD (2013) Fiber and functional gastrointestinal disorders. Am J Gastroenterol 108:718–727
Favier CF, Vaughan EE, de Vos WM, Akkermans ADL (2002) Molecular monitoring of succession of bacterial communities in human neonates. Appl Environ Microbiol 68:219–226
Fleming SE, Choi SY, Fitch MD (1991) Absorption of short-chain fatty acids from the rat cecum in vivo. J Nutr 121:1787–1797
Flint HJ, Scott KP, Louis P, Duncan SH (2012) The role of the gut microbiota in nutrition and health. Nat Rev Gastroenterol Hepatol 9:577–589

Frank DN, St Amand AL, Feldman RA, Boedeker EC, Harpaz N, Pace NR (2007) Molecular-phylogenetic characterization of microbial community imbalances in human inflammatory bowel diseases. Proc Natl Acad Sci USA 104:13780–13785

Fukuda S, Toh H, Hase K, Oshima K, Nakanishi Y, Yoshimura K, Tobe T, Clarke JM, Topping DL, Suzuki T, Taylor TD, Itoh K, Kikuchi J, Morita H, Hattori M, Ohno H (2011a) Bifidobacteria can protect from enteropathogenic infection through production of acetate. Nature 469:543–547

Fukuda S, Toh H, Hase K, Oshima K, Nakanishi Y, Yoshimura K, Tobe T, Clarke JM, Topping DL, Suzuki T, Taylor TD, Itoh K, Kikuchi J, Morita H, Hattori M, Ohno H (2011b) Bifidobacteria can protect from enteropathogenic infection through production of acetate. Nature 469:543–547

Fung KY, Cosgrove L, Clarke J, Lockett T, Morita T, Hase K, Topping DL (2014) Large bowel short chain fatty acid production, immune status and immune cell function: implications for chronic inflammatory disease. Nova Science Publishers, New York

Fung KY, Ooi CC, Lewanowitsch T, Tan S, Tan HT, Lim TK, Lin Q, Williams DB, Lockett TJ, Cosgrove LJ, Chung MC, Head RJ (2012) Identification of potential pathways involved in induction of apoptosis by butyrate and 4-benzoylbutyrate in HT29 colorectal cancer cells. J Proteome Res 11:6019–6029

Furusawa Y, Obata Y, Fukuda S, Endo TA, Nakato G, Takahashi D, Nakanishi Y, Uetake C, Kato K, Kato T, Takahashi M, Fukuda NN, Murakami S, Miyauchi E, Hino S, Atarashi K, Onawa S, Fujimura Y, Lockett T, Clarke JM, Topping DL, Tomita M, Hori S, Ohara O, Morita T, Koseki H, Kikuchi J, Honda K, Hase K, Ohno H (2013) Commensal microbe-derived butyrate induces the differentiation of colonic regulatory T cells. Nature 504:446–450

Galecka M, Szachta P, Bartnicka A, Lykowska-Szuber L, Eder P, Schwiertz A (2013) *Faecalibacterium prausnitzii* and Crohn's disease—is there any connection? Pol J Microbiol 62:91–95

Gallo RL, Hooper LV (2012) Epithelial antimicrobial defence of the skin and intestine. Nat Rev Immunol 12:503–516

Ganapathy V, Gopal E, Miyauchi S, Prasad PD (2005) Biological functions of SLC5A8, a candidate tumour suppressor. Biochem Soc Trans 33:237–240

Garrett WS (2015) Cancer and the microbiota. Science 348:80–86

Gevers D, Kugathasan S, Denson LA, Vazquez-Baeza Y, van Treuren W, Ren B, Schwager E, Knights D, Song SJ, Yassour M, Morgan XC, Kostic AD, Luo C, Gonzalez A, McDonald D, Haberman Y, Walters T, Baker S, Rosh J, Stephens M, Heyman M, Markowitz J, Baldassano R, Griffiths A, Sylvester F, Mack D, Kim S, Crandall W, Hyams J, Huttenhower C, Knight R, Xavier RJ (2014) The treatment-naive microbiome in new-onset Crohn's disease. Cell Host Microbe 15:382–392

Gibson GR, Probert HM, Van Loo J, Rastall RA, Roberfroid MB (2004) Dietary modulation of the human colonic microbiota: updating the concept of prebiotics. Nutr Res Rev 17:259–275

Gibson GR, Roberfroid MB (1995) Dietary modulation of the human colonic microbiota—introducing the concept of prebiotics. J Nutr 125:1401–1412

Goldman SM, Kamel F, Ross GW, Jewell SA, Marras C, Hoppin JA, Umbach DM, Bhudhikanok GS, Meng C, Korell M, Comyns K, Hauser RA, Jankovic J, Factor SA, Bressman S, Lyons KE, Sandler DP, Langston JW, Tanner CM (2014) Peptidoglycan recognition protein genes and risk of Parkinson's disease. Mov Disord 29:1171–1180

Grembecka M (2015) Sugar alcohols-their role in the modern world of sweeteners: a review. Eur Food Res Technol 241:1–14

Groschwitz KR, Hogan SP (2009) Intestinal barrier function: molecular regulation and disease pathogenesis. J Allergy Clin Immunol 124:3–20 (quiz 21-2)

Group NHW, Peterson J, Garges S, Giovanni M, McInnes P, Wang L, Schloss JA, Bonazzi V, McEwen JE, Wetterstrand KA, Deal C, Baker CC, di Francesco V, Howcroft TK, KARP RW, Lunsford RD, Wellington CR, Belachew T, Wright M, Giblin C, David H, Mills M, Salomon R, Mullins C, Akolkar B, Begg L, Davis C, Grandison L, Humble M, Khalsa J,

Little AR, Peavy H, Pontzer C, Portnoy M, Sayre MH, Starke-Reed P, Zakhari S, Read J, Watson B, Guyer M (2009) The NIH human microbiome project. Genome Res 19:2317–2323

Haenen D, Zhang J, Souza da Silva C, Bosch G, van der Meer IM, van Arkel J, van den Borne JJ, Perez Gutierrez O, Smidt H, Kemp B, Muller M, Hooiveld GJ (2013) A diet high in resistant starch modulates microbiota composition, SCFA concentrations, and gene expression in pig intestine. J Nutr 143:274–283

Halmos EP, Christophersen CT, Bird AR, Shepherd SJ, Gibson PR, Muir JG (2015) Diets that differ in their FODMAP content alter the colonic luminal microenvironment. Gut 64:93–100

Hamer HM, Jonkers DMAE, Vanhoutvin SALW, Troost FJ, Rijkers G, de Bruine A, Bast A, Venema K, Brummer RJM (2010) Effect of butyrate enemas on inflammation and antioxidant status in the colonic mucosa of patients with ulcerative colitis in remission. Clin Nutr 29: 738–744

Harig JM, Soergel KH, Komorowski RA, Wood CM (1989) Treatment of diversion colitis with short-chain fatty-acid irrigation. N Engl J Med 320:23–28

Haro C, Garcia-Carpintero S, Alcala-Diaz JF, Gomez-Delgado F, Delgado-Lista J, Perez-Martinez P, Zuniga OAR, Quintana-Navarro GM, Landa BB, Clemente JC, Lopez-Miranda J, Camargo A, Perez-Jimeanez F (2016) The gut microbial community in metabolic syndrome patients is modified by diet. J Nutr Biochem 27:27–31

Heller F, Florian P, Bojarski C, Richter J, Christ M, Hillenbrand B, Mankertz J, Gitter AH, Burgel N, Fromm M, Zeitz M, Fuss I, Strober W, Schulzke JD (2005) Interleukin-13 is the key effector Th2 cytokine in ulcerative colitis that affects epithelial tight junctions, apoptosis, and cell restitution. Gastroenterology 129:550–564

Heneker T, Edwards CA, Preston T (1998) Digestion and colonic fermentation of maize porridge using a $13CO_2$ breath test. South African Congress of Gastroenterology

Hijazi Z, Molla AM, Al-Habashi H, Muawad WM, Molla AM, Sharma PN (2004) Intestinal permeability is increased in bronchial asthma. Arch Dis Child 89:227–229

Hoogerwerf WA, Tsao SC, Devuyst O, Levine SA, Yun CH, Yip JW, Cohen ME, Wilson PD, Lazenby AJ, Tse CM, Donowitz M (1996) NHE2 and NHE3 are human and rabbit intestinal brush-border proteins. Am J Physiol 270:G29–G41

Hou JK, Abraham B, El-Serag H (2011) Dietary intake and risk of developing inflammatory bowel disease: a systematic review of the literature. Am J Gastroenterol 106:563–573

Huttenhower C, Gevers D, Knight R, Abubucker S, Badger JH, Chinwalla AT, Creasy HH, Earl AM, Fitzgerald MG, Fulton RS, Giglio MG, Hallsworth-Pepin K, Lobos EA, Madupu R, Magrini V, Martin JC, Mitreva M, Muzny DM, Sodergren EJ, Versalovic J, Wollam AM, Worley KC, Wortman JR, Young SK, Zeng QD, Aagaard KM, Abolude OO, Allen-Vercoe E, Alm EJ, Alvarado L, Andersen GL, Anderson S, Appelbaum E, Arachchi HM, Armitage G, Arze CA, Ayvaz T, Baker CC, Begg L, Belachew T, Bhonagiri V, Bihan M, Blaser MJ, Bloom T, Bonazzi V, Brooks JP, Buck GA, Buhay CJ, Busam DA, Campbell JL, Canon SR, Cantarel BL, Chain PSG, Chen IMA, Chen L, Chhibba S, Chu K, Ciulla DM, Clemente JC, Clifton SW, Conlan S, Crabtree J, Cutting MA, Davidovics NJ, Davis CC, Desantis TZ, Deal C, Delehaunty KD, Dewhirst FE, Deych E, Ding Y, Dooling DJ, Dugan SP, Dunne WM, Durkin AS, Edgar RC, Erlich RL, Farmer CN, Farrell RM, Faust K, Feldgarden M, Felix VM, Fisher S, Fodor AA, Forney LJ, Foster L, di Francesco V, Friedman J, Friedrich DC, Fronick CC, Fulton LL, Gao HY, Garcia N, Giannoukos G, Giblin C, Giovanni MY, Goldberg JM, Goll J, Gonzalez A, Griggs A et al (2012) Structure, function and diversity of the healthy human microbiome. Nature 486:207–214

Inan MS, Rasoulpour RJ, Yin L, Hubbard AK, Rosenberg DW, Giardina C (2000) The luminal short-chain fatty acid butyrate modulates NF-kappaB activity in a human colonic epithelial cell line. Gastroenterology 118:724–734

Inoue K, Shirai T, Ochiai H, Kasao M, Hayakawa K, Kimura M, Sansawa H (2003) Blood-pressure-lowering effect of a novel fermented milk containing gamma-aminobutyric acid (GABA) in mild hypertensives. Eur J Clin Nutr 57:490–495

Iwanaga T, Kishimoto A (2015) Cellular distributions of monocarboxylate transporters: a review. Biomed Res 36:279–301

James SL, Christophersen CT, Bird AR, Conlon MA, Rosella O, Gibson PR, Muir JG (2015) Abnormal fibre usage in UC in remission. Gut 64:562–570

Jenkins CLD, Lewis D, Bushell R, Belobrajdic DP, Bird AR (2011) Chain length of cereal fructans isolated from wheat stem and barley grain modulates in vitro fermentation. J Cereal Sci 53:188–191

Jimenez E, Marin ML, Martin R, Odriozola JM, Olivares M, Xaus J, Fernandez L, Rodriguez JM (2008) Is meconium from healthy newborns actually sterile? Res Microbiol 159:187–193

Keenan MJ, Zhou J, Hegsted M, Pelkman C, Durham HA, Coulon DB, Martin RJ (2015) Role of resistant starch in improving gut health, adiposity, and insulin resistance. Adv Nutr 6:198–205

Khan KJ, Ullman TA, Ford AC, Abreu MT, Abadir A, Marshall JK, Talley NJ, Moayyedi P (2011) Antibiotic therapy in inflammatory bowel disease: a systematic review and meta-analysis. Am J Gastroenterol 106:661–673

Kimberg DV, Field M, Johnson J, Henderson A, Gershon E (1971) Stimulation of intestinal mucosal adenyl cyclase by cholera enterotoxin and prostaglandins. J Clin Invest 50:1218–1230

Kleessen B, Hartmann L, Blaut M (2007a) Oligofructose and long-chain inulin: influence on the gut microbial ecology of rats associated with a human faecal flora. Br J Nutr 86:291

Kleessen B, Schwarz S, Boehm A, Fuhrmann H, Richter A, Henle T, Krueger M (2007b) Jerusalem artichoke and chicory inulin in bakery products affect faecal microbiota of healthy volunteers. Br J Nutr 98:540–549

Kleessen B, Stoof G, Proll J, Schmiedl D, Noack J, Blaut M (1997) Feeding resistant starch affects fecal and cecal microflora and short-chain fatty acids in rats. J Anim Sci 75:2453–2462

Koenig JE, Spor A, Scalfone N, Fricker AD, Stombaugh J, Knight R, Angenent LT, Ley RE (2011) Succession of microbial consortia in the developing infant gut microbiome. Proc Natl Acad Sci USA 108:4578–4585

Komiyama Y, Andoh A, Fujiwara D, Ohmae H, Araki Y, Fujiyama Y, Mitsuyama K, Kanauchi O (2011) New prebiotics from rice bran ameliorate inflammation in murine colitis models through the modulation of intestinal homeostasis and the mucosal immune system. Scand J Gastroenterol 46:40–52

Lanza E, Yu BB, Murphy G, Albert PS, Caan B, Marshall JR, Lance P, Paskett ED, Weissfeld J, Slattery M, Burt R, Iber F, Shike M, Kikendall JW, Brewer BK, Schatzkin A, Polyp Prevention Trial Study Group (2007) The polyp prevention trial-continued follow-up study: no effect of a low fat, high-fiber, high-fruit, and—vegetable diet on adenoma recurrence eight years after randomization. Cancer Epidemiol Biomark Prev 16:1745–1752

le Blay G, Michel C, Blottiere HM, Cherbut C (1999) Enhancement of butyrate production in the rat caecocolonic tract by long-term ingestion of resistant potato starch. Br J Nutr 82:419–426

le Chatelier E, Nielsen T, Qin JJ, Prifti E, Hildebrand F, Falony G, Almeida M, Arumugam M, Batto JM, Kennedy S, Leonard P, Li JH, Burgdorf K, Grarup N, Jorgensen T, Brandslund I, Nielsen HB, Juncker AS, Bertalan M, Levenez F, Pons N, Rasmussen S, Sunagawa S, Tap J, Tims S, Zoetendal EG, Brunak S, Clement K, Dore J, Kleerebezem M, Kristiansen K, Renault P, Sicheritz-Ponten T, de Vos WM, Zucker JD, Raes J, Hansen T, Bork P, Wang J, Ehrlich SD, Pedersen O, Meta HITC (2013) Richness of human gut microbiome correlates with metabolic markers. Nature 500:541

le Leu R, Young G, Hu Y, Winter J, Conlon M (2013) Dietary red meat aggravates dextran sulfate sodium-induced colitis in mice whereas resistant starch attenuates inflammation. Dig Dis Sci 58:3475–3482

le Leu RK, Winter JM, Christophersen CT, Young GP, Humphreys KJ, Hu Y, Gratz SW, Miller RB, Topping DL, Bird AR, Conlon MA (2015) Butyrylated starch intake can prevent red meat-induced O6-methyl-2-deoxyguanosine adducts in human rectal tissue: a randomised clinical trial. Br J Nutr 114:220–230

le Poul E, Loison C, Struyf S, Springael JY, Lannoy V, Decobecq ME, Brezillon S, Dupriez V, Vassart G, van Damme J, Parmentier M, Detheux M (2003a) Functional characterization of human receptors for short chain fatty acids and their role in polymorphonuclear cell activation. J Biol Chem 278:25481–25489 le Poul E, Loison C, Struyf S, Springael JY, Lannoy V, Decobecq ME, Brezillon S, Dupriez V, Vassart G, van Damme J, Parmentier M, Detheux M (2003b) Functional characterization of human receptors for short chain fatty acids and their role in polymorphonuclear cell activation. J Biol Chem 278:25481–25489

Leach JD (2007) Evolutionary perspective on dietary intake of fibre and colorectal cancer. Eur J Clin Nutr 61:140–142

Lecona E, Olmo N, Turnay J, Santiago-Gomez A, de Silanes IL, Gorospe M, Lizarbe MA (2008) Kinetic analysis of butyrate transport in human colon adenocarcinoma cells reveals two different carrier-mediated mechanisms. Biochem J 409:311–320

Lee WJ, Hase K (2014) Gut microbiota-generated metabolites in animal health and disease. Nat Chem Biol 10:416–424

Leitch EC, Walker AW, Duncan SH, Holtrop G, Flint HJ (2007a) Selective colonization of insoluble substrates by human faecal bacteria. Environ Microbiol 9:667–679

Leitch ECM, Walker AW, Duncan SH, Holtrop G, Flint HJ (2007b) Selective colonization of insoluble substrates by human faecal bacteria. Environ Microbiol 9:667–679

Lewis K, Lutgendorff F, Phan V, Soderholm JD, Sherman PM, McKay DM (2010) Enhanced translocation of bacteria across metabolically stressed epithelia is reduced by butyrate. Inflamm Bowel Dis 16:1138–1148

Ley RE, Turnbaugh PJ, Klein S, Gordon JI (2006) Microbial ecology—human gut microbes associated with obesity. Nature 444:1022–1023

Lin MY, de Zoete MR, Van Putten JP, Strijbis K (2015) Redirection of epithelial immune responses by short-chain fatty acids through inhibition of histone deacetylases. Front Immunol 6:554

Louis P (2012) Does the human gut microbiota contribute to the etiology of autism spectrum disorders? Dig Dis Sci 57:1987–1989

Luceri C, Femia AP, Fazi M, di Martino C, Zolfanelli F, Dolara P, Tonelli F (2016) Effect of butyrate enemas on gene expression profiles and endoscopic/histopathological scores of diverted colorectal mucosa: a randomized trial. Dig Liver Dis 48:27–33

Lukito W, Malik SG, Surono IS, Wahlqvist ML (2015) From 'lactose intolerance' to 'lactose nutrition'. Asia Pac J Clin Nutr 24:S1–S8

Machiels K, Joossens M, Sabino J, De PV, Arijs I, Eeckhaut V, Ballet V, Claes K, Van IF, Verbeke K, Ferrante M, Verhaegen J, Rutgeerts P, Vermeire S (2013) A decrease of the butyrate-producing species *Roseburia hominis* and *Faecalibacterium prausnitzii* defines dysbiosis in patients with ulcerative colitis. Gut 63:1275–1283

Machiels K, Joossens M, Sabino J, de Preter V, Arijs I, Eeckhaut V, Ballet V, Claes K, Van Immerseel F, Verbeke K, Ferrante M, Verhaegen J, Rutgeerts P, Vermeire S (2014) A decrease of the butyrate-producing species Roseburia hominis and Faecalibacterium prausnitzii defines dysbiosis in patients with ulcerative colitis. Gut 63:1275–1283

Manichanh C, Rigottier-Gois L, Bonnaud E, Gloux K, Pelletier E, Frangeul L, Nalin R, Jarrin C, Chardon P, Marteau P, Roca J, Dore J (2006) Reduced diversity of faecal microbiota in Crohn's disease revealed by a metagenomic approach. Gut 55:205–211

Martin FPJ, Dumas ME, Wang YL, Legido-Quigley C, Yap IKS, Tang HR, Zirah S, Murphy GM, Cloarec O, Lindon JC, Sprenger N, Fay LB, Kochhar S, van Bladeren P, Holmes E, Nicholson JK (2007) A top-down systems biology view of microbiome-mammalian metabolic interactions in a mouse model. Mol Syst Biol 3:112

Martin LJM, Dumon HJW, Champ MMJ (1998) Production of short-chain fatty acids from resistant starch in a pig model. J Sci Food Agric 77:71–80

Martinez I, Kim J, Duffy PR, Schlegel VL, Walter J (2010) Resistant starches types 2 and 4 have differential effects on the composition of the fecal microbiota in human subjects. PLoS One 5: e15046

Maslowski KM, Vieira AT, Ng A, Kranich J, Sierro F, Yu D, Schilter HC, Rolph MS, Mackay F, Artis D, Xavier RJ, Teixeira MM, Mackay CR (2009) Regulation of inflammatory responses by gut microbiota and chemoattractant receptor GPR43. Nature 461:1282–1286

Mattila E, Uusitalo-Seppala R, Wuorela M, Lehtola L, Nurmi H, Ristikankare M, Moilanen V, Salminen K, Seppala M, Mattila PS, Anttila VJ, Arkkila P (2012) Fecal transplantation, through colonoscopy, is effective therapy for recurrent Clostridium difficile infection. Gastroenterology 142:490–496

McOrist AL, Miller RB, Bird AR, Keogh JB, Noakes M, Topping DL, Conlon MA (2011) Fecal butyrate levels vary widely among individuals but are usually increased by a diet high in resistant starch. J Nutr 141:883–889

Meddings JB, Jarand J, Urbanski SJ, Hardin J, Gall DG (1999) Increased gastrointestinal permeability is an early lesion in the spontaneously diabetic BB rat. Am J Physiol 276:G951–G957

Michielan A, D'Inca R (2015) Intestinal permeability in inflammatory bowel disease: pathogenesis, clinical evaluation, and therapy of leaky gut. Mediators Inflamm 2015:628157

Miller Jones J (2014) CODEX-aligned dietary fiber definitions help to bridge the 'fiber gap'. Nutr J 13:34

Muir JG, Rose R, Rosella O, Liels K, Barrett JS, Shepherd SJ, Gibson PR (2009) Measurement of short-chain carbohydrates in common Australian vegetables and fruits by high-performance liquid chromatography (HPLC). J Agric Food Chem 57:554–565

Mulle JG, Sharp WG, Cubells JF (2013) The gut microbiome: a new frontier in autism research. Curr Psychiatry Rep 15:337

Murphy N, Norat T, Ferrari P, Jenab M, Bueno-de-Mesquita B, Skeie G, Dahm CC, Overvad K, Olsen A, Tjonneland A, Clavel-Chapelon F, Boutron-Ruault MC, Racine A, Kaaks R, Teucher B, Boeing H, Bergmann MM, Trichopoulou A, Trichopoulos D, Lagiou P, Palli D, Pala V, Panico S, Tumino R, Vineis P, Siersema P, van Duijnhoven F, Peeters PHM, Hjartaker A, Engeset D, Gonzalez CA, Sanchez MJ, Dorronsoro M, Navarro C, Ardanaz E, Quiros JR, Sonestedt E, Ericson U, Nilsson L, Palmqvist R, Khaw KT, Wareham N, Key TJ, Crowe FL, Fedirko V, Wark PA, Chuang SC, Riboli E (2012) Dietary fibre intake and risks of cancers of the colon and rectum in the European Prospective Investigation into Cancer and Nutrition (EPIC). PLoS One 7:e39361

Musch MW, Bookstein C, Xie Y, Sellin JH, Chang EB (2001) SCFA increase intestinal Na absorption by induction of NHE3 in rat colon and human intestinal C2/bbe cells. Am J Physiol Gastrointest Liver Physiol 280:G687–G693

Naseer MI, Bibi F, Alqahtani MH, Chaudhary AG, Azhar EI, Kamal MA, Yasir M (2014) Role of gut microbiota in obesity, type 2 diabetes and Alzheimer's disease. CNS Neurol Disord-Drug Targets 13:305–311

Nemani K, Ghomi RH, Mccormick B, Fan XD (2015) Schizophrenia and the gut-brain axis. Prog Neuropsychopharmacol Biol Psychiatry 56:155–160

Nicholson JK, Holmes E, Kinross J, Burcelin R, Gibson G, Jia W, Pettersson S (2012) Host-gut microbiota metabolic interactions. Science 336:1262–1267

Nicholson JK, Holmes E, Wilson ID (2005) Gut microorganisms, mammalian metabolism and personalized health care. Nat Rev Microbiol 3:431–438

Norat T, Bingham S, Ferrari P, Slimani N, Jenab M, Mazuir M, Overvad K, Olsen A, Tjonneland A, Clavel F, Boutron-Ruault MC, Kesse E, Boeing H, Bergmann MM, Nieters A, Linseisen J, Trichopoulou A, Trichopoulos D, Tountas Y, Berrino F, Palli D, Panico S, Tumino R, Vineis P, Bueno-de-Mesquita HB, Peeters PHM, Engeset D, Lund E, Skeie G, Ardanaz E, Gonzalez C, Navarro C, Quiros JR, Sanchez MJ, Berglund G, Mattisson I, Hallmans G, Palmqvist R, Day NE, Khaw KT, Key TJ, San Joaquin M, Hemon B, Saracci R, Kaaks R, Riboli E (2005) Meat, fish, and colorectal cancer risk: the European prospective investigation into cancer and nutrition. J Natl Cancer Inst 97:906–916

O'Callaghan NJ, Toden S, Bird AR, Topping DL, Fenech M, Conlon MA (2012) Colonocyte telomere shortening is greater with dietary red meat than white meat and is attenuated by resistant starch. Clin Nutr 31:60–64

O'Keefe SJD, Chung D, Mahmoud N, Sepulveda AR, Manafe M, Arch J, Adada H, van der Merwe T (2007) Why do African Americans get more colon cancer than native Africans? J Nutr 137:175S–182S

Ou JH, Delany JP, Zhang M, Sharma S, O'Keefe SJD (2012) Association between low colonic short-chain fatty acids and high bile acids in high colon cancer risk populations. Nutr Cancer Int J 64:34–40

Pal S, Murugan C, Clarke JM, Young GP, Ramakrishna BS, Lockett T (2013) Esterified starch as a treatment for acute gastroenteritis—a double blind randomized controlled trial. Gastroenterology 144:S904

Park Y, Hunter DJ, Spiegelman D, Bergkvist L, Berrino F, Van den Brandt PA, Buring JE, Colditz GA, Freudenheim JL, Fuchs CS, Giovannucci E, Goldbohm RA, Graham S, Harnack L, Hartman AM, Jacobs DR, Kato I, Krogh V, Leitzmann MF, Mccullough ML, Miller AB, Pietinen P, Rohan TE, Schatzkin A, Willett WC, Wolk A, Zeleniuch-Jacquotte A, Zhang SMM, Smith-Warner SA (2005) Dietary fiber intake and risk of colorectal cancer—a pooled analysis of prospective cohort studies. Jama-J Am Med Assoc 294:2849–2857

Parsons BN, Wigley P, Simpson HL, Williams JM, Humphrey S, Salisbury AM, Watson AJ, Fry SC, O'Brien D, Roberts CL, O'Kennedy N, Keita AV, Soderholm JD, Rhodes JM, Campbell BJ (2014) Dietary supplementation with soluble plantain non-starch polysaccharides inhibits intestinal invasion of Salmonella Typhimurium in the chicken. PLoS One 9:e87658

Peeters M, Geypens B, Claus D, Nevens H, Ghoos Y, Verbeke G, Baert F, Vermeire S, Vlietinck R, Rutgeerts P (1997) Clustering of increased small intestinal permeability in families with Crohn's disease. Gastroenterology 113:802–807

Perrier C, Corthesy B (2011) Gut permeability and food allergies. Clin Exp Allergy 41:20–28

Piche T (2014) Tight junctions and IBS–the link between epithelial permeability, low-grade inflammation, and symptom generation? Neurogastroenterol Motil 26:296–302

Ploger S, Stumpff F, Penner GB, Schulzke JD, Gabel G, Martens H, Shen Z, Gunzel D, Aschenbach JR (2012) Microbial butyrate and its role for barrier function in the gastrointestinal tract. Ann NY Acad Sci 1258:52–59

Png CW, Linden SK, Gilshenan KS, Zoetendal EG, McSweeney CS, Sly LI, McGuckin MA, Florin THJ (2010) Mucolytic bacteria with increased prevalence in IBD mucosa augment in vitro utilization of mucin by other bacteria. Am J Gastroenterol 105:2420–2428

Puddu A, Sanguineti R, Montecucco F, Viviani GL (2014) Evidence for the gut microbiota short-chain fatty acids as key pathophysiological molecules improving diabetes. Mediators Inflamm 2014:162021

Pullan RD, Thomas GA, Rhodes M, Newcombe RG, Williams GT, Allen A, Rhodes J (1994) Thickness of adherent mucus gel on colonic mucosa in humans and its relevance to colitis. Gut 35:353–359

Qin JJ, Li RQ, Raes J, Arumugam M, Burgdorf KS, Manichanh C, Nielsen T, Pons N, Levenez F, Yamada T, Mende DR, Li JH, Xu JM, Li SC, Li DF, Cao JJ, Wang B, Liang HQ, Zheng HS, Xie YL, Tap J, Lepage P, Bertalan M, Batto JM, Hansen T, le Paslier D, Linneberg A, Nielsen HB, Pelletier E, Renault P, Sicheritz-Ponten T, Turner K, Zhu HM, Yu C, Li ST, Jian M, Zhou Y, Li YR, Zhang XQ, Li SG, Qin N, Yang HM, Wang J, Brunak S, Dore J, Guarner F, Kristiansen K, Pedersen O, Parkhill J, Weissenbach J, Bork P, Ehrlich SD, Wang J, Meta HITC (2010) A human gut microbial gene catalogue established by metagenomic sequencing. Nature 464:59–65

Qin JJ, Li YR, Cai ZM, Li SH, Zhu JF, Zhang F, Liang SS, Zhang WW, Guan YL, Shen DQ, Peng YQ, Zhang DY, Jie ZY, Wu WX, Qin YW, Xue WB, Li JH, Han LC, Lu DH, Wu PX, Dai YL, Sun XJ, Li ZS, Tang AF, Zhong SL, Li XP, Chen WN, Xu R, Wang MB, Feng Q, Gong MH, Yu J, Zhang YY, Zhang M, Hansen T, Sanchez G, Raes J, Falony G, Okuda S, Almeida M, Lechatelier E, Renault P, Pons N, Batto JM, Zhang ZX, Chen H, Yang RF, Zheng WM, Li SG, Yang HM, Wang J, Ehrlich SD, Nielsen R, Pedersen O, Kristiansen K, Wang J (2012) A metagenome-wide association study of gut microbiota in type 2 diabetes. Nature 490:55–60

Raghupathy P, Ramakrishna BS, Oommen SP, Ahmed TS, Priyaa T, Dziura J, Young GP, Binder HJ (2006) Amylase-resistant starch as adjunct to oral rehydration therapy in children with diarrhea. J Pediatr Gastroenterol Nutr 42:362–368

Rajendran VM, Binder HJ (1994) Apical membrane Cl-butyrate exchange: mechanism of short chain fatty acid stimulation of active chloride absorption in rat distal colon. J Membr Biol 141:51–58
Rajilic-Stojanovic M, Shanahan F, Guarner F, de Vos WM (2013) Phylogenetic analysis of dysbiosis in ulcerative colitis during remission. Inflamm Bowel Dis 19:481–488
Ramakrishna BS, Nance SH, Roberts-Thomson IC, Roediger WE (1990) The effects of enterotoxins and short-chain fatty acids on water and electrolyte fluxes in ileal and colonic loops in vivo in the rat. Digestion 45:93–101
Ramakrishna BS, Venkataraman S, Srinivasan P, Dash P, Young GP, Binder HJ (2000) Amylase-resistant starch plus oral rehydration solution for cholera. N Engl J Med 342:308–313
Ramirez-Farias C, Slezak K, Fuller Z, Duncan A, Holtrop G, Louis P (2009) Effect of inulin on the human gut microbiota: stimulation of Bifidobacterium adolescentis and Faecalibacterium prausnitzii. Br J Nutr 101:541–550
Ridaura VK, Faith JJ, Rey FE, Cheng JY, Duncan AE, Kau AL, Griffin NW, Lombard V, Henrissat B, Bain JR, Muehlbauer MJ, Ilkayeva O, Semenkovich CF, Funai K, Hayashi DK, Lyle BJ, Martini MC, Ursell LK, Clemente JC, Van Treuren W, Walters WA, Knight R, Newgard CB, Heath AC, Gordon JI (2013) Gut microbiota from twins discordant for obesity modulate metabolism in mice. Science 341:1079
Roberfroid MB (2005) Introducing inulin-type fructans. Br J Nutr 93:S13–S25
Roberfroid MB (2007) Inulin-type fructans: functional food ingredients. J Nutr 137:2493S–2502S
Roberts CL, Keita AV, Duncan SH, O'Kennedy N, Soderholm JD, Rhodes JM, Campbell BJ (2010) Translocation of Crohn's disease *Escherichia coli* across M-cells: contrasting effects of soluble plant fibres and emulsifiers. Gut 59:1331–1339
Roberts CL, Keita AV, Parsons BN, Prorok-Hamon M, Knight P, Winstanley C, O'Kennedy N, Soderholm JD, Rhodes JM, Campbell BJ (2013) Soluble plantain fibre blocks adhesion and M-cell translocation of intestinal pathogens. J Nutr Biochem 24:97–103
Said HM (2011) Intestinal absorption of water-soluble vitamins in health and disease. Biochem J 437:357–372
Sakaguchi S, Yamaguchi T, Nomura T, Ono M (2008) Regulatory T cells and immune tolerance. Cell 133:775–787
Savage DC (1977) Microbial ecology of gastrointestinal-tract. Annu Rev Microbiol 31:107–133
Schamberger GP, Diez-Gonzalez F (2002) Selection of recently isolated colicinogenic *Escherichia coli* strains inhibitory to *Escherichia coli* O157: H7. J Food Prot 65:1381–1387
Scheperjans F, Aho V, Pereira PAB, Koskinen K, Paulin L, Pekkonen E, Haapaniemi E, Kaakkola S, Eerola-Rautio J, Pohja M, Kinnunen E, Murros K, Auvinen P (2015) Gut microbiota are related to Parkinson's disease and clinical phenotype. Mov Disord 30:350–358
Scheppach W (1996) Treatment of distal ulcerative colitis with short-chain fatty acid enemas. a placebo-controlled trial. German-Austrian SCFA Study Group. Dig Dis Sci 41:2254–2259
Schieber AM, Lee YM, Chang MW, Leblanc M, Collins B, Downes M, Evans RM, Ayres JS (2015) Disease tolerance mediated by microbiome *E. coli* involves inflammasome and IGF-1 signaling. Science 350:558–563
Schneeman BO (1986) Dietary fiber—physical and chemical-properties, methods of analysis, and physiological-effects. Food Technol 40:104–110
Segal I, Edwards CA, Walker ARP (2000) Continuing low colon cancer incidence in African populations. Am J Gastroenterol 95:859–860
Sellon RK, Tonkonogy S, Schultz M, Dieleman LA, Grenther W, Balish E, Rennick DM, Sartor RB (1998) Resident enteric bacteria are necessary for development of spontaneous colitis and immune system activation in interleukin-10-deficient mice. Infect Immun 66:5224–5231
Singh N, Gurav A, Sivaprakasam S, Brady E, Padia R, Shi H, Thangaraju M, Prasad PD, Manicassamy S, Munn DH, Lee JR, Offermanns S, Ganapathy V (2014) Activation of Gpr109a, receptor for niacin and the commensal metabolite butyrate, suppresses colonic inflammation and carcinogenesis. Immunity 40:128–139
Slavin JL (2001) Dietary fiber and colon cancer. Marcel Dekker Inc, New York

Smith K, McCoy KD, Macpherson AJ (2007) Use of axenic animals in studying the adaptation of mammals to their commensal intestinal microbiota. Semin Immunol 19:59–69

Smith PM, Howitt MR, Panikov N, Michaud M, Gallini CA, Bohlooly Y, Glickman JN, Garrett WS (2013) The microbial metabolites, short-chain fatty acids, regulate colonic Treg cell homeostasis. Science 341:569–573

Sokol H, Pigneur B, Watterlot L, Lakhdari O, Bermudez-Humaran LG, Gratadoux JJ, Blugeon S, Bridonneau C, Furet JP, Corthier G, Grangette C, Vasquez N, Pochart P, Trugnan G, Thomas G, Blottiere HM, Dore J, Marteau P, Seksik P, Langella P (2008) *Faecalibacterium prausnitzii* is an anti-inflammatory commensal bacterium identified by gut microbiota analysis of Crohn disease patients. Proc Natl Acad Sci USA 105:16731–16736

Soliman ML, Rosenberger TA (2011) Acetate supplementation increases brain histone acetylation and inhibits histone deacetylase activity and expression. Mol Cell Biochem 352:173–180

Steinhart AH, Hiruki T, Brzezinski A, Baker JP (1996) Treatment of left-sided ulcerative colitis with butyrate enemas: a controlled trial. Aliment Pharmacol Ther 10:729–736

Stephen AM (1994) Whole grains—impact of consuming whole grains on physiological-effects of dietary fiber and starch. Crit Rev Food Sci Nutr 34:499–511

Subramanya S, Ramakrishna BS, Binder HJ, Farthing MJ, Young GP (2006) Evaluation of oral rehydration solution by whole-gut perfusion in rats: effect of osmolarity, sodium concentration and resistant starch. J Pediatr Gastroenterol Nutr 43:568–575

Subramanya SB, Rajendran VM, Srinivasan P, Kumar NSN, Ramakrishna BS, Binder HJ (2007) Differential regulation of cholera toxin-inhibited Na-H exchange isoforms by butyrate in rat ileum. Am J Physiol Gastrointest. Liver Physiol 293:G857–G863

Suzuki T, Yoshida S, Hara H (2008) Physiological concentrations of short-chain fatty acids immediately suppress colonic epithelial permeability. Br J Nutr 100:297–305

Swidsinski A, Loening-Boecke V, Herber A (2009) Mucosal flora in Crohn's disease and ulcerative colitis—an overview. J Physiol Pharmacol 60(suppl 6):61–71

Takahashi K, Nishida A, Fujimoto T, Fujii M, Shioya M, Imaeda H, Inatomi O, Bamba S, Andoh A, Sugimoto M (2016) Reduced abundance of butyrate-producing bacteria species in the fecal microbial community in Crohn's disease. Digestion 93:59–65

Tan J, Mckenzie C, Potamitis M, Thorburn AN, Mackay CR, Macia L (2014) The role of short-chain fatty acids in health and disease. Adv Immunol 121:91–119

Tao R, de Zoeten EF, Ozkaynak E, Chen C, Wang L, Porrett PM, Li B, Turka LA, Olson EN, Greene MI, Wells AD, Hancock WW (2007) Deacetylase inhibition promotes the generation and function of regulatory T cells. Nat Med 13:1299–1307

Tennant DR (2014) Potential intakes of total polyols based on UK usage survey data. Food Addit Contam Part A-Chem Anal Control Eposure Risk Assess 31:574–586

Thangaraju M, Cresci GA, Liu K, Ananth S, Gnanaprakasam JP, Browning DD, Mellinger JD, Smith SB, Digby GJ, Lambert NA, Prasad PD, Ganapathy V (2009) GPR109A is a G-protein-coupled receptor for the bacterial fermentation product butyrate and functions as a tumor suppressor in colon. Cancer Res 69:2826–2832

Thibault R, Blachier F, Darcy-Vrillon B, De CP, Bourreille A, Segain JP (2010) Butyrate utilization by the colonic mucosa in inflammatory bowel diseases: a transport deficiency. Inflamm Bowel Dis 16:684–695

Thibault R, de Coppet P, Daly K, Bourreille A, Cuff M, Bonnet C, Mosnier JF, Galmiche JP, Shirazi-Beechey S, Segain JP (2007) Down-regulation of the monocarboxylate transporter 1 is involved in butyrate deficiency during intestinal inflammation. Gastroenterology 133:1916–1927

Thorburn AN, Mckenzie CI, Shen S, Stanley D, Macia L, Mason LJ, Roberts LK, Wong CH, Shim R, Robert R, Chevalier N, Tan JK, Marino E, Moore RJ, Wong L, McConville MJ, Tull DL, Wood LG, Murphy VE, Mattes J, Gibson PG, Mackay CR (2015) Evidence that asthma is a developmental origin disease influenced by maternal diet and bacterial metabolites. Nat Commun 6:7320

Thornton JR, Dryden A, Kelleher J, Losowsky MS (1987) Super-efficient starch absorption—a risk factor for colonic neoplasia. Dig Dis Sci 32:1088–1091

Toden S, Bird AR, Topping DL, Conlon MA (2007) High red meat diets induce greater numbers of colonic DNA double-strand breaks than white meat in rats: attenuation by high-amylose maize starch. Carcinogenesis 28:2355–2362
Toden S, Lockett TJ, Topping DL, Scherer BL, Watson EJ, Southwood JG, Clarke JM (2014) Butyrylated starch affects colorectal cancer markers beneficially and dose-dependently in genotoxin-treated rats. Cancer Biol Ther 15:1515–1523
Tolhurst G, Heffron H, Lam YS, Parker HE, Habib AM, Diakogiannaki E, Cameron J, Grosse J, Reimann F, Gribble FM (2012) Short-chain fatty acids stimulate glucagon-like peptide-1 secretion via the G-protein-coupled receptor FFAR2. Diabetes 61:364–371
Topping D (2007) Cereal complex carbohydrates and their contribution to human health. J Cereal Sci 46:220–229
Topping DL, Bird AR (1999) Foods, nutrients and digestive health. Aust J Nutr Diet 56:S22–S34
Topping DL, Clifton PM (2001) Short-chain fatty acids and human colonic function: roles of resistant starch and nonstarch polysaccharides. Physiol Rev 81:1031–1064
Topping DL, Illman RJ (1986) Bacterial fermentation in the human large-bowel—time to change from the roughage model of dietary fiber. Med J Austral 144:307–309
Tosh SM, Brummer Y, Wolever TMS, Wood PJ (2008) Glycemic response to oat bran muffins treated to vary molecular weight of beta-glucan. Cereal Chem 85:211–217
Trowell HC, Burkitt DP (1987) The development of the concept of dietary fiber. Mol Aspects Med 9:7–15
Tungland BC, Meyer D (2002) Nondigestible oligo—and polysaccharides (Dietary fiber): their physiology and role in human health and food. Compr Rev Food Sci Food Saf 1:73–92
Tuohy KM, Conterno L, Gasperotti M, Viola R (2012) Up-regulating the human intestinal microbiome using whole plant foods, polyphenols, and/or fiber. J Agric Food Chem 60:8776–8782
Turnbaugh PJ, Ley RE, Mahowald MA, Magrini V, Mardis ER, Gordon JI (2006) An obesity-associated gut microbiome with increased capacity for energy harvest. Nature 444:1027–1031
Vadivel G, Prasad PD, Thangaraju M, Martin PM, Singh N (2014) Butyrate-mediated protection against colonic inflammation and colon carcinogenesis: role of butyrate transporters and butyrate receptors. Food sources, functions and health benefits, Butyrate
Van den Abbeele P, Gerard P, Rabot S, Bruneau A, El Aidy S, Derrien M, Kleerebezem M, Zoetendal EG, Smidt H, Verstraete W, Van de Wiele T, Possemiers S (2011) Arabinoxylans and inulin differentially modulate the mucosal and luminal gut microbiota and mucin-degradation in humanized rats. Environ Microbiol 13:2667–2680
Van der Merwe B, Erasmus C, Taylor JRN (2001) African maize porridge: a food with slow in vitro starch digestibility. Food Chem 72:347–353
Vieira ELM, Leonel AJ, Sad AP, Beltrao NRM, Costa TF, Ferreira TMR, Gomes-Santos AC, Faria AMC, Peluzio MCG, Cara DC, Alvarez-Leite JI (2012) Oral administration of sodium butyrate attenuates inflammation and mucosal lesion in experimental acute ulcerative colitis. J Nutr Biochem 23:430–436
Vinolo MA, Rodrigues HG, Hatanaka E, Hebeda CB, Farsky SH, Curi R (2009) Short-chain fatty acids stimulate the migration of neutrophils to inflammatory sites. Clin Sci (Lond) 117:331–338
Vinolo MA, Rodrigues HG, Hatanaka E, Sato FT, Sampaio SC, Curi R (2011a) Suppressive effect of short-chain fatty acids on production of proinflammatory mediators by neutrophils. J Nutr Biochem 22:849–855
Vinolo MA, Rodrigues HG, Nachbar RT, Curi R (2011b) Regulation of inflammation by short chain fatty acids. Nutrients 3:858–876
Vivinus-Nebot M, Frin-Mathy G, Bzioueche H, Dainese R, Bernard G, Anty R, Filippi J, Saint-Paul MC, Tulic MK, Verhasselt V, Hebuterne X, Piche T (2014) Functional bowel symptoms in quiescent inflammatory bowel diseases: role of epithelial barrier disruption and low-grade inflammation. Gut 63:744–752
Wahlqvist ML (2015) Lactose nutrition in lactase nonpersisters. Asia Pac J Clin Nutr 24:S21–S25

Walker ARP, Walker BF, Walker AJ (1986) Fecal pH, dietary fiber intake, and proneness to colon cancer in 4 South-African populations. Br J Cancer 53:489–495

Walker AW, Ince J, Duncan SH, Webster LM, Holtrop G, Ze X, Brown D, Stares MD, Scott P, Bergerat A, Louis P, Mcintosh F, Johnstone AM, Lobley GE, Parkhill J, Flint HJ (2011) Dominant and diet-responsive groups of bacteria within the human colonic microbiota. ISME J 5:220–230

Wang HB, Wang PY, Wang X, Wan YL, Liu YC (2012) Butyrate enhances intestinal epithelial barrier function via up-regulation of tight junction protein Claudin-1 transcription. Dig Dis Sci 57:3126–3135

Wang L, Christophersen CT, Sorich MJ, Gerber JP, Angley MT, Conlon MA (2011) Low relative abundances of the mucolytic bacterium *Akkermansia muciniphila* and *Bifidobacterium* spp. in feces of children with autism. Appl Environ Microbiol 77:6718–6721

West NP, Christophersen CT, Pyne DB, Cripps AW, Conlon MA, Topping DL, Kang S, McSweeney CS, Fricker PA, Aguirre D, Clarke JM (2013) Butyrylated starch increases colonic butyrate concentration but has limited effects on immunity in healthy physically active individuals. Exerc Immunol Rev 19:102–119

Windey K, de Preter V, Verbeke K (2012) Relevance of protein fermentation to gut health. Mol Nutr Food Res 56:184–196

Windey K, Houben E, Deroover L, Verbeke K (2015) Contribution of colonic fermentation and fecal water toxicity to the pathophysiology of lactose-intolerance. Nutrients 7:7505–7522

Winter J, Nyskohus L, Young GP, Hu Y, Conlon MA, Bird AR, Topping DL, Le Leu RK (2011) Inhibition by resistant starch of red meat-induced promutagenic adducts in mouse colon. Cancer Prev Res 4:1920–1928

Wong JMW, de Souza R, Kendall CWC, Emam A, Jenkins DJA (2006) Colonic health: fermentation and short chain fatty acids. J Clin Gastroenterol 40:235–243

Xu H, McCoy A, Li J, Zhao Y, Ghishan FK (2015) Sodium butyrate stimulates NHE8 expression via its role on activating NHE8 basal promoter activity. Am Physiol-Gastrointest Liver Physiol 309:G500–G505

Yatsunenko T, Rey FE, Manary MJ, Trehan I, Dominguez-Bello MG, Contreras M, Magris M, Hidalgo G, Baldassano RN, Anokhin AP, Heath AC, Warner B, Reeder J, Kuczynski J, Caporaso JG, Lozupone CA, Lauber C, Clemente JC, Knights D, Knight R, Gordon JI (2012) Human gut microbiome viewed across age and geography. Nature 486:222

Yetukuri L, Ekroos K, Vidal-Puig A, Oresic M (2008) Informatics and computational strategies for the study of lipids. Mol Biosyst 4:121–127

Yu YN, Fang JY (2015) Gut microbiota and colorectal cancer. Gastrointest Tumors 2:26–32

Ze X, Duncan SH, Louis P, Flint HJ (2012) Ruminococcus bromii is a keystone species for the degradation of resistant starch in the human colon. ISME J 6:1535–1543

Zeissig S, Burgel N, Gunzel D, Richter J, Mankertz J, Wahnschaffe U, Kroesen AJ, Zeitz M, Fromm M, Schulzke JD (2007) Changes in expression and distribution of claudin 2, 5 and 8 lead to discontinuous tight junctions and barrier dysfunction in active Crohn's disease. Gut 56:61–72

Chapter 3
Exploring the Bioactive Landscape of the Gut Microbiota to Identify Metabolites Underpinning Human Health

Páraic Ó Cuív, Sriti Burman, Sian Pottenger and Mark Morrison

1 Introduction

North America, Europe and Australasia have amongst the highest incidences of chronic gastrointestinal and metabolic diseases including inflammatory bowel diseases (IBD), colorectal cancer (CRC) and obesity (Molodecky et al. 2012; Stevens et al. 2012; Bray et al. 2013). Although once considered rare in large parts of the world the incidences of IBD and obesity in particular have also been steadily increasing in Asia, South America and the Middle East (Ng et al. 2014; Kaplan 2015). These diseases are associated with considerable socioeconomic costs; for example, the estimated costs to the global economy from obesity approaches US$2 trillion per annum, which equates to 2.8 % of global gross domestic product (Dobbs et al. 2014). Thus, there is an urgent need to develop more effective preventative and therapeutic strategies to ameliorate the impacts of these diseases.

Genomic studies have revealed that IBD, CRC and obesity are underpinned by specific host genetic susceptibilities that are considered to be necessary but often not sufficient for disease to develop (Jostins et al. 2012; Peters et al. 2015; Locke et al. 2015), and it is now recognised that environmental factors and lifestyle choices also affect disease risk. Epidemiological studies also suggest that host genetic, environmental factors and lifestyle choices either alone or in combination does not fully explain disease risk implying that other risk factors remain to be identified. With that context, the human gastrointestinal tract harbours a diverse microbial community (gut microbiota) that provides a range of ecological and metabolic functions relevant to host health and well-being (reviewed by Backhed et al. 2005). Human- and animal-based studies have now also identified the gut microbiota as an important risk factor in the aetiology of chronic gut diseases. First,

P. Ó Cuív (✉) · S. Burman · S. Pottenger · M. Morrison
The University of Queensland Diamantina Institute, The University of Queensland, Translational Research Institute, Woolloongabba QLD 4102, Australia
e-mail: p.ocuiv@uq.edu.au

D.J. Beale et al. (eds.), *Microbial Metabolomics*,
DOI 10.1007/978-3-319-46326-1_3

human studies have revealed that the microbiota varies between healthy and diseased individuals (e.g. Turnbaugh et al. 2009; Qin et al. 2010; Nakatsu et al. 2015) and these variations are associated with changes in the disease state (Cotillard et al. 2013; Nakatsu et al. 2015; De Cruz et al. 2015). Second, germ-free animals are protected from disease but become susceptible following microbiota transfer (Turnbaugh et al. 2006; Zackular et al. 2013; Schaubeck et al. 2016). Third, both human- and animal-based studies have revealed these diseases are responsive to interventions that modulate the activity of the gut microbiota including antibiotics (Zackular et al. 2013; Murphy et al. 2013; Schaubeck et al. 2016), diet (Donohoe et al. 2014; Quince et al. 2015), probiotics (Kadooka et al. 2010; Bassaganya-Riera et al. 2012) and faecal microbiota transfers (Suskind et al. 2015).

Even though IBD, CRC and obesity are a heterogeneous group of diseases, they are all characterised by an activated inflammatory response. Nuclear factor-kappa B (NF-κB) is a master regulator of gut epithelial integrity and inflammation, and activation of the NF-κB signalling pathway plays a key role in driving the inflammatory response during the onset and progression of these diseases. Consistent with this, the NF-κB signalling pathway is a validated therapeutic target for the treatment of IBD (Atreya et al. 2008) (Fig. 1), and it is also a recognised therapeutic target for CRC (Sakamoto and Maeda 2010), and for obesity and its co-morbidities (Donath 2014; Esser et al. 2015). The NF-κB pathway is particularly well recognised as a therapeutic target for IBD, however, many of the current therapeutics are only partially effective and/or have significant side effects. For instance, glucocorticosteroids can affect linear growth and bone health in paediatric subjects; methotrexate can cause hepatotoxicity, and as a teratogen, the treatment of female subjects is complicated; salicylates are associated with an increased risk of bleeding. Similarly, the newer biologics (e.g. anti-TNFα factors) are expensive, increase the risk of infection and suffer from a loss of response. Interestingly, the gut microbiota plays a central role in modulating the host immune response and specific gut microbes have been shown to possess potent NF-κB suppressive capabilities that can ameliorate the inflammatory response (Ménard et al. 2004; Sokol et al. 2008; Heuvelin et al. 2009; Petrof et al. 2009; Eeckhaut et al. 2012; Khokhlova et al. 2012; Kaci et al. 2013). This suggests that exploiting gut microbe-derived NF-κB suppressive bioactives may provide new opportunities to maintain host health. In this Chapter, we examine our current understanding of the host-microbiota interaction and outline strategies to identify and characterise the NF-κB suppressive capabilities of the gut microbiota. In particular, we propose that an integrated approach combining culture-dependent and independent approaches with a more mechanistic dissection of the microbiota provided by improved cultivation techniques, high-throughput functional screens and metabolomic and genetic dissections is necessary to transform our understanding of gut health and support the development of new preventative and therapeutic strategies.

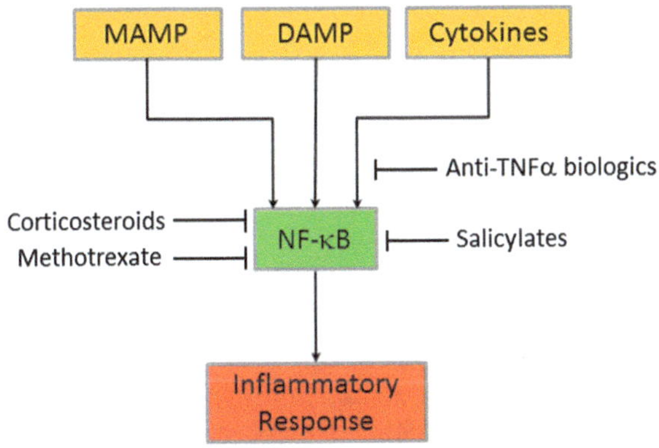

Fig. 1 The NF-κB pathway as a validated drug target for the treatment of chronic gut diseases. The NF-κB pathway can be activated by several mechanisms including microbe-associated molecular patterns (MAMP; e.g. via lipopolysaccharide, flagellin from the gut microbiota), damage associated molecular patterns (DAMP; e.g. via extracellular detection of normally intracellular proteins) or cytokines. Targeting of the NF-κB pathway for the treatment of chronic gut diseases is best recognised for IBD with glucocorticosteroids (corticosteroids), methotrexate, salicylates (e.g. mesalazine, sulfasalazine) and anti-TNFα biologics interfering with pathway signalling. However, this pathway is also increasingly targeted for CRC, and obesity and its co-morbidities

2 The Human Holobiont: An Emergent Paradigm of Human Health

The publication of the human genome sequence was a seminal milestone in our history. Published with much excitement in 2001, it promised new insights and understanding of what it means to be human (Venter et al. 2001; Lander et al. 2001). Initial estimates of the number of protein-coding genes deemed necessary to explain the biological and phenotypic complexity characteristic of humans varied widely, however, there was considerable surprise when it was revealed that the human genome is comprised of as few as 25,000 genes (International Human Genome Sequencing Consortium 2004). Humans are not autonomous and following a period of introspection it was increasingly recognised that our associated microbiota provides a range of functions relevant to health and disease. Thus, in its aftermath, there was an increasing call to sequence our second genome—that of the human microbiota (Davies 2001; Relman and Falkow 2001). This international effort to sequence the human microbiome has principally been led by the Human Microbiome Project (HMP) funded by the US National Institutes of Health (Peterson et al. 2009) and the MetaHIT Project (Ehrlich 2010) funded by the European Union, with additional coordination of other global efforts mediated through the International Human Microbiome Consortium.

Humans and their associated microbiota co-exist as a symbiotic multispecies assemblage termed a "holobiont" that is defined as a physical association between a host and its associated microbiota for significant portions of their life history (Bordenstein and Theis 2015). The emergence of the holobiont concept has dramatically altered our perception of human health—where the role of microbes was traditionally viewed from the perspective of infectious diseases—to one where the microbiota is viewed as an integral component that contributes essential functionalities relevant to the fitness of the holobiont. The assembly of a holobiont is a dynamic process that impacts both the host and microbiota (Gilbert 2014). For instance, the human gut provides a wide variety of ecological niches that are characterised by a constant temperature, oxygen tension, humidity and nutrient supply. This supports colonisation by a numerically abundant and diverse microbiota that in return helps prevent colonisation by potential pathogens, detoxifies harmful compounds, produces essential nutrients and catalyses the biotransformation of dietary substrate so they can be utilised by the host (Fig. 2). The "hologenome" then is comprised of the genetic potential encoded by the host's genome and their associated microbiota (microbiome) and can thus be considered as an extension of the host genotype itself. Notably, the hologenome is dynamic in terms of its composition with the potential to change more rapidly than the host genome alone via gene acquisition or loss which also confers a greater adaptive potential to the holobiont (Quercia et al. 2014).

The holobiont concept provides a new paradigm for a more holistic understanding of the aetiology of chronic gut diseases. For instance, some of the genetic

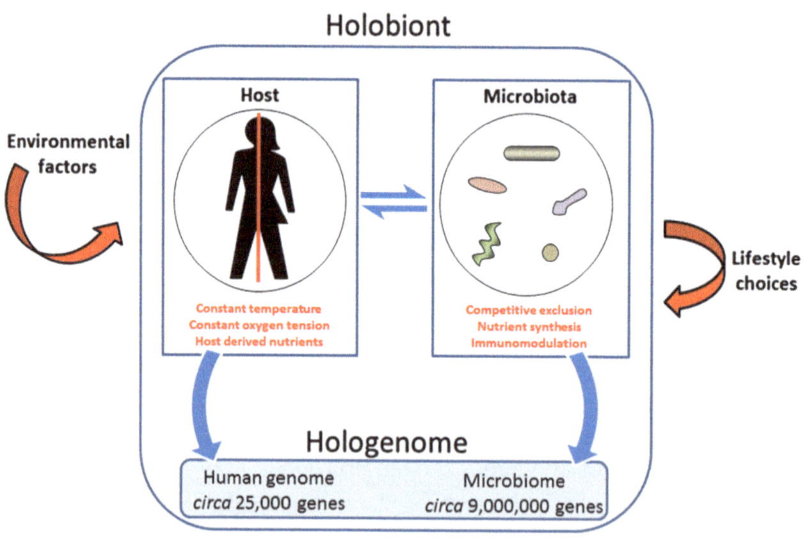

Fig. 2 The emergent holobiont model of human health. The holobiont assembly impacts the fitness of the host and its associated microbiota and is characterised by specific host–microbiota interdependencies. The hologenome is comprised of the genetic capacity of both the human genome and microbiome. The holobiont provides a new model to examine the impact of environmental factors and lifestyle choices on host health and disease risk

susceptibility loci for these diseases also affect the ability of specific microbes to colonise the gut (reviewed by Spor et al. 2011) suggesting that the contribution of genetic susceptibility and microbiota composition to disease risk may be intrinsically linked. In addition, the holobiont has provided a framework on which the impact of environmental factors and lifestyle choices on health and disease risk can be dissected and this has informed the development of new strategies to rationally modulate the holobiont phenotype to improve host health (Zeevi et al. 2015). The holobiont may also represent an optimum biological system to bioprospect for novel NF-κB suppressive bioactives as the gut microbiota has co-evolved with the development of the host mucosal immune system. In particular, we hypothesise that select microbes produce bioactives that actively suppress the NF-κB-mediated immune response perhaps as an essential capability to allow for successful colonisation and persistence. These NF-κB suppressive bioactives may have specific attributes that are relevant to the development of new therapeutics including high bioactivity, bioavailability and target site specificity, as well as stability in the gut environment. Thus, these bioactives could potentially be used directly or serve as lead molecules for the development of novel NF-κB suppressive therapeutics. Alternatively, determining the mechanism by which they exert their suppressive effects could help to identify new cellular targets that could be drugged by existing or new therapeutics. Taken together the identification and characterisation of these bioactives may help realise new opportunities to prevent or treat chronic gut diseases.

3 Insights into the Structure: Function Capacity of the Human Gut Microbiota

Our understanding and appreciation of the diversity and functional capacity of the gut microbiota are largely based on distinct yet complementary culture-dependent and culture-independent analyses of the gut environment (Fig. 3). Many of the current reference strains used in gut microbiota research were first isolated in the mid-twentieth century following the advent of techniques in anaerobic microbiology (for a historical perspective see Rajilić-Stojanović and de Vos 2014). However, it has long been recognised that the vast majority of gut microbes are resistant to cultivation as revealed by the discordance between microscopic counts of microbial cells and those recovered using traditional laboratory based cultivation. This phenomenon was first described in aquatic environments and termed "the great plate count anomaly" (reviewed by Staley and Konopka 1985). Instead, advances in molecular biology and DNA sequencing technology culminated in the establishment of culture-independent approaches to study the microbiota, based largely on 16S rRNA gene community profiling and metagenomics. A crucial discovery was that the microbial 16S rRNA gene could be used as a molecular clock to infer phylogeny and provide an estimate of microbial diversity (Woese and Fox 1977).

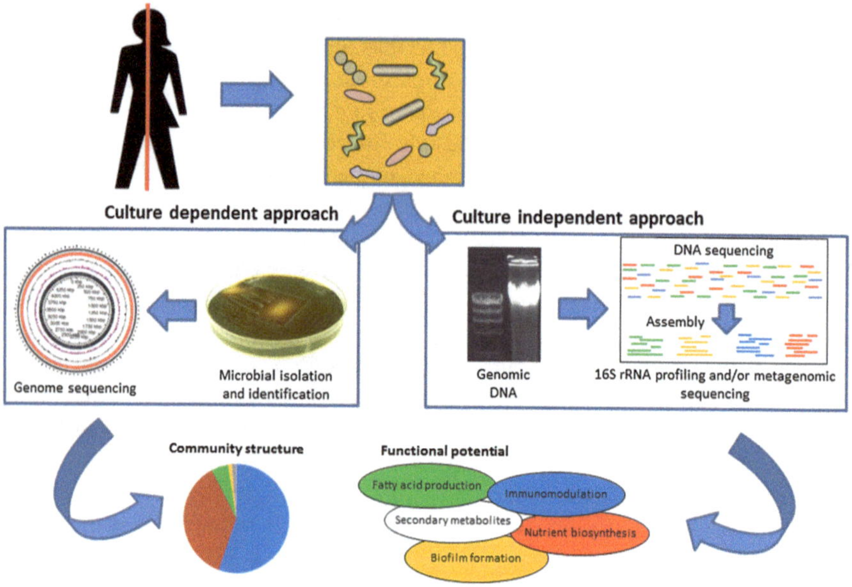

Fig. 3 The analysis of the gut microbiota by culture-dependent and culture-independent approaches. For culture-dependent approaches gut microbes are ideally recovered from gut samples (e.g. faeces, biopsy tissue) as axenic cultures. The phylogeny and functional potential of the isolates can then be assessed by 16S rRNA gene and/or genome sequencing. The functional characteristics of the isolates can be assessed by phenotypic profiling. For culture-independent approaches bulk DNA is typically recovered directly from gut samples and the DNA is then used for 16S rRNA gene profiling and/or metagenomic sequencing. The diversity and functional capacity of the microbiota can be assessed from the resultant sequence data

The 16S rRNA gene is approximately 1550 bp in length and has a divergence rate of 1–2 % per 50 million years (Ochman et al. 1999). The gene is comprised of conserved and (hyper)variable regions and this architecture has been exploited in culture-independent studies to assess microbial diversity. Here, near full-length or subsections of the 16S rRNA gene are amplified by polymerase chain reaction (PCR) using primers targeting the conserved regions and the intervening variable regions are used to infer phylogeny (Klindworth et al. 2013). The length of the 16S rRNA gene sequence can affect phylogenetic assignment (Kim et al. 2011; Franzén et al. 2015) and it does not provide any information on the functional potential of the taxa identified. However, the development of metagenomic approaches, facilitating the sequencing of bulk DNA recovered from microbial communities has now provided new opportunities to both assess microbial diversity through sequencing of defined phylogenetic marker genes (Sunagawa et al. 2013) and the functional capacity of the microbiome (Qin et al. 2010; Li et al. 2014a). In a landmark study Qin et al. (2010) examined the microbiome of 124 subjects and determined that it is comprised of a genetic pool of up to 3.3 million non-redundant genes that is as much as 150× that of the human genome. In practical terms the functional activity of an individual's gut microbiome is supported by ~500,000 non-redundant genes. In line with an earlier estimate (Yang et al. 2009), a subsequent study by Li et al. (2014a) identified over 9.8 million non-redundant genes in the human gut microbiome. It is likely that the number of non-redundant genes remains underestimated, however, metagenomics may have reached a point of diminishing returns where a greater effort has to be expended to provide an even deeper insight into the gene repertoire of the gut microbiota. Together, both 16S rRNA-based profiling and metagenomic sequencing have provided a unique insight into the gut microbiome and revealed that the vast majority of gut microbes remain uncultured (reviewed by Rajilic-Stojanovic et al. 2007; Rajilić-Stojanović and de Vos 2014).

The adult human gut microbiota is comprised of viruses, bacteria, archaea and eukaryotes with the number of microbial cells inhabiting the adult human gut outnumbering host cells by an order of magnitude (Savage 1977). The gut environment is characterised by a host driven top-down pressure on the microbiota that selects for a community of distantly related microbes with similar functional capabilities ensuring redundancy of microbial processes essential for the host. In contrast, intra-microbiota competition results in a bottom-up pressure that selects for functional specialisation. Consequently, the structure of the gut microbiota is characterised by distinct inter-subject variability although the core functional capabilities of the microbiota (e.g. short chain fatty acid (SCFA) production, vitamin biosynthesis) are largely conserved (Turnbaugh et al. 2009; Lozupone et al.2012). The diversity and functional attributes of the bacterial and archaeal communities in the human gut is best understood. The human gut microbiota is dominated by bacteria affiliated with the phyla Firmicutes and Bacteroidetes with smaller numbers of other phyla including Actinobacteria, Fusobacteria, Proteobacteria and Verrucomicrobia also present (Rajilic Stojanovic et al. 2007; Lozupone et al. 2012). The diversity of the microbiota becomes increasingly complex at deeper phylogenetic levels and the human gut can harbour up to several hundred individual strains that vary substantially between individuals (Greenblum et al. 2015; Yassour et al. 2016). The gut archaea have a low abundance and are comprised of methanogenic and non-methanogenic archaea (Rieu-Lesme et al. 2005; Gill et al. 2006; Nam et al. 2008; Oxley et al. 2010; Ó Cuív et al. 2011a). The methanogenic gut archaea are dominated by strains affiliated with *Methanobrevibacter* and *Methanosphaera* spp. although the diversity of human gut methanogenic archaea may be underestimated (Gill et al. 2006; Nam et al. 2008; Scanlan et al. 2008; Mihajlovski et al. 2008; Ó Cuív et al. 2011a). Despite the substantial inter-subject variability, the healthy gut microbiota has been shown to be comprised of a core microbiota that is widely shared between individuals and that includes some of the most abundant members of the microbiota (Tap et al. 2009; Qin et al. 2010; Jalanka-Tuovinen et al. 2011; Sekelja et al. 2011; Rajilic-Stojanovic et al. 2012; Martínez et al. 2013; Li et al. 2013), and an accessory microbiota that is less widely shared and typically comprised of low abundance taxa that are nonetheless metabolically active (Peris-Bondia et al. 2011).

Both culture-dependent and culture-independent approaches have helped to identify important differences to the structure-function activity between the gut microbiota of healthy individuals and those with chronic gut diseases. For instance, perturbations of signature bacterial species from the core microbiota have been associated with chronic gut diseases (Qin et al. 2010; Nakatsu et al. 2015). In CD, the abundance of specific core bacteria differs from the healthy gut (Kang et al. 2010; Mondot et al. 2011; Prideaux et al. 2013; De Cruz et al. 2015), these differences are coincident with the onset of active disease (De Cruz et al. 2015) and a restoration of their abundance may support remission (Dey et al. 2013; De Cruz et al. 2015). Furthermore, in vivo and in vitro-based experiments have demonstrated that structure-function differences between the healthy and diseased gut microbiota are associated with variations in biological activities that are relevant to gut health. While the immunoregulatory capacity of the microbiota remains largely unknown several representative isolates from the core microbiota produce bioactive factors that can suppress NF-κB activation (Ménard et al. 2004; Sokol et al. 2008; Heuvelin et al. 2009; Khokhlova et al. 2012; Quevrain et al. 2016), modulate the balance and/or activity of regulatory and effector T-cell populations (Atarashi et al. 2011; Atarashi et al. 2013; Qiu et al. 2013; Li et al. 2014b) and restore barrier function (Martin et al. 2015), thus attenuating the host inflammatory response and helping to maintain gut homeostasis. Based on these observations, members of the core gut microbiota have been proposed as "next-generation" probiotics for the treatment of chronic gut diseases (Neef and Sanz 2013). There continues to be a growing appreciation of the NF-κB suppressive capabilities of individual members of the core microbiota and other gut bacteria. However, while some of the NF-κB suppressive factors produced by gut bacteria have been identified, in many instances they remain to be determined (Kelly et al. 2004; Lakhdari et al. 2011; Kaci et al. 2011; Santos Rocha et al. 2012; Kaci et al. 2013).

Advancements in DNA sequencing technologies continue apace and the cost and speed at which sequence data can be produced and annotated continues to dramatically improve (Loman et al. 2012; Land et al. 2015). However, despite the wealth of gut microbiome associated sequence data now available in the public databases, the overwhelming majority of gene products have not been functionally characterised. Indeed, it is estimated that up to 75 % of protein families are assigned to uncharacterised orthologous groups and novel families (Qin et al. 2010; Ellrott et al. 2010; The Human Microbiome Project Consortium 2012). This challenge is further compounded by the fact that DNA sequence data are typically annotated using automated pipelines with little manual curation resulting in the introduction and propagation of annotation errors, and ultimately spurious function prediction (Schnoes et al. 2009; Promponas et al. 2015). It is widely acknowledged that the ability to functionally dissect the gut microbiome has not kept pace with DNA sequencing technology and it is notable that the functions of over one-third of the gene complement of the model organism and best characterised gut bacterium *Escherichia coli* (*E. coli*) K-12 remain undetermined (Hu et al. 2009). This shortcoming is increasingly being addressed (Nichols et al. 2011; Meng et al. 2012; Paradis-Bleau et al. 2014; Rajagopala et al. 2014) supported largely by *E. coli*'s ease of propagation and its amenability to genetic dissection. In contrast, the vast majority of gut microbes are fastidious anaerobes that are not known to be amenable to genetic dissection and hence their genetic potential remains cryptic. This has led to suggestions that an increased effort must be expended to functionally characterise existing gene sets as this will provide new insights into the microbial factors supporting gut health or driving disease (Roberts 2004; Galperin and Koonin 2010; Anton et al. 2014; Joice et al. 2014).

Based on these collective observations, we contend that new advances in microbial isolation coupled with parallel developments in functional characterisation and dissection approaches will provide the best opportunities to develop streamlined strategies to identify NF-κB suppressive and other types of bioactives produced by gut microbes. In particular, considering the complexity of the gut microbiota these strategies must be cost-effective, scalable and amenable to automation, and the following sections provide an overview of each of these aspects.

4 Bringing the Microbiome to Life: Culture-Dependent Analysis of the Gut Microbiota

It is indisputable that the development of new approaches to isolate and propagate fastidious gut microbes has not kept pace with those of culture-independent approaches. In particular, microbial culturing is widely perceived to be a time and labour intensive process and much information can now be provided without having to isolate individual microbes (Table 1). Nonetheless, both approaches are complementary and in some instances culture-dependent approaches provide the best opportunity to dissect the functional capacity of the microbiota. For instance, culture-dependent approaches allow specific axenic isolates to be directly linked with NF-κB suppressive capabilities and, moreover, they provide a valuable resource to test experimental hypotheses (e.g. Koch's postulates).

The vast majority of gut microbes are strict anaerobes and require an environment with a low redox potential in which to grow. The history of isolating and cultivating fastidious gut microbes extends from the late-nineteenth century (Rajilić-Stojanović and de Vos 2014). Many of the techniques used in contemporary laboratories were developed and adapted by Hungate (1969) and colleagues (Eller et al. 1971; Macy et al. 1972; Bryant 1972; Balch et al. 1979) and have been used to isolate and propagate facultative anaerobic microbes (e.g. *E. coli*), microaerophilic microbes (e.g. *Lactobacillus* spp.), aerotolerant anaerobic microbes (e.g. *Bacteroides* spp.) and obligate anaerobic microbes (e.g. *Clostridium* spp.) (Virginia Polytechnic Institute and State University Anaerobe Laboratory 1975; Dowell et al. 1981). These techniques can be readily established and remain relevant today although the ability to isolate and propagate fastidious anaerobic microbes has been advanced by the development of anaerobic chambers for microbiological culturing

Table 1 Culture-dependent analysis of the gut microbiota—opportunities and challenges

Culture-dependent approach	Advantage	Disadvantage
Microbial isolation	• Enables experimental hypotheses to be evaluated (e.g. Koch's postulates) • Provides a resource for further experimentation	• Time-consuming and labour intensive
Genomic characterisation	• Enables the functional potential of an isolate to be assessed • The 16S rRNA gene sequence can be associated with specific functional genes • Intraspecies genetic variability can be assessed where multiple isolates are available	• Genomic data can be provided by culture-independent means • Genome annotations can result in a high number of genes of unknown function
Functional characterisation	• Facilitates phenotypic profiling (e.g. metabolic, physiological characteristics) • Functional attributes can be linked with the 16S rRNA gene and/or genomic content	• Limited ability to genetically dissect microbial isolates

which further reduce the risk of inadvertent oxygen contamination and allow many standard techniques (e.g. spread plates, streak plates) to be used to isolate target microbes.

The distinct ecological niches present along the human gut can be challenging to replicate in a laboratory environment particularly as the nutritional requirements of many target microbes are unknown. The use of "habitat simulating" media has been widely used to circumvent this challenge and typically includes sterile aqueous extracts of faecal or rumen digesta, in addition to sources of amino acids, carbohydrates and other nutrients (Eller et al. 1971; Barcenilla et al. 2000; McSweeney et al. 2005; Lagier et al. 2012), leading to the isolation of phylogenetically diverse gut microbes including bacteria that have specific host dependencies such as *Akkermansia muciniphila* (Derrien et al. 2004) and the obligate symbiont segmented filamentous bacterium (Schnupf et al. 2015). Although habitat simulating media often support the growth of subdominant populations, their enrichment and isolation is often complicated because of rapid overgrowth by fast growing, numerically abundant microbes. More selective media have been developed for the isolation of specific gut taxa including *Bacteroides* spp. (Livingston et al. 1978), *Bifidobacterium* spp. (Ferraris et al. 2010) and *Enterococcus* spp. (Isenberg et al. 1970) by identifying specific nutritional dependencies, and promoters/inhibitors of growth (e.g. antibiotics, bile salts, sodium azide). Alternatively, subdominant populations can be enriched by selecting for a specific phenotype (e.g. spore formation) and this has enabled taxonomically novel microbes to be directly recovered on nutrient-rich habitat simulating media (Atarashi et al. 2013; Browne et al. 2016).

Accordingly, based on these, our own (Ó Cuív et al. 2011b, 2015) and other (Rettedal et al. 2014; Ma et al. 2014) observations, many "uncultured" microbes grow reproducibly well in vitro when isolated as axenic cultures. Thus, many more novel gut microbes could be recovered if the practical considerations involved with screening large numbers of microbial isolates under strict anaerobic conditions could be overcome.

To improve the throughput of microbial isolation, Stevenson et al. (2004) developed an approach called "Plate wash PCR" to recover axenic isolates of previously uncultured bacteria from agricultural soil and the guts of wood-feeding termites. Briefly, an inoculum is plated in duplicate on solid medium and following growth the colonies are re-suspended *en masse* from one of the replicate plates and the sample extracted DNA is screened using specific PCR primers. By this approach, a broad range of growth parameters can be rapidly screened to determine conditions supporting the growth of target taxa. Once identified, colonies from the matching replica plate are grown in multiwell plates and screened with specific primers to identify the target isolate. Plate wash PCR was successfully used to isolate a Lachnospiraceae affiliated bacterium that inhibits colonisation of the murine gut by *Clostridium difficile* VPI 10463 (Reeves et al. 2012), and it has been adapted to support the isolation of human gut bacteria affiliated with the HMP's most-wanted taxa using a microfluidic platform (Ma et al. 2014). Goodman et al. (2011) described a similar approach but determined the diversity of microbial isolates recovered on the replica culture plate by 16S rRNA-based microbial profiling. In addition, to further improve the throughput of the isolation process, a most probable number (MPN) approach was used to create, in 384 well plates, personalised archived culture collections of axenic isolates directly from faecal samples without picking individual colonies. The MPN approach is based on extinction culturing, whereby diluting microbial cells so that ≤ 1 culturable cell is used as an inoculum supports the production of axenic cultures (Button et al. 1993). This favours the isolation of the most abundant rather than the fastest growing or most culturable microbes and the MPN method has also been used to produce axenic cultures of previously uncultured rumen bacteria (Kenters et al. 2011). Rettedal et al. (2014) also used 16S rRNA profiling to profile gut bacteria recovered on a broad range of solid growth media. Then, by a process termed cultivation-based multiplex phenotyping, they combined growth on solid medium with antibiotic selection and 16S rRNA profiling to selectively target and recover target bacteria including members of the HMP's most-wanted taxa (Fodor et al. 2012). Recently, Browne et al. (2016) applied a similar approach to isolate spore forming bacteria from the human gut.

Separately, Raoult and colleagues (Lagier et al. 2012) coined the term "culturomics" and demonstrated that increasing the throughput of microbial isolation greatly extended the number of cultured isolates from the human gut. By this approach, 32,500 colonies representing 340 bacterial species and including 31 previously unidentified species were obtained using 212 culture conditions and three human faecal samples. Culturomics was also shown to be superior to culture-independent approaches in its ability to detect bacteria that were below the detection threshold of 16S rRNA profiling approaches (Lagier et al. 2012; Dubourg et al. 2013). It is notable that these studies were performed using methodologies that could be readily established in a standard microbiological laboratory (e.g. the use of anaerobic jars to produce microaerobic or anaerobic conditions). It is likely that the use of an anaerobic chamber could have further increased the recovery of fastidious obligate anaerobes, however, manipulating large numbers of isolates in multiwell plates and a confined environment is challenging. Interestingly, Raoult and colleagues (La Scola et al. 2014; Dione et al. 2015) discovered that the addition of antioxidants to the growth medium permitted the growth of strict anaerobic bacteria under atmospheric conditions. This observation could revolutionise our ability to isolate fastidious gut bacteria particularly if it can be verified that their growth and metabolic activity is similar under aerobic and anaerobic conditions, and it complements advancements in automated colony picking robotic platforms that are capable of operating in an anaerobic chamber.

The wealth of sequence data now available for gut microbes has also helped to direct the isolation of gut microbes. For instance, Pope et al. (2011) described the successful isolation of an uncultured bacterium affiliated with the Succinivibrionaceae from foregut digesta samples collected from Tammar wallabies. Here, metagenomic data were used to partially reconstruct and model the bacterium's metabolism and physiological features, and then tailored culture conditions were developed to direct the axenic cultivation of the bacterium by a process termed metagenome directed isolation. Bomar et al. (2011) similarly used metatranscriptomic data to direct the isolation of an abundant *Rikenella*-like bacterium from the gut of a medicinal leech. Recently, Oberhardt et al. (2015) developed a web-based platform that uses a database of microbe-medium combinations to predict media for microbes based on their 16S rRNA sequence. The exploitation of sequence data to help bring the microbiome to life is a vital development as much of these data languishes mostly unused in online databases.

The throughput of microbe identification has also been expedited by developments in matrix-assisted laser desorption/ionisation-time of flight mass spectrometry (MALDI-TOF-MS) based analyses. The early classification of microbes was primarily based on physiological and morphological characteristics (Virginia Polytechnic Institute and State University Anaerobe Laboratory 1975), however, the development of 16S rRNA-based phylogenetics allowed the genetic relatedness of these isolates to be determined. The gold standard of 16S rRNA-based phylogeny taxonomy is based on the production of near full-length gene sequences that are used to infer relatedness (Kim et al. 2011; Franzén et al.2015). The identification of microbial isolates is typically achieved by low-throughput Sanger sequencing of the 16S rRNA gene, however, due to its low rate of divergence, it is widely recognised that the 16S rRNA gene is limited in its ability to provide phylogenetic resolution of the microbiota at lower phylogenetic levels. Other genes can also be used as phylogenetic markers [e.g. *gyrB, rpoB* (Sunagawa et al. 2013; Fish et al. 2013)] but these are less well established and not routinely used. Instead, MALDI-TOF-MS-based analyses now provide an alternative and in many respects a superior means to identify microbial isolates. The ability to identify specific isolates is typically based on the mass patterns of ribosomal or other abundant housekeeping proteins and is determined by reference to a database of spectra produced using representative isolates. This approach is particularly valuable in providing a cost-effective rapid and sensitive assessment of intraspecies variability without any prior knowledge of the strains being tested, although the ability to distinguish between very closely related strains can be challenging (Sandrin et al. 2013). We anticipate that MALDI-TOF-MS-based identification of microbial isolates will increasingly supplant 16S rRNA gene-based identification as the reference databases become more comprehensive and the technology more robust and affordable.

In summation, meaningful progress has been made to increase the efficiency and throughput of microbial isolation and these have increased the diversity of gut microbes that are available in international biorepositories (e.g. Biodefense and Emerging Infections Research Resources Repository (BEI Resources), Deutsche Sammlung von Mikroorganismen und Zellkulturen (DSMZ) GmbH) or that are held in private laboratory culture collections. Continued advances in automated microbial isolation and identification will further expedite these efforts and support a more mechanistic dissection of the gut microbiota in the maintenance of gut homeostasis and the prevention of chronic gut diseases, although important challenges remain. Much of our understanding of the diversity and functional capability of the gut microbiota is based on analyses of faecal associated microbiota, which can be collected in a non-invasive manner and up to 54 % of the faecal mass is comprised of microbial biomass (Stephen and Cummings 1980; Rose et al. 2015) thus providing copious material for experimental interrogation. However, it has been long recognised that the faecal and mucosa associated microbiota differ (Zoetendal et al. 2002; Ott et al. 2004; Lepage et al. 2005; Eckburg et al. 2005) and it is now also recognised that the mucosa associated microbiota is also characterised by a distinct biogeography (Obata et al. 2010; Aguirre de Carcer et al. 2011; Pedron et al. 2012; Sonnenberg et al. 2012; Zhang et al. 2014) that likely reflects different ecological niches driven by variations in nutrient availability, oxygen tension, pH and immune activation (reviewed by Donaldson et al. 2015). The aetiology of several chronic gut diseases is characterised by site-specific differences with CD predominantly affecting the ileum and proximal colon, and UC and CRC predominantly affecting the distal colon. Thus, the spatial distribution of the gut microbiota may have implications for our understanding of host–microbe interactions and their relationship to health and disease, and future efforts should seek to preferentially culture gut microbes from sites relevant to disease.

5 Dissecting the Functional Potential of the Gut Microbiota: Advances in In Vitro Approaches to Identify NF-κB Suppressive Gut Microbes

In 2014 it was reported that over 1000 cultured gut microbial species had been described in the scientific literature and this number continues to increase rapidly due to the new advances in microbial cultivation techniques (Rajilić-Stojanović and de Vos 2014). The NF-κB suppressive activities of the vast majority of existing isolates have not been assessed but taken together with the increasing rate of microbial isolation there is a need for improved functional screening strategies to effectively identify these strains. Strategies to identify immunomodulatory microbes should address three key criteria. First, the assays should be biologically relevant, sensitive and specific, facilitating the identification of virulent or cytotoxic microbes at an early point in the screening process. Second, the assays should allow the extent of immunomodulatory activity to be quantified and the host pathways affected to be readily identified and dissected to determine the target of the bioactive. Third, the assays should be cost-effective and robust, easy to perform and amenable to scaling to an automated high-throughput format. Historically, the use of well-established cell lines to identify NF-κB suppressive microbial isolates broadly fulfils these criteria.

The NF-κB pathway has been extensively characterised and transcription can be activated via two alternate pathways, called the canonical and non-canonical pathways. These pathways can be activated either independently (e.g. TNFα/IL-1β activates the canonical pathway, B-cell activating factor activates the non-canonical pathway) or in tandem (e.g. CD40L/Lipopolysaccharide activate both pathways). Many gut bacteria are considered to be pathobionts—symbionts that are capable of acting as pathogens under certain environmental conditions—and are capable of stimulating an immune response. Consequently, the ability of gut bacteria to supress NF-κB activation is often initially assessed using peripheral blood mononuclear cells (PBMC) as several studies have reported that peripheral blood cells predict the in vivo immunomodulatory potential of different bacteria (Foligne et al. 2007; Sokol et al. 2008). Alternatively, NF-κB suppressive capability can be assessed using peripheral blood derived cell lines (e.g. human monocyte-like THP-1 cell line, murine RAW macrophage cell line) stimulated with a specific NF-κB pathway agonist. These cell lines have rapid and reproducible growth characteristics and they express a broad range of Toll-like receptors (TLR) [e.g. THP-1 cells expresses all TLRs including the surface TLRs (i.e. TLR1/2, TLR2, TLR4, TLR5 and TLR6/2)]. These characteristics can be used to identify microbes that modulate NF-κB activity, or that express virulence or cytotoxic factors, in a high-throughput manner.

Despite the usefulness of immune cell lines the ability of gut microbes to suppress NF-κB activity is typically assessed using intestinal epithelial cell culture lines. Numerous epithelial cell lines are widely used by researchers, however, the HT-29, Caco-2 and T84 cell lines and their derivatives are amongst the most widely used to assess immunomodulatory activity. Gut epithelial cells are constantly exposed to microbial factors and are thus broadly unresponsive to stimulation by the healthy gut microbiota. Consistent with this, HT-29, Caco-2 and T84 cells express a subset of functional TLRs (e.g. TLR2, TLR3, TLR4 and TLR5) (Cario et al. 2000; Melmed et al. 2003; Lakhdari et al. 2010), and the cell surface receptors are predominantly expressed basolaterally. Several gut epithelial cell lines carrying NF-κB reporter genes including secreted embryonic alkaline phosphatase (Lakhdari et al. 2010), luciferase (Kaci et al. 2011) and green fluorescent protein (Mastropietro et al. 2015) have been described. Using reporter cell lines, Blottière and colleagues at the Institute National de la Recherche Agronomique (France) have led efforts to identify gut bacteria and metagenomic clones capable of modulating NF-κB expression using high-throughput screening approaches (e.g. Lakhdari et al. 2010, 2011; Santos Rocha et al. 2012). In the most exhaustive study to date, Lakhdari et al. (2011) used a series of immune and intestinal epithelial reporter cells to determine the NF-κB suppressive capabilities of 49 strains of well-described gut bacteria. Interestingly, thirteen NF-κB suppressive strains were identified although their activity was cell line-dependent (one isolate suppressed NF-κB activation in HT-29 cells whereas the other twelve isolates suppressed activation in Caco-2 cells) suggesting that the responsiveness may be affected by the genotype of the cell lines.

While cancer derived intestinal epithelial cell lines may provide biological insights relevant to CRC, a major criticism is that they typically lack the genetic susceptibilities relevant to IBD and obesity. For instance, IBD is associated with over 160 genetic susceptibility loci (McGovern et al. 2010; Jostins et al. 2012) and is characterised by disease heterogeneity with differences in location, severity and extent that may change over time. Host genetics can also influence therapeutic responsiveness and CD carriers of the *nod2* mutation are more likely to be refractory to glucocorticosteroid treatment although they can be effectively treated by TNFα biologics (Niess et al. 2012). While primary cells can be used as an alternative to immortal cell lines, they have a finite life span which typically precludes long-term study. Also, the diversity in cell lineages found in the gut epithelium (e.g. epithelial, goblet, enteroendocrine, Paneth cells) is not reflected in homogenous primary cells or immortal cell lines. This issue has been addressed by recent advances in gut epithelial culture methods from human and laboratory animals which have resulted in the generation of "mini-guts" from intestinal samples containing adult, human embryonic or inducible stem cells that retain the phenotype of the tissue of origin. Mini-guts produced from embryonic or induced pluripotent stem cells are termed induced intestinal organoids while those produced from adult stem cells are termed enteroids (small intestinal) or colonoids (colonic) (Stelzner et al. 2012). Induced intestinal organoid cultures take longer to establish and retain a foetal phenotype and consequently enteroids/colonoids are considered to be a more representative model for human disease. Enteroids/colonoids are derived from intestinal samples containing adult stem cells following cultivation in the presence of growth factors and ultimately form three-dimensional cultures containing differentiated epithelial cells (Sato and Clevers 2013; VanDussen et al. 2015; Mahe et al. 2015). The cultured cells can be grown as spheroids with the apical membrane facing a single internal lumen compartment or, alternatively, they can be grown as monolayers in a transwell system. These cell cultures can also be stably maintained through repeated rounds of propagation and freezing thus recapitulating the main elements of cancer cell culture lines and providing a superior in vitro model to assess NF-κB suppressive capabilities. For instance, the impact of NF-κB suppressive bioactives on individual epithelial cell subtypes could be assessed by fluorescence-activated cell sorting using antibodies targeting the NF-κB complex and lineage specific markers. Enteroids/colonoids can be generated from animals carrying reporter genes or, alternatively, Schwank et al. (2013) reported that the CRISPR/Cas9 system could be used to edit organoid genome sequences. Along with new developments in CRISPR/Cas9-based large fragment deletions and insertions (Wang et al. 2015; Zhang et al. 2015a), this may provide new opportunities to produce patient-specific reporter cell lines. Together, these developments offer new opportunities to identify and dissect disease-specific pathways as well as assess their responsiveness to different therapeutics.

6 Metabolomic-Based Strategies to Identify NF-κB Suppressive Bioactives

The healthy gut microbiota produces a diverse array of factors including proteins (Rieu et al. 2014), peptides (Kaci et al. 2011; Quevrain et al. 2016), polysaccharide-peptidoglycans (Matsumoto et al. 2009) and secondary metabolites (Bansal et al. 2010; Gonzalez-Sarrias et al. 2010; Lim et al. 2015; Lee et al. 2015) that are capable of suppressing NF-κB, revealing this capability is characterised by a high degree of functional redundancy. Metabolomic approaches have played a central role in the identification of these factors although they have been challenged by the sheer diversity of metabolites produced by gut microbes. In addition, many of these metabolites are produced at low concentrations and include novel metabolites that are not represented in existing databases, further hindering identification efforts. Nonetheless, effective bioassay guided fractionation strategies that typically involve successive fractionation coupled with functional assays to track the fraction(s) retaining suppressive activity can be devised to identify NF-κB suppressive bioactive factors (Fig. 4).

Microbes in the healthy gut environment are physically separated from epithelial cells by a mucus layer and bioactive factors must be capable of traversing this barrier to reach their cellular target. Many NF-κB bioactives are secreted into the extracellular milieu and the first stage of the screening process involves the preparation of a cell-free supernatant fraction of spent medium that can be assessed for suppressive activity. The supernatant fraction of most fastidious gut microbes is likely to contain SCFA which are produced by anaerobes as an end product of fermentation, and are amongst the most abundant metabolites produced. Acetate, propionate and butyrate are produced at the highest concentrations with other SCFA

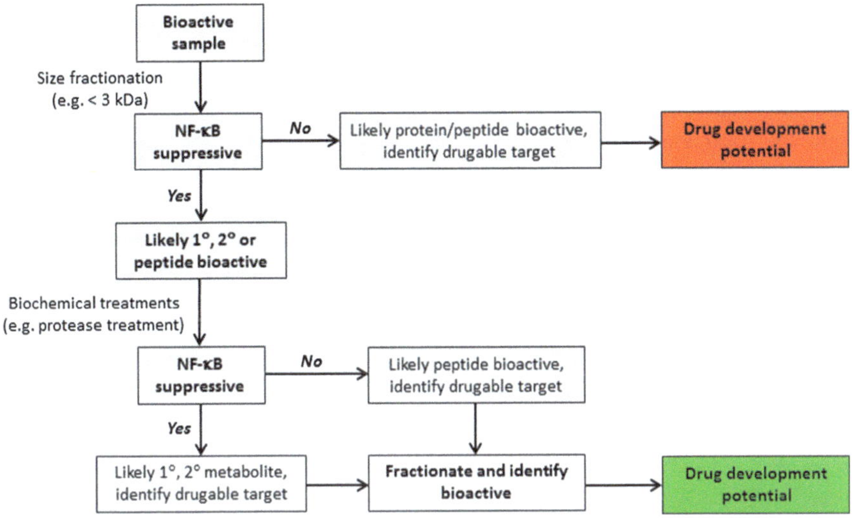

Fig. 4 Bioassay guided fractionation strategy to enrich and purify NF-κB suppressive bioactives. The bioactive fractions are successively fractionated and the NF-κB suppressive activity in the fractions is assessed after each treatment. Fractions enriched in primary and secondary metabolites (1° and 2° respectively), and peptides are typically produced by size fractionation. Fractions >3 kDa are typically not considered to be suitable for drug development but may help to identify new drugable targets. Metabolites and peptide bioactives can be further fractionated using biochemical treatments (e.g. protease, denaturant, thermal treatment). Peptide bioactives can be further fractionated and used to identify new drugable targets or as lead molecules for drug development. The 1° and 2° metabolites can be further fractionated (e.g. solid phase extractions, HPLC based fractionation) to identify the bioactive factor. Secondary metabolites can also be used to identify new drugable targets or as lead molecules for drug development

produced at lower concentrations. SCFA bind to G-protein coupled receptors including GPR41 and GPR43 (acetate, butyrate, propionate), GPR109A (butyrate) and OLFR78 (lactate, propionate). These receptors are found on a range of cells including immune and epithelial cells (Karaki et al. 2008; Pluznick et al. 2013; Tazoe et al. 2009; Thangaraju et al. 2009; Vinolo et al. 2011). SCFA are amongst the most bioactive metabolites produced by the microbiota and they affect a variety of cellular process including NF-κB activity (Inan et al. 2000; Tedelind et al. 2007; Lakhdari et al. 2011). To identify culture supernatants possessing non-SCFA suppressors of NF-κB activity, the SCFA concentrations in the spent culture supernatant are typically first determined and the ability of similar concentrations of SCFA to suppress NF-κB activation is then assessed (Lakhdari et al. 2011).

A variety of fractionation strategies have been described to differentiate between bioactives with specific biochemical characteristics. For instance, small peptides and secondary metabolites are considered to be more conducive to drug development as they are more likely to possess desirable characteristics and be less costly to produce (Uhlig et al. 2014; Fosgerau and Hoffmann 2015; Harvey et al. 2015). These bioactives can often be readily separated from larger macromolecules on the basis of size and a simple 3 kDa molecular weight cut-off filter allows peptides up to 27 amino acids long (assuming an average amino acid size of 110 Da) to be easily separated from larger molecules. Bioactive secondary metabolites and peptides can subsequently be distinguished using routine (e.g. protease, denaturing, thermal) treatments and the sample can subsequently be further fractionated (e.g. solid phase extractions, HPLC based fractionation) to further reduce the complexity of the samples. This approach has been used effectively to identify NF-κB suppressive peptides produced by *Faecalibacterium prausnitzii* (Quevrain et al. 2016). However, the identification of secondary metabolites can be more challenging and it can be necessary to fractionate large sample volumes to identify the metabolites of interest, although this process can be expedited if isogenic mutant or non-suppressive strains can be processed in parallel (Donia et al. 2014). Once sufficiently enriched and concentrated the bioactives can be identified using specialist metabolomic methodologies and equipment.

7 Genetic-Based Strategies to Identify NF-κB Suppressive Bioactives

Much of our understanding of the functional capacity of the microbial world has been provided by the genetic dissection of clinically and agriculturally relevant bacteria that are only distantly related to the microbes that typically inhabit the human gut. The ability to conclusively link genes and function is a central challenge in elucidating the functional potential of the microbiota. However, with limited exceptions (Rey et al. 2013; Ichimura et al. 2013), few molecular tools have been described for the characterisation of gut microbes. The vast majority of the currently available microbial isolates are not known to be amenable to genetic transformation although (meta)genomics has revealed evidence of extensive lateral gene transfer within the gut microbiome. To address this challenge, we recently developed an innovative approach termed metaparental mating that expedites the directed isolation of genetically tractable gut bacteria from mixed microbial communities (Ó Cuív et al. 2015). The metaparental mating approach is based on the well-established biparental mating approach (Simon et al. 1983; Simon et al. 1986) and uses RP4 (RK2)-mediated bacterial conjugation and a broad host range mobilisable shuttle vector. Metaparental mating has several advantages over alternative natural (i.e. transduction, transformation) or contrived (e.g. electroporation, sonoporation) genetic transformation approaches. First, RP4-based conjugation is very promiscuous and has been shown to mediate the transfer of DNA to a diverse range of bacteria (Whitehead and Hespell 1990; Picardeau 2008; Tolonen et al. 2009; Dominguez and O'Sullivan 2013) and also to archaea (Dodsworth et al. 2010), fungi (Nishikawa et al. 1990) and animal cells (Waters 2001). Second, the metaparental mating can be performed under anaerobic conditions and stably transformed recipients can be recovered by selection of a vector encoded marker. In addition, as the antibiotic resistance phenotype of the recipients may not be known, the laboratory *E. coli* ST18 donor strain can be efficiently counter selected without antibiotics by nutritional auxotrophy (i.e. the omission of δ-aminolevulinic acid from the selection medium). Third, the RP4-based conjugation can be readily scaled and automated (Clarke et al. 2005) to increase the throughput of the meta-parental mating mediated isolation process.

We used the metaparental mating approach to specifically target bacteria affiliated with the Firmicutes as these comprise the majority of the human gut microbial core although they are underrepresented in microbial culture collections. Furthermore, few of these bacteria, and in particular those affiliated with the Clostridia, have been genetically characterised although many strains are capable of modulating the host immune response (Sokol et al. 2008; Ivanov et al. 2009; Atarashi et al. 2011, 2013; Li et al. 2014b; Quevrain et al. 2016). In support of this effort we developed a series of modular vectors termed pEHR5 that can be conjugated from an *E. coli* host to a pool of potential recipients. As the efficiency of conjugation can be affected by the size of the vectors, their modular architecture helps minimise their overall size. In addition, it allows individual modules to be easily exchanged ensuring that the base vectors are flexible and can be readily re-purposed. Similar modular vectors have been used in a broad range of non-*E. coli* hosts to support protein expression and the construction of fluorescently labelled bacterial strains (Herrero et al. 1990; Charpentier et al. 2004; Fodor et al. 2004; Heap et al. 2009; Dammeyer et al. 2013; Wright et al. 2015). By this approach, we recovered a broad suite of axenic fastidious gut bacteria affiliated with the Firmicutes that were stably transformed with pEHR5-based vectors. In addition, we demonstrated that the metaparental mating approach and the pEHR vectors can be used for heterologous protein expression by constructing fluorescently labelled gut bacteria (Ó Cuív et al. 2015).

The pEHR5 vector system is freely available to the research community without the need for a restrictive material transfer agreement and it offers a basis for the development of a uniform and streamlined set of molecular tools for the isolation and functional genetic characterisation of fastidious microbes. Nonetheless, the metaparental mating approach can plausibly be applied with any RP4 mobilisable vector bearing an appropriate resistance marker(s) and origin(s) of replication, thus allowing genetically tractable bacteria to be recovered from complex microbial communities. We confirmed this hypothesis by using the narrow host range vector pJQ200sk(+) (Quandt and Hynes 1993) to demonstrate that *E. coli* transconjugants bearing pJQ200sk(+) could be selectively recovered from an anaerobic enrichment from human faeces by metaparental mating (Fig. 5). In addition, we used the vector pGusA*mob* [(Girbal et al. 2003), pGusA modified to carry an *oriT*] to target the recovery of Firmicutes affiliated bacteria and isolated transconjugants affiliated with *Blautia hathewayi*, *Streptococcus pleomorphus* and *Anaerococcus vaginalis* on M10-based medium. We have now also demonstrated that the pEHR vectors can be cured using standard molecular techniques to yield naïve strains (Pottenger and Ó Cuív, Unpublished data).

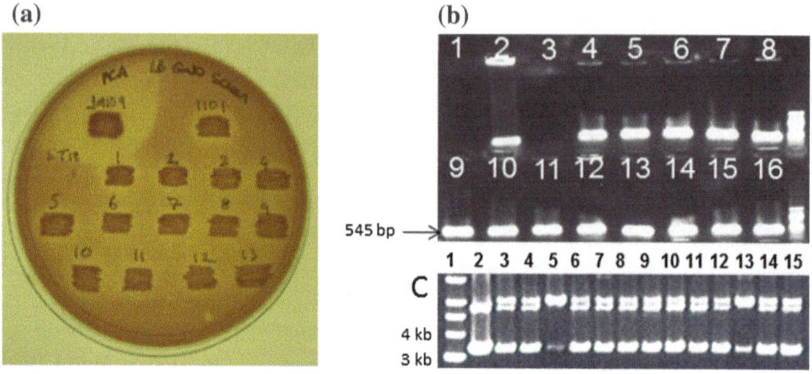

Fig. 5 **a** Identification of *E. coli* transconjugants carrying pJQ200sk(+) (5.4 kb) recovered by metaparental mating. The *E. coli* transconjugants were recovered on LB medium and replica plated onto MacConkey Agar supplemented with 30 μg.ml^{-1} gentamicin sulphate to differentiate between the different laboratory and commensal strains. The laboratory strain *E. coli* JM109 carrying pJQ200sk(+) was characterised by a clear zone around the patched culture consistent with its inability to ferment lactose. In contrast, the recent human gut isolate *E. coli* PC1101 was capable of fermenting lactose and the 13 transconjugants recovered exhibited a similar phenotype. As expected, *E. coli* ST18 did not grow on MacConkey agar due to its nutritional requirement for δ-aminolevulinic acid. **b** The identity of the transconjugants was confirmed by PCR using *E. coli* specific primers (Sabat et al. 2000). Successful confirmation is indicated by a 545 bp product and, as expected, *E. coli* ST18 carrying pJQ200sk(+) (*Lane 2*) and all of the transconjugants (*Lanes 4–16*) produced a product of the correct size. In contrast, no products were observed for PCR lacking DNA template (*Lane 1*) or containing *Campylobacter jejuni* DNA template (*Lane 3*). **c** The presence and integrity of the plasmid vector was assessed by agarose gel electrophoresis. Plasmid vector prepared from each of the 13 transconjugants (*Lanes 3–15*) exhibited similar mobility to plasmid DNA prepared from *E. coli* JM109 (*Lane 2*) confirming that they were stably transformed

The metaparental mating approach and the pEHR vector series are significant developments for the genetic dissection of the gut microbiota by forward and reverse genetic approaches. For instance, it is now known that many gut bacteria carry putative biosynthetic gene clusters for secondary metabolites (Letzel et al. 2013; Donia et al. 2014; Cimermancic et al. 2014; Donia and Fischbach 2015; Hadjithomas et al. 2015), some of which may encode for NF-κB suppressive bioactives. The specific gene clusters underpinning the production of NF-κB suppressive bioactives could potentially be identified by comparative genomics of suppressive and non-suppressive strains, however, reverse genetic approaches can now also plausibly be applied to specifically disrupt target genes and conclusively confirm their role in the production of specific bioactives (Donia et al. 2014). Consistent with this, mutagenesis strategies based on homologous recombination (Al-Hinai et al. 2012; Heap et al. 2012; Faulds-Pain and Wren 2013) and the Ll. *ltrB* group II intron (Chen et al. 2005; Heap et al. 2007; Tolonen et al. 2009) have been described for a diverse range bacteria affiliated with the Firmicutes. While the specific factors underpinning NF-κB suppressive activity have been identified in some cases (e.g. Rieu et al. 2014; Quevrain et al. 2016) in most instances, they remain cryptic and forward genetic approaches including transposon mutagenesis (Liu et al. 2013; Ichimura et al. 2013; Zhang et al. 2015b) and in vivo transposomics (Vidal et al. 2009; Veeranagouda et al. 2012) have been successfully developed for fastidious bacteria. While forward genetic approaches can be applied with fastidious gut microbes they are constrained by the number of mutant clones that have to be screened to achieve good coverage of the genome. For example, assuming a genome size of 4 Mb and an average gene size of 753 bp (Li et al. 2014a), over 12,000 mutants would have to be screened to achieve 90 % coverage of the genome (Clarke and Carbon 1976).

An alternative approach involves the construction of medium/large insert gene libraries (e.g. plasmid, cosmid, fosmid, BAC libraries) that are screened for NF-κB suppressive activity in a suitable microbial host. Using this approach, approximately 230 clones would have to be screened to achieve 90 % coverage of the genome assuming a genome size of 4 Mb and an average insert size of 40 kb. Gene libraries generally assume that all of the genetic elements supporting the immunomodulatory activity are linked, expressed and functional in the microbial host. *E. coli* has relaxed requirements for promoter recognition and this approach has been exploited to identify metagenomic fosmids derived from human gut microbiota that are capable of suppressing/activating NF-κB (Lakhdari et al. 2010; Cohen et al. 2015). Nonetheless, as few as 40 % of heterologous genes are expressed in *E. coli* (Aakvik et al. 2011) and new cloning vectors have been developed that have extended the host range of large insert vectors (e.g. Aakvik et al. 2009; Kakirde et al. 2011). Currently, the replication range of these vectors is mostly limited to proteobacteria and examples of vectors for more distantly related phyla, especially the Firmicutes, are limited (Hain et al. 2008; Liu et al. 2009).

8 Bioprospecting for NF-κB Suppressive Bioactives: *Faecalibacterium prausnitzii* as a Case Study

The butyrate producing gut bacterium, *F. prausnitzii*, comprises part of the core microbiota in healthy adult humans and is ubiquitously found in the gut of mammals and insects (Foglesong et al. 1984; Bjerrum et al. 2006; Castillo et al. 2007; Qin et al. 2010; Nava and Stappenbeck 2011; Miquel et al. 2013; Oikonomou et al. 2013). This suggests that *F. prausnitzii* plays a critical role in host metabolism and physiology and consequently it is widely considered to be a model gut bacterium with relevance to health and disease. In that context, much progress has been made in identifying the true metabolic potential of *F. prasunitzii* and its contribution to health and well being (Sokol et al. 2008; Quevrain et al. 2016). Swidsinski et al. (2008) first reported a reduced population of *F. prausnitzii* in CD subjects and Sokol et al. (2008) subsequently demonstrated a low abundance of *F. prausnitzii* in ileal biopsies from CD subjects at the time of surgery was associated with recurrence six months postoperatively, and that the abundance at six months was consistently lower in subjects with recurrent disease in comparison to those in remission. In support of this observation a longitudinal study with an Australian CD cohort examined the mucosa associated microbial communities in subjects undergoing ileal resection and determined that patients who were in remission 6 months postoperatively had a higher population of *F. prausnitzii* and other members of the Firmicutes at surgery (De Cruz et al. 2015). Notably, changes in the abundance of *F. prausnitzii* in CD subjects have been reported in different ethnic populations (Prideaux et al. 2013) and perturbations have also been reported in other inflammatory and metabolic disorders like ulcerative colitis, coeliac disease, juvenile spondyloarthritis and type 2 diabetes (Sokol et al. 2009; De Palma et al. 2010; Remely et al. 2014; Gill et al. 2015) suggesting that it plays an important role in maintaining gut homeostasis.

The immunomodulatory potential of *F. prausnitzii* A2-165 was first identified by Sokol et al. (2008) who demonstrated that the bacterium exerted an anti-inflammatory effect in PBMCs by inducing IL-10 and suppressing IL-12 and INFγ secretion. They also showed that spent culture supernatant but not sterile medium, UV-killed *F. prausnitzii* or cellular fractions were able to block the activation of NF-κB and reduce IL-8 secretion in Caco-2 cells. Butyrate exerts physiological and anti-inflammatory effects in the gut (Canani et al. 2011; Ploger et al. 2012), however, the presence of butyrate in the spent culture supernatants did not suppress NF-κB activation in Caco-2 cells suggesting that other bioactive factors were responsible for the anti-inflammatory effects (Sokol et al. 2008). Critically, *F. prausnitzii* whole cells as well as filter-sterilised culture supernatant could attenuate the overall severity of trinitrobenzene sulphonic acid induced colitis in BALB/c mice by both a gut-dependent and gut-independent route. Separate studies have also supported these observations and revealed that *F. prausnitzii* and/or its supernatant can induce T_{reg} proliferation (Qiu et al. 2013; Martin et al. 2014), modulate T-cell responses (Rossi et al. 2016) and improve gut barrier function (Carlsson et al. 2013; Martin et al. 2015; Laval et al. 2015), thus also contributing to the suppression of inflammation. Together, these observations indicated that the anti-inflammatory activity could be largely attributed to a secreted bioactive.

In addition to butyrate, it is now known that *F. prausnitzii* produces a range of distinct immunomodulatory bioactives relevant to host health including peptides and secondary metabolites. Using a peptidomic approach, Quevrain et al. (2016) identified 7 peptides derived from a 15 kDa protein termed MAM (Microbial Anti-inflammatory Molecule) that is phylogenetically narrowly distributed. Intracellular expression of the MAM protein in human epithelial cells suppressed NF-κB activation in a specific and dose-dependent manner possibly by affecting IκκB function. Furthermore, *Lactococcus lactis* expressing MAM was capable of ameliorating dinitrobenzene sulphonic acid induced colitis in BALB/c mice. The mechanism of action of the MAM protein remains to be determined including whether its NF-κB suppressive activity is mediated by the intact protein and/or its derived peptides, and how these are delivered to the cell. In addition to MAM, *F. prausnitzii* produces a range of (precursor) anti-inflammatory secondary metabolites. Using a gnotobiotic mouse model, Miquel et al. (2015) revealed that the protective effect of *F. prausnitzii* following colonisation was associated with the presence of salicylic acid and shikimic acid in gut and serum metabolomic profiles. Salicylic acid is a precursor of 5-aminosalycilic acid and is capable of suppressing IL-8 secretion from TNFα stimulated HT-29 cells. In contrast, shikimic acid is not capable of suppressing IL-8 secretion from TNFα stimulated HT-29 cells, however, this molecule is a precursor of anti-inflammatory aromatic compounds including salicylic acid and 3,4-oxo-eisopropylideneshikimic acid (Xing et al. 2013).

These observations underline the role played by *F. prausnitzii* in the maintenance of gut homeostasis and reveal the evolution of a variety of strategies to affect specific aspects of gut function and the immune response. Consistent with this, *F. prausnitzii* has been suggested as a candidate next generation probiotic for the treatment of gut inflammatory diseases (Sokol et al. 2008; Neef and Sanz 2013). Critically, the characterisation of *F. prausnitzii* has provided a template by which the contribution of other microbes to gut health can be examined. It should be noted that although *F. prausnitzii* A2-165 has been known to suppress NF-κB since 2008 and its genome was sequenced in 2009, the specific bioactives supporting this activity remained unidentified until 2015, highlighting the limited capacity of -omic approaches to identify novel functional capabilities. We have now used meta-parental mating to isolate genetically tractable strains of *F. prausnitzii* (Ó Cuív et al. 2015), and we anticipate that these will further expedite the functional dissection of this important gut bacterium.

9 Concluding Remarks and Future Perspective

The healthy gut microbiota plays a vital role in helping to maintain gut homeostasis and preventing the onset of chronic gut disease. Surprisingly, little is known about the essential functionalities that underlie this capability and how they might be exploited to develop more effective therapeutic interventions. The rapid advances in DNA sequencing technologies continue to provide an unprecedented insight into structure-functional activity of the gut microbiome. In contrast, the development of complementary approaches including microbial culturing, functional assays, metabolomics and genetic technologies have not kept pace with these developments. This has hindered efforts to realise the functional potential of the microbiota, however, the successful metabolomic dissection of *F. prausnitzii* will encourage and inform the development of improved methodologies for other gut bacteria. Similarly, new advances in microbial culturing and genetic techniques will provide new opportunities to support a more mechanistic dissection of these functionalities. We anticipate that the effective integration of these disparate yet complementary approaches will afford the best opportunity to effectively bioprospect the gut microbiota and support the discovery of novel bioactives, and the development of new therapeutics.

Acknowledgements We gratefully acknowledge the financial support of the University of Queensland Diamantina Institute to PÓC, SP and MM. SB is the recipient of an Australian Postgraduate Award. The Translational Research Institute is supported by a grant from the Australian Government.

Competing interest

The authors declare no competing interest.

References

Aakvik T, Degnes KF, Dahlsrud R et al (2009) A plasmid RK2-based broad-host-range cloning vector useful for transfer of metagenomic libraries to a variety of bacterial species. FEMS Microbiol Lett 296:149–158

Aakvik T, Lale R, Liles M et al (2011) Metagenomic libraries for functional screening. In: Handbook of molecular microbial ecology I. Wiley, Hoboken

Aguirre de Carcer D, Ó Cuív P, Wang T et al (2011) Numerical ecology validates a biogeographical distribution and gender-based effect on mucosa-associated bacteria along the human colon. ISME J 5:801–809

Al-Hinai MA, Fast AG, Papoutsakis ET (2012) Novel system for efficient isolation of *Clostridium* double-crossover allelic exchange mutants enabling markerless chromosomal gene deletions and DNA integration. Appl Environ Microbiol 78:8112–8121

Anton BP, Kasif S, Roberts RJ et al (2014) Objective: biochemical function. Frontiers Genet 5:210

Atarashi K, Tanoue T, Oshima K et al (2013) Treg induction by a rationally selected mixture of Clostridia strains from the human microbiota. Nature 500:232–236

Atarashi K, Tanoue T, Shima T et al (2011) Induction of colonic regulatory T cells by indigenous *Clostridium* species. Science 331:337–341

Atreya I, Atreya R, Neurath MF (2008) NF-κB in inflammatory bowel disease. J Intern Med 263:591–596

Backhed F, Ley RE, Sonnenburg JL et al (2005) Host-bacterial mutualism in the human intestine. Science 307:1915–1920

Balch WE, Fox GE, Magrum LJ et al (1979) Methanogens: reevaluation of a unique biological group. Microbiol Rev 43:260–296

Bansal T, Alaniz RC, Wood TK et al (2010) The bacterial signal indole increases epithelial-cell tight-junction resistance and attenuates indicators of inflammation. Proc Natl Acad Sci U S A 107:228–233

Barcenilla A, Pryde SE, Martin JC et al (2000) Phylogenetic relationships of butyrate-producing bacteria from the human gut. Appl Environ Microbiol 66:1654–1661

Bassaganya-Riera J, Viladomiu M, Pedragosa M et al (2012) Immunoregulatory mechanisms underlying prevention of colitis-associated colorectal cancer by probiotic bacteria. PLoS ONE 7:e34676

Bjerrum L, Engberg RM, Leser TD et al (2006) Microbial community composition of the ileum and cecum of broiler chickens as revealed by molecular and culture-based techniques. Poult Sci 85:1151–1164

Bomar L, Maltz M, Colston S et al (2011) Directed culturing of microorganisms using metatranscriptomics. MBio 2:e00012–11

Bordenstein SR, Theis KR (2015) Host biology in light of the microbiome: ten principles of holobionts and hologenomes. PLoS Biol 13:e1002226

Bray F, Ren JS, Masuyer E et al (2013) Global estimates of cancer prevalence for 27 sites in the adult population in 2008. Int J Cancer 132:1133–1145

Browne HP, Forster SC, Anonye BO et al (2016) Culturing of 'unculturable' human microbiota reveals novel taxa and extensive sporulation. Nature (advance online publication)

Bryant MP (1972) Commentary on the Hungate technique for culture of anaerobic bacteria. Am J Clin Nutr 25:1324–1328

Button DK, Schut F, Quang P et al (1993) Viability and isolation of marine bacteria by dilution culture: theory, procedures, and initial results. Appl Env Microbiol 59:881–891

Canani RB, Costanzo MD, Leone L et al (2011) Potential beneficial effects of butyrate in intestinal and extraintestinal diseases. World J Gastroenterol 17:1519–1528

Cario E, Rosenberg IM, Brandwein SL et al (2000) Lipopolysaccharide activates distinct signaling pathways in intestinal epithelial cell lines expressing toll-like receptors. J Immunol 164:966–972

Carlsson AH, Yakymenko O, Olivier I et al (2013) *Faecalibacterium prausnitzii* supernatant improves intestinal barrier function in mice DSS colitis. Scand J Gastroenterol 48:1136–1144

Castillo M, Skene G, Roca M et al (2007) Application of 16S rRNA gene-targetted fluorescence in situ hybridization and restriction fragment length polymorphism to study porcine microbiota along the gastrointestinal tract in response to different sources of dietary fibre. FEMS Microbiol Ecol 59:138–146

Charpentier E, Anton AI, Barry P et al (2004) Novel cassette-based shuttle vector system for gram-positive bacteria. Appl Environ Microbiol 70:6076–6085

Chen Y, Mcclane BA, Fisher DJ et al (2005) Construction of an alpha toxin gene knockout mutant of *Clostridium perfringens* type A by use of a mobile group II intron. Appl Environ Microbiol 71:7542–7547

Cimermancic P, Medema MH, Claesen J et al (2014) Insights into secondary metabolism from a global analysis of prokaryotic biosynthetic gene clusters. Cell 158:412–421

Clarke L, Carbon J (1976) A colony bank containing synthetic Col El hybrid plasmids representative of the entire *E. coli* genome. Cell 9:91–99

Clarke P, Ó Cuív P, O'Connell M (2005) Novel mobilizable prokaryotic two-hybrid system vectors for high-throughput protein interaction mapping in *Escherichia coli* by bacterial conjugation. Nucleic Acids Res 33:e18

Cohen LJ, Kang HS, Chu J et al (2015) Functional metagenomic discovery of bacterial effectors in the human microbiome and isolation of commendamide, a GPCR G2A/132 agonist. Proc Natl Acad Sci USA 112:E4825–E4834

Cotillard A, Kennedy SP, Kong LC et al (2013) Dietary intervention impact on gut microbial gene richness. Nature 500:585–588

Dammeyer T, Timmis KN, Tinnefeld P (2013) Broad host range vectors for expression of proteins with (Twin-) Strep-tag, His-tag and engineered, export optimized yellow fluorescent protein. Microb Cell Fact 12:49

Davies J (2001) In a map for human life, count the microbes, too. Science 291:2316

De Cruz P, Kang S, Wagner J et al (2015) Association between specific mucosa-associated microbiota in Crohn's disease at the time of resection and subsequent disease recurrence: a pilot study. J Gastroenterol Hepatol 30:268–278

De Palma G, Nadal I, Medina M et al (2010) Intestinal dysbiosis and reduced immunoglobulin-coated bacteria associated with coeliac disease in children. BMC Microbiol 10:63

Derrien M, Vaughan EE, Plugge CM et al (2004) *Akkermansia muciniphila* gen. nov., sp. nov., a human intestinal mucin-degrading bacterium. Int J Syst Evol Microbiol 54:1469–1476

Dey N, Soergel DA, Repo S et al (2013) Association of gut microbiota with post-operative clinical course in Crohn's disease. BMC Gastroenterol 13:131

Dione N, Khelaifia S, La Scola B et al (2015) A quasi-universal medium to break the aerobic/anaerobic bacterial culture dichotomy in clinical microbiology. Clin Microbiol Infect

Dobbs R, Sawers C, Thompson F et al (2014) Overcoming obesity: an initial economic analysis. McKinsey Global Institute, Europe

Dodsworth JA, Li L, Wei S et al (2010) Interdomain conjugal transfer of DNA from bacteria to archaea. Appl Environ Microbiol 76:5644–5647

Dominguez W, O'sullivan DJ (2013) Developing an efficient and reproducible conjugation-based gene transfer system for bifidobacteria. Microbiology 159:328–338

Donaldson GP, Lee SM, Mazmanian SK (2015) Gut biogeography of the bacterial microbiota. Nat Rev Micro 14:20–32 (advance online publication)
Donath MY (2014) Targeting inflammation in the treatment of type 2 diabetes: time to start. Nat Rev Drug Discov 13:465–476
Donia MS, Cimermancic P, Schulze CJ et al (2014) A systematic analysis of biosynthetic gene clusters in the human microbiome reveals a common family of antibiotics. Cell 158:1402–1414
Donia MS, Fischbach MA (2015) Small molecules from the human microbiota. Science 349:1254766
Donohoe DR, Holley D, Collins LB et al (2014) A gnotobiotic mouse model demonstrates that dietary fiber protects against colorectal tumorigenesis in a microbiota-and butyrate-dependent manner. Cancer Discov 4:1387–1397
Dowell VR, Hawkins TM, Control CFD (1981) Laboratory methods in anaerobic bacteriology. U. S. Department of Health and Human Services, Public Health Service, Centers for Disease Control, USA
Dubourg G, Lagier JC, Armougom F et al (2013) The gut microbiota of a patient with resistant tuberculosis is more comprehensively studied by culturomics than by metagenomics. Eur J Clin Microbiol Infect Dis 32:637–645
Eckburg PB, Bik EM, Bernstein CN et al (2005) Diversity of the human intestinal microbial flora. Science 308:1635–1638
Eeckhaut V, Machiels K, Perrier C et al (2012) *Butyricicoccus pullicaecorum* in inflammatory bowel disease. Gut 62:1745–1752
Ehrlich SD (2010) Metagenomics of the intestinal microbiota: potential applications. Gastroenterol Clin Biol 34(Supplement 1):S23–S28
Eller C, Crabill MR, Bryant MP (1971) Anaerobic roll tube media for nonselective enumeration and isolation of bacteria in human feces. Appl Microbiol 22:522–529
Ellrott K, Jaroszewski L, Li W et al (2010) Expansion of the protein repertoire in newly explored environments: human gut microbiome specific protein amilies. PLoS Comput Biol 6:e1000798
Esser N, Paquot N, Scheen AJ (2015) Anti-inflammatory agents to treat or prevent type 2 diabetes, metabolic syndrome and cardiovascular disease. Expert Opin Investig Drugs 24:283–307
Faulds-Pain A, Wren BW (2013) Improved bacterial mutagenesis by high-frequency allele exchange, demonstrated in *Clostridium difficile* and *Streptococcus suis*. Appl Environ Microbiol 79:4768–4771
Ferraris L, Aires J, Waligora-Dupriet AJ et al (2010) New selective medium for selection of bifidobacteria from human feces. Anaerobe 16:469–471
Fish JA, Chai B, Wang Q et al (2013) FunGene: the functional gene pipeline and repository. Front Microbiol 4:291
Fodor AA, Desantis TZ, Wylie KM et al (2012) The "most wanted" taxa from the human microbiome for whole genome sequencing. PLoS ONE 7:e41294
Fodor BD, Kovacs AT, Csaki R et al (2004) Modular broad-host-range expression vectors for single-protein and protein complex purification. Appl Environ Microbiol 70:712–721
Foglesong MA, Cruden DL, Markovetz AJ (1984) Pleomorphism of fusobacteria isolated from the cockroach hindgut. J Bacteriol 158:474–480
Foligne B, Nutten S, Grangette C et al (2007) Correlation between in vitro and in vivo immunomodulatory properties of lactic acid bacteria. World J Gastroenterol 13:236–243
Fosgerau K, Hoffmann T (2015) Peptide therapeutics: current status and future directions. Drug Discov Today 20:122–128
Franzén O, Hu J, Bao X et al (2015) Improved OTU-picking using long-read 16S rRNA gene amplicon sequencing and generic hierarchical clustering. Microbiome 3:1–14
Galperin MY, Koonin EV (2010) From complete genome sequence to "complete" understanding? Trends Biotechnol 28:398–406
Gilbert SF (2014) A holobiont birth narrative: the epigenetic transmission of the human microbiome. Front Genet 5:282
Gill SR, Pop M, Deboy RT et al (2006) Metagenomic analysis of the human distal gut microbiome. Science 312:1355–1359

Gill T, Asquith M, Rosenbaum JT et al (2015) The intestinal microbiome in spondyloarthritis. Curr Opin Rheumatol 27:319–325

Girbal L, Mortier-Barriere I, Raynaud F et al (2003) Development of a sensitive gene expression reporter system and an inducible promoter-repressor system for *Clostridium acetobutylicum*. Appl Environ Microbiol 69:4985–4988

Gonzalez-Sarrias A, Larrosa M, Tomas-Barberan FA et al (2010) NF-kappaB-dependent anti-inflammatory activity of urolithins, gut microbiota ellagic acid-derived metabolites, in human colonic fibroblasts. Br J Nutr 104:503–512

Goodman AL, Kallstrom G, Faith JJ et al (2011) Extensive personal human gut microbiota culture collections characterized and manipulated in gnotobiotic mice. Proc Natl Acad Sci USA 108:6252–6257

Greenblum S, Carr R, Borenstein E (2015) Extensive strain-level copy-number variation across human gut microbiome species. Cell 160:583–594

Hadjithomas M, Chen IM, Chu K et al (2015) IMG-ABC: a knowledge base to fuel discovery of biosynthetic gene clusters and novel secondary metabolites. MBio 6:e00932

Hain T, Otten S, von both U et al (2008) Novel bacterial artificial chromosome vector pUvBBAC for use in studies of the functional genomics of *Listeria* spp. Appl Environ Microbiol 74:1892–1901

Harvey AL, Edrada-Ebel R, Quinn RJ (2015) The re-emergence of natural products for drug discovery in the genomics era. Nat Rev Drug Discov 14:111–129

Heap JT, Ehsaan M, Cooksley CM et al (2012) Integration of DNA into bacterial chromosomes from plasmids without a counter-selection marker. Nucleic Acids Res 40:e59–e59

Heap JT, Pennington OJ, Cartman ST et al (2007) The ClosTron: a universal gene knock-out system for the genus *Clostridium*. J Microbiol Methods 70:452

Heap JT, Pennington OJ, Cartman ST et al (2009) A modular system for *Clostridium* shuttle plasmids. J Microbiol Methods 78:79–85

Herrero M, De Lorenzo V, Timmis KN (1990) Transposon vectors containing non-antibiotic resistance selection markers for cloning and stable chromosomal insertion of foreign genes in gram negative bacteria. J Bacteriol 172:6557–6567

Heuvelin E, Lebreton C, Grangette C et al (2009) Mechanisms involved in alleviation of intestinal inflammation by *Bifidobacterium breve* soluble factors. PLoS ONE 4:e5184

Hu P, Janga SC, Babu M et al (2009) Global functional atlas of *Escherichia coli* encompassing previously uncharacterized proteins. PLoS Biol 7:e96

Hungate RE (1969) A roll tube method for cultivation of strict anaerobes. In: Norris JR, Ribbons DW (eds) Methods in microbiology. Academic Press, Cambridge

Ichimura M, Uchida K, Nakayama-Imaohji H et al (2013) Mariner-based transposon mutagenesis for *Bacteroides* species. J Basic Microbiol 54:558–567

Inan MS, Rasoulpour RJ, Yin L et al (2000) The luminal short-chain fatty acid butyrate modulates NF-kappaB activity in a human colonic epithelial cell line. Gastroenterology 118:724–734

International Human Genome Sequencing Consortium (2004) Finishing the euchromatic sequence of the human genome. Nature 431:931–945

Isenberg HD, Goldberg D, Sampson J (1970) Laboratory studies with a selective *Enterococcus* medium. Appl Microbiol 20:433–436

Ivanov II, Atarashi K, Manel N et al (2009) Induction of intestinal Th17 cells by segmented filamentous bacteria. Cell 139:485

Jalanka-Tuovinen J, Salonen A, Nikkilä J et al (2011) Intestinal microbiota in healthy adults: temporal analysis reveals individual and common core and relation to intestinal symptoms. PLoS ONE 6:e23035

Joice R, Yasuda K, Shafquat A et al (2014) Determining microbial products and identifying molecular targets in the human microbiome. Cell Metab 20:731–741

Jostins L, Ripke S, Weersma RK et al (2012) Host-microbe interactions have shaped the genetic architecture of inflammatory bowel disease. Nature 491:119–124

Kaci G, Goudercourt D, Dennin V et al (2013) Anti-inflammatory properties of *Streptococcus salivarius* a commensal bacterium of the oral cavity and digestive tract. Appl Environ Microbiol 80:928–934

Kaci G, Lakhdari O, Dore J et al (2011) Inhibition of the NF-kappaB pathway in human intestinal epithelial cells by commensal *Streptococcus salivarius*. Appl Environ Microbiol 77:4681–4684

Kadooka Y, Sato M, Imaizumi K et al (2010) Regulation of abdominal adiposity by probiotics (*Lactobacillus gasseri* SBT2055) in adults with obese tendencies in a randomized controlled trial. Eur J Clin Nutr 64:636–643

Kakirde KS, Wild J, Godiska R et al (2011) Gram negative shuttle BAC vector for heterologous expression of metagenomic libraries. Gene 475:57–62

Kang S, Denman SE, Morrison M et al (2010) Dysbiosis of fecal microbiota in Crohn's disease patients as revealed by a custom phylogenetic microarray. Inflamm Bowel Dis 16:2034–2042

Kaplan GG (2015) The global burden of IBD: from 2015 to 2025. Nat Rev Gastroenterol Hepatol 12:720–727

Karaki S, Tazoe H, Hayashi H et al (2008) Expression of the short-chain fatty acid receptor, GPR43, in the human colon. J Mol Histol 39:135–142

Kelly D, Campbell JI, King TP et al (2004) Commensal anaerobic gut bacteria attenuate inflammation by regulating nuclear-cytoplasmic shuttling of PPAR-gamma and RelA. Nat Immunol 5:104–112

Kenters N, Henderson G, Jeyanathan J et al (2011) Isolation of previously uncultured rumen bacteria by dilution to extinction using a new liquid culture medium. J Microbiol Methods 84:52–60

Khokhlova EV, Smeianov VV, Efimov BA et al (2012) Anti-inflammatory properties of intestinal *Bifidobacterium* strains isolated from healthy infants. Microbiol Immunol 56:27–39

Kim M, Morrison M, Yu Z (2011) Evaluation of different partial 16S rRNA gene sequence regions for phylogenetic analysis of microbiomes. J Microbiol Methods 84:81–87

Klindworth A, Pruesse E, Schweer T et al (2013) Evaluation of general 16S ribosomal RNA gene PCR primers for classical and next-generation sequencing-based diversity studies. Nucleic Acids Res 41:e1–e1

La Scola B, Khelaifia S, Lagier JC et al (2014) Aerobic culture of anaerobic bacteria using antioxidants: a preliminary report. Eur J Clin Microbiol Infect Dis 33:1781–1783

Lagier JC, Armougom F, Million M et al (2012) Microbial culturomics: paradigm shift in the human gut microbiome study. Clin Microbiol Infect 18:1185–1193

Lakhdari O, Cultrone A, Tap J et al (2010) Functional metagenomics: a high throughput screening method to decipher microbiota-driven NF-κB modulation in the human gut. PLoS ONE 5: e13092

Lakhdari O, Tap J, Béguet-crespel F et al (2011) Identification of NF-κB modulation capabilities within human intestinal commensal bacteria. J Biomed Biotechnol

Land M, Hauser L, Jun SR et al (2015) Insights from 20 years of bacterial genome sequencing. Funct Integr Genom 15:141–161

Lander ES, Linton LM, Birren B et al (2001) Initial sequencing and analysis of the human genome. Nature 409:860–921

Laval L, Martin R, Natividad JN et al (2015) *Lactobacillus rhamnosus* CNCM I-3690 and the commensal bacterium *Faecalibacterium prausnitzii* A2-165 exhibit similar protective effects to induced barrier hyper-permeability in mice. Gut Microbes 6:1–9

Lee SY, Jeong JJ, Le TH et al (2015) Ocotillol, a majonoside R2 metabolite, ameliorates 2,4,6-trinitrobenzenesulfonic acid-induced colitis in mice by restoring the balance of Th17/Treg cells. J Agric Food Chem 63:7024–7031

Lepage P, Seksik P, Sutren M et al (2005) Biodiversity of the mucosa-associated microbiota is stable along the distal digestive tract in healthy individuals and patients with IBD. Inflamm Bowel Dis 11:473–480

Letzel AC, Pidot SJ, Hertweck C (2013) A genomic approach to the cryptic secondary metabolome of the anaerobic world. Nat Prod Rep 30:392–428

Li K, Bihan M, Methe BA (2013) Analyses of the stability and core taxonomic memberships of the human microbiome. PLoS ONE 8:e63139

Li J, Jia H, Cai X et al (2014a) An integrated catalog of reference genes in the human gut microbiome. Nat Biotechnol 32:834–841

Li YN, Huang F, Cheng HJ et al (2014b) Intestine-derived *Clostridium leptum* induces murine tolerogenic dendritic cells and regulatory T cells in vitro. Hum Immunol 75:1232–1238

Lim SM, Jeong JJ, Kang GD et al (2015) Timosaponin AIII and its metabolite sarsasapogenin ameliorate colitis in mice by inhibiting NF-kappaB and MAPK activation and restoring Th17/Treg cell balance. Int Immunopharmacol 25:493–503

Liu H, Bouillaut L, Sonenshein AL et al (2013) Use of a mariner-based transposon mutagenesis system to isolate *Clostridium perfringens* mutants deficient in gliding motility. J Bacteriol 195:629–636

Liu X, Peng D, Luo Y et al (2009) Construction of an *Escherichia coli* to *Bacillus thuringiensis* shuttle vector for large DNA fragments. Appl Microbiol Biotechnol 82:765–772

Livingston SJ, Kominos SD, Yee RB (1978) New medium for selection and presumptive identification of the *Bacteroides fragilis* group. J Clin Microbiol 7:448–453

Locke AE, Kahali B, Berndt SI et al (2015) Genetic studies of body mass index yield new insights for obesity biology. Nature 518:197–206

Loman NJ, Constantinidou C, Chan JZM et al (2012) High-throughput bacterial genome sequencing: an embarrassment of choice, a world of opportunity. Nat Rev Micro 10:599–606

Lozupone CA, Stombaugh JI, Gordon JI et al (2012) Diversity, stability and resilience of the human gut microbiota. Nature 489:220–230

Ma L, Kim J, Hatzenpichler R et al (2014) Gene-targeted microfluidic cultivation validated by isolation of a gut bacterium listed in human microbiome project's most wanted taxa. Proc Natl Acad Sci USA 111:9768–9773

Macy JM, Snellen JE, Hungate RE (1972) Use of syringe methods for anaerobiosis. Am J Clin Nutrition 25:1318–1323

Mahe MM, Sundaram N, Watson CL et al (2015) Establishment of human epithelial enteroids and colonoids from whole tissue and biopsy. J Vis Exp. doi:10.3791/52483

Martin R, Chain F, Miquel S et al (2014) The commensal bacterium *Faecalibacterium prausnitzii* is protective in DNBS-induced chronic moderate and severe colitis models. Inflamm Bowel Dis 20:417–430

Martin R, Miquel S, Chain F et al (2015) *Faecalibacterium prausnitzii* prevents physiological damages in a chronic low-grade inflammation murine model. BMC Microbiol 1:67

Martínez I, Muller CE, Walter J (2013) Long-term temporal analysis of the human fecal microbiota revealed a stable core of dominant bacterial species. PLoS ONE 8:e69621

Mastropietro G, Tiscornia I, Perelmuter K et al (2015) HT-29 and Caco-2 reporter cell lines for functional studies of nuclear factor kappa b activation. Mediators Inflamm 2015:13

Matsumoto S, Hara T, Nagaoka M et al (2009) A component of polysaccharide peptidoglycan complex on Lactobacillus induced an improvement of murine model of inflammatory bowel disease and colitis-associated cancer. Immunology 128:e170–e180

McGovern DPB, Gardet A, Torkvist L et al (2010) Genome-wide association identifies multiple ulcerative colitis susceptibility loci. Nat Genet 42:332–337

McSweeney CS, Denman SE, Mackie RI (2005) Rumen bacteria. In: Makkar HPS, McSweeney CS (eds) Methods in gut microbial ecology for ruminants. Springer, Dordrecht

Melmed G, Thomas LS, Lee N et al (2003) Human intestinal epithelial cells are broadly unresponsive to Toll-like receptor 2-dependent bacterial ligands: implications for host-microbial interactions in the gut. J Immunol 170:1406–1415

Ménard S, Candalh C, Bambou JC et al (2004) Lactic acid bacteria secrete metabolites retaining anti-inflammatory properties after intestinal transport. Gut 53:821–828

Meng J, Kanzaki G, Meas D et al (2012) A genome-wide inducible phenotypic screen identifies antisense RNA constructs silencing *Escherichia coli* essential genes. FEMS Microbiol Lett 329:45–53

Mihajlovski A, Alric M, Brugère JF (2008) A putative new order of methanogenic archaea inhabiting the human gut, as revealed by molecular analyses of the mcrA gene. Res Microbiol 159:516–521

Miquel S, Leclerc M, Martin R et al (2015) Identification of metabolic signatures linked to anti-inflammatory effects of *Faecalibacterium prausnitzii*. MBio 6

Miquel S, Martin R, Rossi O et al (2013) *Faecalibacterium prausnitzii* and human intestinal health. Curr Opin Microbiol 16:255–261

Molodecky NA, Soon IS, Rabi DM et al (2012) Increasing incidence and prevalence of the inflammatory bowel diseases with time, based on systematic review. Gastroenterol 142:46–54. e42; quiz e30.

Mondot S, Kang S, Furet JP et al (2011) Highlighting new phylogenetic specificities of Crohn's disease microbiota. Inflamm Bowel Dis 17:185–192

Murphy EF, Cotter PD, Hogan A et al (2013) Divergent metabolic outcomes arising from targeted manipulation of the gut microbiota in diet-induced obesity. Gut 62:220–226

Nakatsu G, Li X, Zhou H et al (2015) Gut mucosal microbiome across stages of colorectal carcinogenesis. Nat Commun 6:8727

Nam YD, Chang HW, Kim KH et al (2008) Bacterial, archaeal, and eukaryal diversity in the intestines of Korean people. J Microbiol 46:491–501

Nava GM, Stappenbeck TS (2011) Diversity of the autochthonous colonic microbiota. Gut Microbes 2

Neef A, Sanz Y (2013) Future for probiotic science in functional food and dietary supplement development. Curr Opin Clin Nutr Metab Care 16:679–687

Ng M, Fleming T, Robinson M et al (2014) Global, regional, and national prevalence of overweight and obesity in children and adults during 1980–2013: a systematic analysis for the Global Burden of Disease Study 2013. Lancet 384:766–781

Nichols RJ, Sen S, Choo YJ et al (2011) Phenotypic landscape of a bacterial cell. Cell 144:143–156

Niess JH, Klaus J, Stephani J et al (2012) NOD2 polymorphism predicts response to treatment in Crohn's disease - first steps to a personalized therapy. Dig Dis Sci 57:879–886

Nishikawa M, Suzuki K, Yoshida K (1990) Structural and functional stability of IncP plasmids during stepwise transmission by trans-kingdom mating: promiscuous conjugation of *Escherichia coli* and *Saccharomyces cerevisiae*. Jpn J Genet 65:323–334

Ó Cuív P, de Aguirre cárcer D, Jones M et al (2011a) The effects from DNA extraction methods on the evaluation of microbial diversity associated with human colonic tissue. Microb Ecol 61:353–362

Ó Cuív P, Klaassens ES, Durkin AS et al (2011b) Draft genome sequence of *Turicibacter sanguinis* PC909, isolated from human feces. J Bacteriol 193:1288–1289

Ó Cuív P, Smith WJ, Pottenger S et al (2015) Isolation of genetically tractable most-wanted bacteria by metaparental mating. Sci Rep 5:13282

Obata T, Goto Y, Kunisawa J et al (2010) Indigenous opportunistic bacteria inhabit mammalian gut-associated lymphoid tissues and share a mucosal antibody-mediated symbiosis. Proc Natl Acad Sci USA 107:7419–7424

Oberhardt MA, Zarecki R, Gronow S et al (2015) Harnessing the landscape of microbial culture media to predict new organism-media pairings. Nat Commun 6

Ochman H, Elwyn S, Moran NA (1999) Calibrating bacterial evolution. Proc Natl Acad Sci USA 96:12638–12643

Oikonomou G, Teixeira AG, Foditsch C et al (2013) Fecal microbial diversity in pre-weaned dairy calves as described by pyrosequencing of metagenomic 16S rDNA. Associations of *Faecalibacterium* species with health and growth. PLoS ONE 8:e63157

Ott SJ, Musfeldt M, Timmis KN et al (2004) In vitro alterations of intestinal bacterial microbiota in fecal samples during storage. Diagn Microbiol Infect Dis 50:237–245

Oxley APA, Lanfranconi MP, Würdemann D et al (2010) Halophilic archaea in the human intestinal mucosa. Environ Microbiol 12:2398–2410

Paradis-Bleau C, Kritikos G, Orlova K et al (2014) A genome-wide screen for bacterial envelope biogenesis mutants identifies a novel factor involved in cell wall precursor metabolism. PLoS Genet 10:e1004056

Pedron T, Mulet C, Dauga C et al (2012) A crypt-specific core microbiota resides in the mouse colon. MBio:3

Peris-Bondia F, Latorre A, Artacho A et al (2011) The active human gut microbiota differs from the total microbiota. PLoS ONE 6:e22448

Peters U, Bien S, Zubair N (2015) Genetic architecture of colorectal cancer. Gut 64:1623–1636

Peterson J, Garges S, Giovanni M et al (2009) The NIH human microbiome project. Genome Res 19:2317–2323

Petrof EO, Claud EC, Sun J et al (2009) Bacteria-free solution derived from *Lactobacillus plantarum* inhibits multiple NF-kappaB pathways and inhibits proteasome function. Inflamm Bowel Dis 15:1537–1547

Picardeau M (2008) Conjugative transfer between *Escherichia coli* and *Leptospira* spp. as a new genetic tool. Appl Environ Microbiol 74:319–322

Ploger S, Stumpff F, Penner GB et al (2012) Microbial butyrate and its role for barrier function in the gastrointestinal tract. Ann NY Acad Sci 1258:52–59

Pluznick JL, Protzko RJ, Gevorgyan H et al (2013) Olfactory receptor responding to gut microbiota-derived signals plays a role in renin secretion and blood pressure regulation. Proc Natl Acad Sci USA 110:4410–4415

Pope PB, Smith W, Denman SE et al (2011) Isolation of Succinivibrionaceae implicated in low methane emissions from Tammar wallabies. Science 333:646–648

Prideaux L, Kang S, Wagner J et al (2013) Impact of ethnicity, geography, and disease on the microbiota in health and inflammatory bowel disease. Inflamm Bowel Dis 19:2906–2918

Promponas VJ, Iliopoulos I, Ouzounis CA (2015) Annotation inconsistencies beyond sequence similarity-based function prediction—phylogeny and genome structure. Stand Genomic Sci 10:108

Qin J, Li R, Raes J et al (2010) A human gut microbial gene catalogue established by metagenomic sequencing. Nature 464:59–65

Qiu X, Zhang M, Yang X et al (2013) *Faecalibacterium prausnitzii* upregulates regulatory T cells and anti-inflammatory cytokines in treating TNBS-induced colitis. J Crohns Colitis 7:e558–e568

Quandt J, Hynes MF (1993) Versatile suicide vectors which allow direct selection for gene replacement in gram-negative bacteria. Gene 127:15–21

Quercia S, Candela M, Giuliani C et al (2014) From lifetime to evolution: timescales of human gut microbiota adaptation. Front Microbiol 5:587

Quevrain E, Maubert MA, Michon C et al (2016) Identification of an anti-inflammatory protein from *Faecalibacterium prausnitzii*, a commensal bacterium deficient in Crohn's disease. Gut 65:415–425

Quince C, Ijaz UZ, Loman N et al (2015) Extensive modulation of the fecal metagenome in children with Crohn's disease during exclusive enteral nutrition. Am J Gastroenterol 110:1718–1729

Rajagopala SV, Sikorski P, Kumar A et al (2014) The binary protein-protein interaction landscape of *Escherichia coli*. Nat Biotechnol 32:285–290

Rajilić-Stojanović M, de Vos WM (2014) The first 1000 cultured species of the human gastrointestinal microbiota. FEMS Microbiol Rev 38:996–1047

Rajilic-Stojanovic M, Heilig HG, Tims S et al (2012) Long-term monitoring of the human intestinal microbiota composition. Environ Microbiol 15:1146–1159

Rajilic-Stojanovic M, Smidt H, de Vos WM (2007) Diversity of the human gastrointestinal tract microbiota revisited. Environ Microbiol 9:2125–2136

Reeves AE, Koenigsknecht MJ, Bergin IL et al (2012) Suppression of *Clostridium difficile* in the gastrointestinal tracts of germfree mice inoculated with a Murine isolate from the family Lachnospiraceae. Infect Immun 80:3786–3794

Relman DA, Falkow S (2001) The meaning and impact of the human genome sequence for microbiology. Trends Microbiol 9:206–208
Remely M, Aumueller E, Jahn D et al (2014) Microbiota and epigenetic regulation of inflammatory mediators in type 2 diabetes and obesity. Benef Microbes 5:33–43
Rettedal EA, Gumpert H, Sommer MOA (2014) Cultivation-based multiplex phenotyping of human gut microbiota allows targeted recovery of previously uncultured bacteria. Nat Commun 5
Rey FE, Gonzalez MD, Cheng J et al (2013) Metabolic niche of a prominent sulfate-reducing human gut bacterium. Proc Natl Acad Sci USA. 110:13582–13587
Rieu-Lesme F, Delbes C, Sollelis L (2005) Recovery of partial 16S rDNA sequences suggests the presence of Crenarchaeota in the human digestive ecosystem. Curr Microbiol 51:317–321
Rieu A, Aoudia N, Jego G et al (2014) The biofilm mode of life boosts the anti-inflammatory properties of *Lactobacillus*. Cell Microbiol 16:1836–1853
Roberts RJ (2004) Identifying protein function—a call for community action. PLoS Biol 2:E42
Rose C, Parker A, Jefferson B et al (2015) The characterization of feces and urine: a review of the literature to inform advanced treatment technology. Crit Rev Environ Sci Technol 45:1827–1879
Rossi O, Van Berkel LA, Chain F et al (2016) *Faecalibacterium prausnitzii* A2-165 has a high capacity to induce IL-10 in human and murine dendritic cells and modulates T cell responses. Sci Rep 6:18507
Sabat G, Rose P, Hickey WJ et al (2000) Selective and sensitive method for PCR amplification of *Escherichia coli* 16S rRNA genes in soil. Appl Environ Microbiol 66:844–849
Sakamoto K, Maeda S (2010) Targeting NF-kappaB for colorectal cancer. Expert Opin Ther Targets 14:593–601
Sandrin TR, Goldstein JE, Schumaker S (2013) MALDI TOF MS profiling of bacteria at the strain level: a review. Mass Spectrom Rev 32:188–217
Santos Rocha C, Lakhdari O, Blottiere HM et al (2012) Anti-inflammatory properties of dairy lactobacilli. Inflamm Bowel Dis 18:657–666
Sato T, Clevers H (2013) Growing self-organizing mini-guts from a single intestinal stem cell: mechanism and applications. Science 340:1190–1194
Savage DC (1977) Microbial ecology of the gastrointestinal tract. Annu Rev Microbiol 31:107–133
Scanlan PD, Shanahan F, Marchesi JR (2008) Human methanogen diversity and incidence in healthy and diseased colonic groups using mcrA gene analysis. BMC Microbiol 8:79
Schaubeck M, Clavel T, Calasan J et al (2016) Dysbiotic gut microbiota causes transmissible Crohn's disease-like ileitis independent of failure in antimicrobial defence. Gut 65:225–237
Schnoes AM, Brown SD, Dodevski I et al (2009) Annotation error in public databases: misannotation of molecular function in enzyme superfamilies. PLoS Comput Biol 5:e1000605
Schnupf P, Gaboriau-Routhiau V, Gros M et al (2015) Growth and host interaction of mouse segmented filamentous bacteria in vitro. Nature 520:99–103
Schwank G, Koo BK, Sasselli V et al (2013) Functional repair of CFTR by CRISPR/Cas9 in intestinal stem cell organoids of cystic fibrosis patients. Cell Stem Cell 13:653–658
Sekelja M, Berget I, Naes T et al (2011) Unveiling an abundant core microbiota in the human adult colon by a phylogroup-independent searching approach. ISME J 5:519–531
Simon R, O'connell M, Labes M et al (1986) Plasmid vectors for the genetic analysis and manipulation of rhizobia and other gram-negative bacteria. Methods Enzymol 118:640–659
Simon R, Priefer U, Puhler A (1983) A broad host range mobilization system for in vivo genetic engineering: transposon mutagenesis in gram negative bacteria. Nat Biotech 1:784
Sokol H, Pigneur B, Watterlot L et al (2008) *Faecalibacterium prausnitzii* is an anti-inflammatory commensal bacterium identified by gut microbiota analysis of Crohn disease patients. Proc Natl Acad Sci USA 105:16731–16736
Sokol H, Seksik P, Furet JP et al (2009) Low counts of *Faecalibacterium prausnitzii* in colitis microbiota. Inflamm Bowel Dis 15:1183–1189
Sonnenberg GF, Monticelli LA, Alenghat T et al (2012) Innate lymphoid cells promote anatomical containment of lymphoid-resident commensal bacteria. Science 336:1321–1325

Spor A, Koren O, Ley R (2011) Unravelling the effects of the environment and host genotype on the gut microbiome. Nat Rev Micro 9:279–290

Staley JT, Konopka A (1985) Measurement of in situ activities of nonphotosynthetic microorganisms in aquatic and terrestrial habitats. Annu Rev Microbiol 39:321–346

Stelzner M, Helmrath M, Dunn JCY et al (2012) A nomenclature for intestinal in vitro cultures. Am Physiol—Gastrointest Liver Physiol 302:G1359–G1363

Stephen AM, Cummings JH (1980) The microbial contribution to human faecal mass. J Med Microbiol 13:45–56

Stevens GA, Singh GM, Lu Y et al (2012) National, regional, and global trends in adult overweight and obesity prevalences. Popul Health Metr 10:22

Stevenson BS, Eichorst SA, Wertz JT et al (2004) New strategies for cultivation and detection of previously uncultured microbes. Appl Environ Microbiol 70:4748–4755

Sunagawa S, Mende DR, Zeller G et al (2013) Metagenomic species profiling using universal phylogenetic marker genes. Nat Meth 10:1196–1199

Suskind DL, Brittnacher MJ, Wahbeh G et al (2015) Fecal microbial transplant effect on clinical outcomes and fecal microbiome in active Crohn's disease. Inflamm Bowel Dis 21:556–663

Swidsinski A, Loening-Baucke V, Vaneechoutte M et al (2008) Active Crohn's disease and ulcerative colitis can be specifically diagnosed and monitored based on the biostructure of the fecal flora. Inflamm Bowel Dis 14:147–161

Tap J, Mondot S, Levenez F et al (2009) Towards the human intestinal microbiota phylogenetic core. Environ Microbiol 11:2574–2584

Tazoe H, Otomo Y, Karaki S et al (2009) Expression of short-chain fatty acid receptor GPR41 in the human colon. Biomed Res 30:149–156

Tedelind S, Westberg F, Kjerrulf M et al (2007) Anti-inflammatory properties of the short-chain fatty acids acetate and propionate: a study with relevance to inflammatory bowel disease. World J Gastroenterol 13:2826–2832

The human microbiome project consortium (2012) Structure, function and diversity of the healthy human microbiome. Nature 486:207–214

Thangaraju M, Cresci GA, Liu K et al (2009) GPR109A is a G-protein-coupled receptor for the bacterial fermentation product butyrate and functions as a tumor suppressor in colon. Cancer Res 69:2826–2832

Tolonen AC, Chilaka AC, Church GM (2009) Targeted gene inactivation in *Clostridium phytofermentans* shows that cellulose degradation requires the family 9 hydrolase Cphy3367. Mol Microbiol 74:1300–1313

Turnbaugh PJ, Hamady M, Yatsunenko T et al (2009) A core gut microbiome in obese and lean twins. Nature 457:480–484

Turnbaugh PJ, Ley RE, Mahowald MA et al (2006) An obesity-associated gut microbiome with increased capacity for energy harvest. Nature 444:1027–1031

Uhlig T, Kyprianou T, Martinelli FG et al (2014) The emergence of peptides in the pharmaceutical business: from exploration to exploitation. EuPA Open Proteomics 4:58–69

Vandussen KL, Marinshaw JM, Shaikh N et al (2015) Development of an enhanced human gastrointestinal epithelial culture system to facilitate patient-based assays. Gut 64:911–920

Veeranagouda Y, Husain F, Wexler HM (2012) Transposon mutagenesis of the anaerobic commensal, *Bacteroides fragilis*, using the EZ:TN5 transposome. FEMS Microbiol Lett 333:94–100

Venter JC, Adams MD, Myers EW et al (2001) The sequence of the human genome. Science 291:1304–1351

Vidal JE, Chen J, Li J et al (2009) Use of an EZ-Tn5-Based Random Mutagenesis System to Identify a Novel Toxin Regulatory Locus in *Clostridium perfringens* Strain 13. PLoS ONE 4: e6232

Vinolo MA, Ferguson GJ, Kulkarni S et al (2011) SCFAs induce mouse neutrophil chemotaxis through the GPR43 receptor. PLoS ONE 6:e21205

Virginia Polytechnic Institute and State University Anaerobe Laboratory (1975) Anaerobe laboratory manual. Virginia Polytechnic Institute and State University, Anaerobe Laboratory, Virginia

Wang B, Li K, Wang A et al (2015) Highly efficient CRISPR/HDR-mediated knock-in for mouse embryonic stem cells and zygotes. Biotechniques 59:201–208

Waters VL (2001) Conjugation between bacterial and mammalian cells. Nat Genet 29:375–376

Whitehead TR, Hespell RB (1990) Heterologous expression of the *Bacteroides ruminicola* xylanase gene in *Bacteroides fragilis* and *Bacteroides uniformis*. FEMS Microbiol Lett 54: 61–65

Whitfield-Cargile CM, Cohen ND, Chapkin RS et al (2016) The microbiota-derived metabolite indole decreases mucosal inflammation and injury in a murine model of NSAID enteropathy. Gut Microbes 7(3):246–261

Woese CR, Fox GE (1977) Phylogenetic structure of the prokaryotic domain: the primary kingdoms. Proc Natl Acad Sci USA 74:5088–5090

Wright O, Delmans M, Stan GB et al (2015) GeneGuard: a modular plasmid system designed for biosafety. ACS Synth Biol 4:307–316

Xing J, You C, Dong K et al (2013) Ameliorative effects of 3,4-oxo-isopropylidene-shikimic acid on experimental colitis and their mechanisms in rats. Int Immunopharmacol 15:524–531

Yang X, Xie L, Li Y et al (2009) More than 9,000,000 unique genes in human gut bacterial community: estimating gene numbers inside a human body. PLoS ONE 4:e6074

Yassour M, Lim MY, Yun HS et al (2016) Sub-clinical detection of gut microbial biomarkers of obesity and type 2 diabetes. Genome Med 8:17

Zackular JP, Baxter NT, Iverson KD et al (2013) The gut microbiome modulates colon tumorigenesis. MBio 4

Zeevi D, Korem T, Zmora N et al (2015) Personalized nutrition by prediction of glycemic responses. Cell 163:1079–1094

Zhang L, Jia R, Palange NJ et al (2015a) Large Genomic Fragment Deletions and Insertions in Mouse Using CRISPR/Cas9. PLoS ONE 10:e0120396

Zhang Y, Grosse-Honebrink A, Minton NP (2015b) A universal mariner transposon system for forward genetic studies in the genus *Clostridium*. PLoS ONE 10:e0122411

Zhang Z, Geng J, Tang X et al (2014) Spatial heterogeneity and co-occurrence patterns of human mucosal-associated intestinal microbiota. ISME J 8:881–893

Zoetendal EG, von Wright A, Vilpponen-Salmela T et al (2002) Mucosa-associated bacteria in the human gastrointestinal tract are uniformly distributed along the colon and differ from the community recovered from feces. Appl Environ Microbiol 68:3401–3407

Chapter 4
Using Metabolomic Approaches to Characterize the Human Pathogen *Leishmania* in Macrophages

Joachim Kloehn, Eleanor C. Saunders and Malcolm J. McConville

1 *Leishmania* and Leishmaniasis

Leishmania spp., are protozoan parasites that belong to the order *Kinetoplastida* (Phylum *Euglenozoa*). The *Kinetoplastida* encompass a large group of flagellated protists including the uniflagellated *Trypanosomatidae*, which all have parasitic life styles, and the biflagellated *Bodonidae* which are typically free living. The family *Trypanosomatidae* comprises the medically important genera *Trypanosoma*, which includes *Trypanosoma brucei* and *Trypanosoma cruzi*, the causative agent of human African trypanosomiasis and Chagas disease, respectively, as well as *Leishmania* spp., which are the cause of a spectrum of diseases collectively termed the leishmaniases (Stuart et al. 2008). *T. brucei*, *T. cruzi*, and *Leishmania* spp., all rely on an insect vector as well as a mammalian host to complete their complex lifecycle. While *T. brucei* and *T. cruzi* are transmitted by the tsetse fly and tri-atomine bugs, respectively, *Leishmania* spp., are transmitted by phlebotomine sandflies.

The pathology of leishmanial diseases range from localized self-healing cutaneous or diffuse cutaneous infections (CL and DCL, respectively) to mucocutaneous leishmaniasis (MCL) and visceral leishmaniasis (VL, also called kala-azar) (Pearson and Sousa 1996). The pathology can be linked to different species—e.g., *Leishmania donovani*, *Leishmania infantum,* and *Leishmania chagasi* cause VL, while *Leishmania major*, *Leishmania mexicana*, *Leishmania amazonensis,* and *Leishmania aethiopica* cause CL or DCL, and *Leishmania braziliensis*, *Leishmania peruviana,*

J. Kloehn · E.C. Saunders · M.J. McConville (✉)
The Department of Biochemistry and Molecular Biology, Bio21 Molecular Science and Biotechnology Institute, University of Melbourne, Parkville, VIC 3010, Australia
e-mail: malcolmm@unimelb.edu.au

M.J. McConville
Metabolomics Australia, Bio21 Molecular Science and Biotechnology Institute, University of Melbourne, Parkville, VIC 3010, Australia

D.J. Beale et al. (eds.), *Microbial Metabolomics*,
DOI 10.1007/978-3-319-46326-1_4

and *Leishmania guyanensis* cause MCL. CL is characterized by open sores around the site of the sandfly bite, which can take years to heal and leave disfiguring scars. Furthermore, CL can recur years after patients have seemingly healed from the initial infection (Marovich et al. 2001; Gangneux et al. 2007). DCL is a more severe form of CL that results in the formation of several hundred nodules or ulcers. The marring lesions and disfiguring scars caused by CL and DCL can result in stigmatization and lead to social exclusion and economic disadvantage (Kassi et al. 2008). MCL results from the dissemination of parasites to mucosal membranes around the mouth and nose and, in some cases, to the genital or optical mucosa (Huna-Baron et al. 2000). This severe form of leishmaniasis can result in devastating destruction and deformation of the face with high risk of secondary infections. VL, or kala-azar, occurs when parasites disseminate to the bone marrow, liver, and spleen, resulting in anemia, fever, weightloss, and enlargement of liver and spleen. If left untreated, VL leads to death in nearly 100 % of cases within a two-year period (WHO.int).

Leishmaniasis is endemic in 88 countries throughout the tropics, subtropics, and the Mediterranean Basin with over 350 million people at risk of infection (WHO.int). Disease prevalence is estimated at 12 million people with more than 2 million new infections occurring annually (WHO.int). Mortalities from VL are increasing worldwide and currently stand at >20,000 deaths annually, making it the second deadliest parasitic disease after malaria (WHO.int; Desjeux 2004; Reithinger 2008; Alvar et al. 2012). The current war in Syria and instability in the Middle East have led to increased incidence and spread of leishmaniasis in the area due to a number of factors including greater refugee migration and an inability to access staff and facilities for diagnosis and treatment (Alawieh et al. 2014).

2 Treatment of Leishmaniasis

Current antileishmanial treatments suffer from one or more major limitations including high toxicity/severe side effects (amphotericin, pentamidine, paromomycin, miltefosine and sitamaquine), the requirement for long-term/parenteral administration (pentavalent antimonials, amphotericin B and paromomycin), high cost (liposomal amphotericin B), variable efficacy (e.g. species specificity, pentavalent antimonials, imidazole and pentamidine), and resistance or likely development of resistance (pentavalent antimonials and miltefosine) (Bouchard et al. 1982; Rangel et al. 1996; Seifert et al. 2003, 2007; Ouellette et al. 2004; Olliaro et al. 2005; Croft et al. 2006; Sindermann and Engel 2006; Bhattacharya et al. 2008; Sundar et al. 2007; Davidson et al. 2009; Moore and Lockwood 2010; Chakravarty and Sundar 2010; Seifert 2011; Freitas-Junior et al. 2012). While combination treatments are being used to reduce the emergence of resistant strains (van Griensven et al. 2010), new drugs which overcome the limitations of the current drugs are needed urgently (Freitas-Junior et al. 2012). Furthermore, despite the fact that humans can generate strong protective immunity against *Leishmania* infection/reinfection (Evans and Kedzierski 2012), no efficacious defined vaccine for preventing human leishmaniasis has been developed to date (Handman 2001; de Oliveira et al. 2009; Kedzierski 2010; McCall et al. 2013; Joshi et al. 2014; Kumar and Engwerda 2014).

3 The Lifecycle of *Leishmania*

Leishmania differentiate through several morphologically and physiologically distinct stages during their complex digenetic lifecycle in their insect vector and animal hosts (Fig. 1). The major developmental stage in the sandfly vector (*Phlebotomus* and *Lutzomyia*) is the motile promastigote, which possesses a single long flagellum that emerges from the anterior flagellar pocket. Sandflies become a vector following the uptake of infected cells or free *Leishmania* parasites (typically low, usually 10–100 parasites) when feeding on a mammalian host (Anjili et al. 2006). Ingested parasites initially differentiate to promastigotes, which undergo a period of rapid proliferation, exploiting the nutrient-rich milieu of the blood meal as it is progressively digested by the sandfly's hydrolases (Pimenta et al. 1997). Following the breakdown of the peritrophic membrane that encapsulates the initial blood meal, the promastigote differentiates through several developmental stages which are distinguished by markedly different replication rates (Gossage et al. 2003; Dostalova and Volf 2012) including the nondividing metacyclic stage that accumulate in the sandfly foregut. This stage exhibits a similar physiology to nondividing (stationary phase) promastigotes (Pro^{stat}) in in vitro culture and appears to be highly virulent and preadapted for life in the mammalian host (Sacks and Perkins 1984, 1985; da Silva and Sacks 1987; Sacks and da Silva 1987; Sacks 1989). The accumulation of large clusters or aggregates of these metacyclic promastigotes at the sandfly mouthparts (specifically the stomodeal valve) causes an alteration in sandfly feeding behavior (e.g. repeated probing of the skin), damage to the stomodeal valve, and regurgitation of the parasite bolus, which may enhance transmission of the parasite to the host (Killick-Kendrick et al. 1977; Beach et al. 1984, 1985; Bates 2007; Rogers and Bates 2007, Schlein et al. 1992; Volf et al. 2004).

The number of *Leishmania* transmitted during a blood meal is typically low (<600 parasites) but can, in some cases, be as high as 100,000 cells (Kimblin et al. 2008). Infectious metacyclics are transmitted along with sandfly saliva and a highly immunogenic polysaccharide gel secreted by the promastigote (Titus and Ribeiro 1988; Bates 2007). Injected promastigotes are initially phagocytosed by polymorphonuclear leucocytes (PMNs) that are rapidly recruited to the sandfly bite site (van Zandbergen et al. 2004). These PMNs undergo apoptosis within a few days and the cellular debris (including released parasites) and/or intact PMNs containing parasites, are phagocytosed by a wave of macrophages that are recruited to the damaged tissue. Macrophages are the primary host cells of *Leishmania* (Handman and Bullen 2002) and it has been proposed that the initial infection of PMNs (in which the

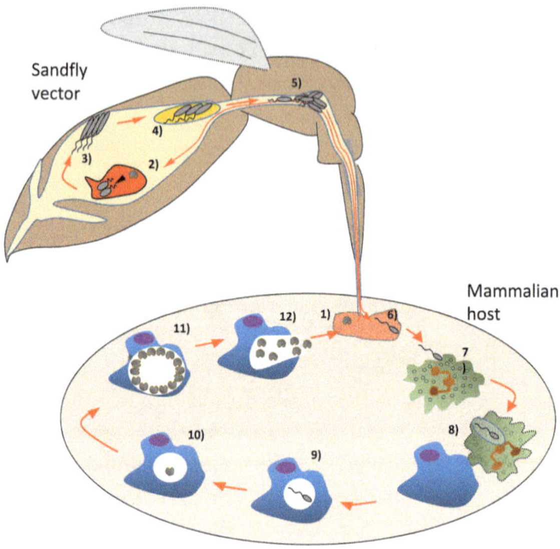

Fig. 1 *Leishmania* undergo a complex digenetic lifecycle developing within the sandfly vector and the mammalian host. Transmission occurs when an infected sandfly takes a blood meal. Amastigotes/infected macrophages are taken up by the sandfly during a blood meal (*1*) (Anjili et al. 2006) and are initially enclosed within a digestive peritrophic sac in the sandfly gut (*red*) (Pimenta et al. 1997). Amastigotes differentiate to flagellated procyclic promastigotes (*2*) which undergo multiple rounds of division before slowing their replication and differentiating into highly motile, but nondividing, nectomonads which escape the peritrophic sac (Gossage et al. 2003; Dostalova and Volf 2012). Nectomonads adhere to the mid-gut wall (*3*) and develop into leptomonads in the thoracic mid-gut (*4*). Leptomonads undergo further rounds of rapid replication and are immobilized in a gel-like matrix which is secreted by this life-cycle stage. Leptomonads differentiate into either nondividing haptomonads, which form a plug by attaching to each other and the stomodeal valve (*5*), or nondividing metacyclics, which are highly infectious and are transmitted during a blood meal (*6*) (Sacks and Perkins 1984, 1985; da Silva and Sacks 1987; Sacks and da Silva 1987; Sacks 1989). In the mammalian host, metacyclic promastigotes are phagocytosed by polymorphic neutrophils (PMN), which are first to reach the site of inflammation following the sandfly bite (*7, 8*) (Laskay et al. 2003; van Zandbergen et al. 2004). PMNs have a short life span and macrophages phagocytose released parasites and/or infected apoptotic PMNs (*8, 9*) (Laskay et al. 2003; van Zandbergen et al. 2004; Ritter et al. 2009). Within the phagolysosome of macrophages, *Leishmania* differentiate into the amastigote stage, adapt to their new milieu (*10*) and start replicating (*11*) (Kaye and Scott 2011). Amastigotes are released when heavily infected macrophages rupture (*12*) allowing infection of other cells (Noronha et al. 2000). The lifecycle begins again when sandflies ingest free amastigotes or infected cells during a blood meal (*1*)

parasites do not replicate) represents a 'Trojan Horse' strategy for targeting macrophages, the preferred host cell for these pathogens (Laskay et al. 2003; van Zandbergen et al. 2004, 2007). Alternatively, extracellular *Leishmania* may be taken up incidentally during the phagocytosis of necrotic PMNs (the 'Trojan rabbit' model) (Ritter et al. 2009). Following uptake by macrophages and the maturation of the vacuole into a phagolysosomal compartment, promastigotes differentiate to small, aflagellate amastigotes which re-enter a proliferative state that is associated with progression of the disease (Kaye and Scott 2011). Further recruitment of naïve macrophages to the site of infection leads to the development of granulomatous lesions that are predominantly composed of infected and uninfected macrophages, as well as other populations of monocytes, PMNs, dendritic cells, and lymphocytes (Amaral et al. 2000; Souza-Lemos et al. 2011). Infected macrophages, or even extracellular amastigotes, can also disseminate to other tissues and establish infection in lymph nodes, mucosal membranes (MCL), or the spleen and liver (VL) (Ridley et al. 1989; McElrath et al. 1988). Subpopulations of *Leishmania*-infected macrophages in lymph nodes are thought to be responsible for long-term persistence of the pathogen, but are inadequately characterized (Dereure et al. 2003).

Macrophages are specialized phagocytic cells that are a vital part of the innate and adaptive immune system. These cells are responsible for internalizing and then killing microbes within the phagolysosome which has a low pH, a complex array of hydrolytic enzymes, and is the target of key microbicidal processes such as the oxidative burst and nitric oxide (NO) synthesis (Russell et al. 2009). However, a number of bacterial (*Coxiella brunetti*, *Listeria monocytogenes*, *Shigella* spp., *Mycobacterium tuberculosis*), fungal (*Cryptococcus neoformans*), and protozoan pathogens (*Leishmania* spp., *T. cruzi* and *Toxoplasma gondii*) actively target macrophages (Dermine and Desjardins 1999). While most of these organisms (including *Leishmania* promastigotes) exhibit mechanisms to escape the phagolysosome or prevent its maturation (Desjardins and Descoteaux 1997; Dermine et al. 2000, 2005; Moradin and Descoteaux 2012), *Leishmania* amastigotes and *Coxiella burnetti* have evolved to withstand the harsh conditions encountered within the mature phagolysosome (Dermine and Desjardins 1999; Thi et al. 2012). In order to survive the acidic pH within the phagolysosome, *Leishmania* amastigotes express proton pumps to maintain a near-neutral intracellular pH (Glaser et al. 1988; Zilberstein et al. 1989; Grigore and Meade 2006). The mechanisms by which *Leishmania* amastigotes withstand the high concentration of proteases within the phagolysosome are poorly understood (Prina et al. 1990). Possibly, *Leishmania* cell surface glycoconjugates play a role in acid resistance and/or *Leishmania* membrane proteins are resistant to proteolysis (Chang and Fong 1983; Prina et al. 1990). Prina et al. (1990) proposed that the enlarged phagolysosomes observed in macrophages infected with *L. mexicana* and related species are beneficial to the parasite as they result in dilution of the lysosomal proteases. Due to the difficulty in studying intracellular amastigotes, most studies rely on the axenic differentiation of promastigotes to amastigotes ($\text{Ama}^{\text{axenic}}$) which can be achieved by lowering the pH of the culture medium and increasing the temperature or through depletion of iron in the culture medium to mimic the conditions of the host macrophage (Saar et al. 1998; Mittra et al. 2013).

4 Characterizing *Leishmania* Metabolism: Insights Offered by '-Omics' Approaches

The paucity of good treatment options and no vaccine means that there is an urgent need to identify new therapeutic targets in *Leishmania*. Many of these targets will ideally be specific to parasite metabolism, targeting pathways critical for the survival and proliferation of the intracellular amastigote stage. Understanding the interplay of host and parasite metabolism is critical in identifying targets that will not affect the host metabolism. Finally, by mapping parasite metabolism it is anticipated that the mode of action (MOA) of current chemotherapies will be revealed and new insights into drug resistance mechanisms gained. The metabolism of *Leishmania* has been studied for several decades using a range of approaches and techniques. Here we discuss these techniques, the insights they have provided into *Leishmania* metabolism as well as their limitations.

5 Genome Studies and *Leishmania* Metabolism Mapping

Genome sequencing has revealed many new potential drug targets in *Leishmania*. To date, the genomes of several *Leishmania* species have been sequenced (Ivens et al. 2005; Peacock et al. 2007; Downing et al. 2011; Real et al. 2013) providing important new insights into the biology of these parasites, including the identification of novel surface proteins, protein kinases, and phosphatases that could be involved in signaling pathways, differentiation and parasite stress responses, as well as many metabolic pathways (Myler et al. 2000; Worthey et al. 2003; Ivens et al. 2005). Genome-wide annotations have led to the development of curated biochemical pathways databases, such as LeishCyc and Kyoto Encyclopedia of Genes and Genomes (KEGG) pathways (Doyle et al. 2009; Saunders et al. 2012). Furthermore, genome-wide flux balance models of *Leishmania* metabolism give unique insight into crucial reactions, predicted auxotrophies and minimal culture media components, potentially lethal or growth-reducing gene deletions, and novel drug target identification (Chavali et al. 2008, 2012; Silva et al. 2015; Subramanian et al. 2015). These tools are particularly useful in providing a means to contextualize and map metabolomics data and are also valuable in hypothesis-driven metabolomic analyses.

However, as is the case with other protozoan genomes, more than half of the genes in the *Leishmania* genomes have not yet been assigned a function (Pawar et al. 2014), highlighting the importance of pairing genomic data with functional gene analysis tools such as metabolomics. Another striking feature to emerge from these analyses is the extraordinarily high level of gene synteny identified by the comparative analysis of *T. brucei*, *T. cruzi,* and *L. major* genomes (El-Sayed et al. 2005). About 6150 genes were found to be shared between these species, while the authors identified about 900 *Leishmania*-specific genes. Furthermore, a comparison of *L. major*, *L. braziliensis,* and *L. infantum* genomes, identified only about 80 genes that are restricted to an individual species despite the early divergence of these species and the distinct clinical manifestation observed during infection with these species (Peacock et al. 2007). This indicates that there are only few species-specific genes which contribute to the different biology of each species, indicating that species and strain-specific differences may result from gene dosage (copy number) (Rogers et al. 2011). Additionally, it should be noted that the genome provides insights into the overall metabolic capacity of *Leishmania,* but allows no conclusion about the significance of metabolic pathways in specific developmental stages and given conditions (e.g. nutrient restriction, drug treatment etc.). For example, enzymes/pathways which may be crucial for survival in the sandfly vector might be dispensable during mammalian host infection and vice versa.

Finally, models of *Leishmania* metabolism may not adequately account for niche/species/stage-specific changes to enzyme compartmentalization which in turn will affect metabolic flux. For example, *Leishmania* and other kinetoplastids maintain an unusual peroxisome-like organelle termed the glycosome (Haanstra et al. 2016). This organelle contains the enzymes of several metabolic pathways including glycolysis, gluconeogenesis, the pentose phosphate pathway (PPP), fatty acid β-oxidation, purine salvage, and pyrimidine synthesis as well as hypothetical proteins with unknown function (Michels et al. 2006; Jamdhade et al. 2015). The localization of enzymes within the glycosome has repercussions on metabolic flux and is thought to provide a means of regulation in the absence of traditional feedback loops. For example, *Leishmania* hexokinase and phosphofructokinase lack conventional activity regulation which could cause the toxic accumulation of hexose phosphates (Bakker et al. 2000; Michels et al. 2006). Indeed, the relocalization of glycolytic enzymes, such as the phosphoglucokinase and the triose phosphate isomerase, to the cytoplasm proved to be fatal for *T. brucei* (Blattner et al. 1998; Helfert et al. 2001). Furthermore, the number and enzymic content of the glycosome changes during the promastigote to amastigote differentiation thereby affecting the metabolic regulation and capacity of the parasite (Cull et al. 2014).

6 Transcriptomic and Proteomic Studies

Unlike the situation in other eukaryotes, protein-encoding genes in *Leishmania* lack introns and are organized and constitutively transcribed as large polycistronic clusters (Myler et al. 2000; Worthey et al. 2003; Ivens et al. 2005). Transcription is not regulated by transcription factors and the polycistronic mRNA is subsequently processed into individual mRNAs by transsplicing and polyadenylation (Ivens et al. 2005). As a result of this unusual mode of gene expression, relative levels of mRNA expression remain remarkably constant across different life-cycle stages under different growth conditions (Leifso et al. 2007). Consequently, transcriptomic studies have not been particularly fruitful in identifying enzymes or metabolic pathways that are activated (or repressed) in amastigotes following their uptake by macrophages, as has been the case in some other bacterial or fungal pathogens (Haile and Papadopoulou 2007). However, there are a limited number of examples where stage-specific changes in protein expression are regulated at the level of mRNA stability. Examples include some of the lysosomal cysteine proteases, as well as some enzymes involved in metabolism and protein synthesis (Saxena et al. 2007). A recent analysis of the *L. major* transcriptome reported that the polycistronic transcripts can be highly heterogeneous in length which may lead to differences in stability, further complicating the interpretation of transcriptomic data (Rastrojo et al. 2013). Lastly, a general limitation of transcriptome analyses is the potentially poor correlation between mRNA and protein levels or between transcript levels and enzymatic activity as observed in other organisms (Miyamoto et al. 2001; Maier et al. 2009; Hoppe 2012; Vogel and Marcotte 2012).

Proteomic approaches have been used to identify differentially expressed proteins in promastigote and amastigote stages, and have given some crucial insights into the metabolism of the two life-cycle stages (Handman et al. 1995; Thiel and Bruchhaus 2001; El Fakhry et al. 2002; Bente et al. 2003; Nugent et al. 2004; Walker et al. 2006; Rosenzweig et al. 2008a, b). Promastigote to amastigote differentiation is associated with a global reduction in protein synthesis but increased expression of some amastigote specific proteins such as A2, an ATP-dependent RNA helicase, the amastin family of proteins, lysosomal cysteine proteinases, some protein chaperones and histones (Handman et al. 1995; Carvalho et al. 2002; Barhoumi et al. 2006; Nasereddin et al. 2010). Comparing the proteome of cultured promastigotes and Ama^{axenic}, Rosenzweig et al. (2008a) reported significant differences in the levels of some metabolic enzymes including increases in enzymes involved in gluconeogenesis, β-oxidation, amino acid catabolism, the tricarboxylic acid (TCA) cycle, and the mitochondrial respiratory chain, suggesting stage-specific remodeling of amastigote metabolism. A number of other studies have identified proteins that are differentially expressed in the distinct life-cycle stages (El Fakhry et al. 2002; Bente et al. 2003; Nugent et al. 2004; Walker et al. 2006). However, the identified differences were surprisingly small and interpretation of these analyses are complicated by the fact that many changes in protein expression are inconsistent across different species and/or the possibility that at some of these changes may be attributed to stage-specific changes in the size or complement of mitochondria and glycosomes in the different developmental stages.

More broadly, while protein levels are commonly considered to provide a measure for the activity of an enzyme/pathway, this correlation is not always observed (Miyamoto et al. 2001). Furthermore, relatively few *Leishmania* proteins have been functionally characterized and there is an increasing number of reports of parasite proteins that have divergent or repurposed roles from those inferred from homology alignments (Oppenheim et al. 2014). For example, *Leishmania* hexokinase functions as a metabolic enzyme in glycosomal glycolysis, as well as heme receptor in the flagellar pocket and mitochondrion, while enolase ‘moonlights’ as a microtubule-binding protein (Krishnamurthy et al. 2005; Quinones et al. 2007;

Vanegas et al. 2007; Collingridge et al. 2010; Tonkin et al. 2015), complicating the interpretation of proteomics analyses.

These studies suggest that *Leishmania* may be more dependent on posttranslational regulatory mechanisms during differentiation or adaptation to changing environmental circumstance than other eukaryotes. Indeed gene families for proteins involved in protein posttranslational modifications (kinases, phosphatases, etc.) are commonly amplified in these parasites. In particular, phosphoproteomic studies have revealed major changes in protein phosphorylation during promastigote to amastigote differentiation (Rosenzweig et al. 2008b; Morales et al. 2010a; Tsigankov et al. 2013, 2014). Interestingly, some of the most abundant proteins to be phosphorylated during differentiation were heat-shock proteins which are otherwise constitutively expressed (Morales et al. 2010a). Rosenzweig et al. (2008b) utilized isobaric tags for relative and absolute quantification/liquid chromatography-mass spectrometry/mass spectrometry (iTRAQ/LC-MS/MS) to characterize a number of posttranslational modifications in *Leishmania* promastigotes and Ama^{axenic}. The *Leishmania* genomes also encode large numbers of protein kinases and phosphatases, several of which have been shown to be essential for differentiation and/or amastigote survival (Ivens et al. 2005; Morales et al. 2010b; Cayla et al. 2014). To date, however, no complete signaling pathways in *Leishmania* (from surface receptor/sensor to downstream effector(s)) have been delineated and the function of most of these posttranslational modifications remains poorly defined. Nonetheless, studies on proteins involved in regulating mitochondrial proteins through the covalent attachment of ubiquitin-like proteins (Gannavaram et al. 2011, 2012) suggest that this will be a rewarding endeavor. In a recent study, Goldman-Pinkovich et al. (2016) described a coordinated arginine deprivation response (ADR) that is activated in response to low arginine levels which are expected to occur in the macrophage phagolysosome. *L. donovani* responded to low external arginine levels by rapidly upregulating the arginine transporter, LdAAP3. This response was dependent on expression of the mitogen activated protein kinase 2 (MAPK2)-dependent signaling pathway (Goldman-Pinkovich et al. 2016). Similarly, stage-specific regulation of expression of the major plasma membrane glucose transporters in *L. mexicana* was found to be regulated by ubiquitination of the cytoplasmic tails of the transporter proteins and their internalization and degradation in the lysosome (Vince et al. 2011).

7 Insights into *Leishmania* Metabolism using Metabolomics

Further advances in our understanding of *Leishmania* adaptive responses in both the sandfly and mammalian hosts will be dependent on the use of complementary approaches, such as metabolomics. Metabolomics is increasingly being used alone or in combination with other '-omics' approaches to identify new or unanticipated metabolic pathways and to characterize metabolic networks in microbial pathogens (Holmes 2010; Creek and Barrett 2014; McConville 2014; McConville et al. 2015; Saunders et al. 2015; Lau et al. 2015; Kim and Creek 2015; Kloehn et al. 2016). The metabolome of a cell/tissue/organism can be considered the major downstream phenotype of changes in the transcriptome and proteome, or the most upstream input into cellular processes from the environment. The later point is particularly relevant when considering pathogens which are highly responsive to changes in the nutrient levels within their specific host niches. Trindade et al. (2016) demonstrated this in a recent study as they identified a major reservoir of *T. brucei* cells in adipose tissue. In comparison with parasites in the bloodstream and central nervous system, the adipose tissue form (ATF) exhibited a distinct metabolism, as it is adapted to its niche by utilizing myristate as a major carbon source through β-oxidation. Furthermore, metabolomics has proven to be particularly useful in drug target discovery research and is expected to become increasingly valuable due to the rapid refinement of existing approaches and the development of new analytical techniques (Rabinowitz et al. 2011). To date in *Leishmania* spp., metabolomics approaches have been employed to characterize mutants, elucidate the MOA of drugs and mechanisms of resistance, and to describe the nutritional requirements and central carbon metabolism of different parasite developmental stages as well as different species (Naderer et al. 2006; De Souza et al. 2006; Creek and Barrett 2014; Vincent et al. 2014; Rojo et al. 2015; Saunders et al. 2015; Arjmand et al. 2016; Westrop et al. 2015). Though beyond the scope of this review, metabolomics has also been used to investigate changes to the host cell upon infection (Lamour et al. 2012; Moreira et al. 2015).

8 Characterization of Genetic Knockouts using Metabolite Profiling

Notwithstanding the technical difficulties of deleting genes in a diploid organism such as *Leishmania*, a significant number of genetically defined mutants have been generated, which have provided new insights into the metabolic requirements of intracellular amastigotes (McConville et al. 2007). To date, metabolomic profiling in *Leishmania* spp., has predominantly been used in a highly targeted fashion to validate the deletion of the desired target gene, complementing detailed molecular biology, biochemistry, and virulence data (Naderer et al. 2006; Saunders et al. 2012; Naderer et al., 2015). For example, Naderer et al. (2006) deleted the gene encoding the gluconeogenic enzyme, fructose 1,6-bisphosphatase (FBPase) in *L. major*. In contrast to wild-type cells, these parasites were unable to utilize gluconeogenic carbon sources (glycerol) to synthesize hexose sugars and mannogen, a unique storage polysaccharide composed of β1,2-linked mannose residues. While the *L. major* ΔFBPase promastigotes grew normally in rich culture medium, they were unable to replicate in macrophages and displayed highly attenuated virulence in mice suggesting that gluconeogenesis is essential for *Leishmania* survival in the macrophage, possibly due to low levels of hexoses in the phagolysosome. Paradoxically, other genetic studies have shown that *Leishmania* spp., are also dependent on the uptake of sugars (Burchmore et al. 2003; Naderer et al. 2010). An abundant source of sugars in the macrophages phagolysosome are likely to be aminosugars (*N*-acetyl-glucosamine and glucosamine) derived from the breakdown of glycosaminoglycans which are internalized by macrophages (Naderer et al. 2010, 2015) which would provide amastigotes with a source of both carbon skeletons and nitrogen groups. The significance of aminosugar uptake and utilization in amastigote metabolism was supported by the finding that deletion of the gene encoding the enzyme glucosamine 6-phosphate deaminase (GND), which converts glucosamine (GlcN) to fructose 6-phosphate, resulted in severe loss of virulence in mice and the capacity to grow in macrophages, while loss of the enzyme N-acetylglucosamine acetyltransferase (GNAT), which converts these parasites to GlcN auxotrophs had no effect on intracellular growth or virulence (Naderer et al. 2015).

In the future, it is anticipated that metabolite profiling will be increasingly used to analyze entire pathways (or nodes) as, for example, demonstrated in its recent application to analysis of the polyamine pathway in *L. amazonensis* (Castilho-Martins et al. 2015). An untargeted (or at least more extensive) metabolite profiling of gene knock-out lines and wild-type cells is particularly recommended in *Leishmania* spp., in order to prevent overlooking unforeseen (side-) effects of the gene depletion. Due to the complexity and interconnectivity of the metabolic network, as well as our still limited understanding of the metabolic network and metabolite-protein interactions, unexpected effects of gene depletions are likely to be relatively common (Hellerstein 2003).

Untargeted metabolomics may also be combined with functional genomics to assign identities to a large number of unknown/hypothetical *Leishmania* genes. Although this approach has not been systematically applied to *Leishmania* spp., such approaches have been useful in identifying pathways and annotating genes in other, more intensely studied organisms such as *Saccharomyces cerevisiae* (Raamsdonk et al. 2001) and *Arabidopsis thaliana* (Fukushima et al. 2014). Indeed, the development of a robust and high-throughput means to characterize mutants will become increasingly important as new technologies, such as the CRISPR (clustered regularly interspaced short palindromic repeat)-Cas9 system (Sollelis et al. 2015; Zhang and Matlashewski 2015), are applied to *Leishmania* making gene disruption easier and feasible in large-scale assays. Extensive annotation of metabolic phenotypes and perturbations will help to refine existing in silico generated metabolic networks and the usefulness of these models for predicting lethality, auxotrophy, and drug susceptibility. When combined with traditional phenotypic data of a given mutant (e.g. growth rate, virulence, viability of each life cycle stage) metabolomics is a powerful way to characterize a mutant and potentially offers a means to ascribe gene function in a high-throughput manner.

9 Studying a Drug's Mode of Action and the Development of Resistance in *Leishmania*

Metabolomics represents a powerful tool to understand both the MOA of drugs as well as how drug resistance occurs (Vincent et al. 2014). In the case of anti-*Leishmania* chemotherapies, this is especially important as the MOA of several frontline drugs remains poorly defined (Croft and Coombs 2003) and the drug efficacy is highly variable dependent on the *Leishmania* species and host genetics (Rangel et al. 1996; Schriefer et al. 2008).

Pentavalent and trivalent antimonials (Sb^{V}, Sb^{III}) have been used to treat VL as well as CL, since the 1960s. Pentavalent antimonials remain effective in 95 % of patients in many areas (Seifert 2011), however, the widespread use of the drugs for several decades has given rise to resistant strains in hyperendemic areas in India (Croft et al. 2006; Ashutosh et al. 2007). Aside from increasing resistance, a major drawback of pentavalent antimonials is their severe cardiotoxicity, which is observed in as many as 10 % of patients (Sundar et al. 1998; Sundar and Chakravarty 2010), and the requirement of parenteral administration as no oral preparation is available. Despite the use of antimonials to treat leishmaniasis (VL and CL) for about 60 years, the compounds' MOA remains unclear (Haldar et al. 2011). Several studies have employed metabolomics in order to determine the effect of drugs on metabolism and understand how resistance occurs (Canuto et al. 2012; t'Kindt et al. 2010; Berg et al. 2013). Most of these studies have employed an untargeted approach to quantify as many metabolites as possible, although the technology employed (capillary electrophoresis-electron spray ionization-time of flight-mass spectrometry (CE-ESI-TOF-MS), LC-MS (Orbitrap)), as well as the *Leishmania* species (*L. infantum*, *L. donovani*), and isolates differ between the publications. t'Kindt et al. (2010) profiled (untreated) antimonial susceptible and resistant clones of *L. donovani*, identifying major differences in phospholipid and sphingolipid metabolism and pools of amino acid/amino acid derivatives. Berg et al. (2013) conducted a more comprehensive analysis, profiling the metabolism of three *L. donovani* clinical isolates that differed in their susceptibility to antimony. Rapidly dividing and nondividing promastigotes (Pro^{log} and Pro^{stat} respectively) were analyzed revealing significant differences in several metabolic pathways including arginine metabolism, cysteine transsulfuration pathway, acylglycines, indole acrylate, glycerophospholipids, and amino acid levels. From these studies it was inferred that drug (Sb^{III}) resistance may be linked to (1) increased protection from drug-induced oxidative stress via greater thiol production and tryptophan degradation, (2) changes to mitochondrial metabolism, (3) increased membrane fluidity (protecting the cell from host-derived oxidants and/or affect drug uptake/export), and (4) accumulation of nonessential amino acids as alternative carbon source for use upon host cell invasion. More recently, Rojo et al. (2015) used a multiplatform approach (LC-MS, CE-MS and gas chromatography (GC)-MS) to increase the diversity of metabolites detected in antimony susceptible and resistant *L. infantum*. Consistent with other metabolomics experiments, a depletion of urea cycle and polyamine biosynthetic pathway intermediates was observed when sensitive promastigotes were treated with Sb^{III} which was not apparent (or reversed) in resistant parasites. Treated parasites also appeared to have a disturbed TCA cycle, with several important intermediates (malate and anaplerotic amino acids, aspartate and glutamate) significantly reduced, as well as changes to parasite membrane composition.

The alkylphosphocoline, miltefosine, was the first oral drug to be licensed for use against leishmaniasis. It is used to treat VL on the Indian subcontinent and CL in South America. Miltefosine was initially developed as an anticancer drug in the 1980s (Eibl and Unger 1990) and was later repurposed as an antileishmanial drug. Drug repurposing/repositioning has been suggested to be an important cost- and time-saving strategy for the identification of new antiparasitic drugs (Andrews et al. 2014). Given that miltefosine is thought to affect lipid metabolism and membrane composition, Vincent et al. (2014) compared miltefosine treated/untreated *L. infantum* promastigotes using a lipidomics approach. Changes to intracellular lipid metabolism were observed as an increase in the abundance of short alkanes, although no change in membrane lipids was detected. These finding contradict earlier studies, which detailed changes to fatty acid metabolism, sterol pathways, and phospholipid levels (Rakotomanga et al. 2004, 2007) although comparison between the two studies is complicated by differences in the experimental design (Vincent et al. 2014). Vincent et al. (2014) also observed an increase in the abundance of selected sugars and DNA damage (released nucleotide fragments). Canuto et al. (2014) performed a more extensive analysis comparing miltefosine-treated/untreated and resistant lines of *L. donovani* while also broadening the metabolite base by employing three different platform technologies (LC-MS, GC-MS and CE-MS). Upon miltefosine treatment, a decrease in the abundance of several key intermediates in the arginine/polyamine pathway (arginine, ornithine and citrulline) was observed in susceptible lines. This suggests that the parasites, due to a decreased trypanothione availability, are particularly susceptible to oxidative stress (potentially induced by miltefosine). In resistant parasites, the abundance of these metabolites, as well as spermidine and intermediates of trypanothione biosynthesis, was increased. The amino acid profile also differed in susceptible and resistant lines. In the resistant line, the abundance of most amino acids was increased, while in the susceptible line many amino acids showed a moderate to strong decrease in abundance upon exposure to the drug suggesting some degree of amino acid starvation. The importance of amino acids as a potential carbon source is detailed below.

Metabolomics has also been used to explore changes in the metabolism of single and combined therapy-resistant (CTR) *L. donovani* lines (Berg et al. 2015). Combination therapies are commonly used in the field with the aim of shortening treatment time, improving efficacy, and delaying the emergence of resistance (Croft and Olliaro 2011). Researchers compared the metabolite profiles, acquired by LC-MS, of lines resistant to amphotericin B-, miltefosine-, antimonial (SB^{III})-, paromomycin-, and combinations thereof, in order to identify unique and common metabolic features. Comparing CTR lines with their singly resistant counterparts revealed that changes in the metabolome were highly varied. For some combinations, the changes were not additive (amphotericin-B/paromomycin, amphotericin/miltefosine and amphotericin-B/antimonial), while in other cases changes to the metabolome were greater than what would be predicted from singly resistant lines (miltefosine/paromomycin and antimonial/paromomycin). Quantitative analysis of amphotericin B/antimonial (AS) and antimonial/paromomycin (SP) resistant lines identified several shared changes in metabolites of proline biosynthesis and the transsulfuration pathway which the researchers suggest is indicative of increased protection against oxidative stress in these CTR lines (proline as a general stress-response metabolite and free radical scavenger, products of transsulfuration pathway feeding into trypanothione biosynthesis). Changes in the abundance of lipid and sterol pathway intermediates were also observed in the tested CTR lines suggesting alterations in membrane composition. Importantly, the researchers sought to validate these conclusions using functional assays (susceptibility to drug-induced and extracellular reactive oxygen stress, genomic DNA damage and membrane fluidity).

As with many other metabolomics studies in *Leishmania*, the bulk of MOA/resistance experiments have been undertaken only in the promastigote stage (Vincent and Barrett 2015). Ultimately, however, any conclusions made about a given drug's MOA or proposed mechanism of resistance need to be verified in the disease causing amastigote stage, such as those generated in vitro ($\text{Ama}^{\text{axenic}}$) or isolated from in vitro infected macrophages ($\text{Ama}^{\text{M}\Phi}$) or from murine lesions ($\text{Ama}^{\text{lesion}}$) (Vincent and Barrett 2015). As detailed in the following sections, each *Leishmania* developmental stage is characterized by unique metabolic features and growth rate which may result in significantly different drug efficacy, MOA, and mechanisms of resistance. Furthermore, the host environment may additionally alter a drug's effectiveness by altering the parasite's environment (e.g. nutrient restriction and stress induced by host microbicidal responses), modulate a drug's availability (e.g. membrane permeability and prodrug catabolism), or indirectly kill the amastigote by inhibiting targets in the macrophage host (e.g. immunomodulatoy effects). Indeed, compound screening experiments have revealed significant differences in the sensitivity of promastigote and amastigote stages to many drugs (De Muylder et al. 2011; De Rycker et al. 2013). These findings indicate that understanding the nutritional environment and metabolism of the host cell is also critically important when delineating a drug's MOA and the parasite's development of resistance. For example, it has been suggested that the accumulation of amino acids in drug-resistant lines preserves the parasite's fitness for the invasion and proliferation in the nutritionally limited environment of the host macrophage by providing an alternative carbon source (Vermeersch et al. 2009; Berg et al. 2013; Canuto et al. 2014). Finally, care needs to be taken when studying field isolates so that detected genetic differences can be aligned with drug susceptibility/resistance measures, virulence, and metabolomics data.

10 Metabolomics Reveals Key Differences Between *Leishmania* Developmental Stages

10.1 Footprinting Approaches

As discussed above, *Leishmania* parasites progress through several different developmental stages during their lifecycle. Many of these developmental stages can be generated in vitro by manipulating temperature and culture conditions allowing detailed studies on parasite metabolism in the absence of the confounding influence of host metabolism (with the caveat mentioned above that in vitro conditions may be an inadequate model for the in vivo environment). A number of studies have measured changes in the culture supernatant (footprinting (Kell et al. 2005)) of these different axenic parasite stages to infer the operation of specific pathways in central carbon metabolism and overall metabolic fluxes (Hart and Coombs 1982; Rainey and MacKenzie 1991; Castilla et al. 1995). Hart and Coombs measured changes in the sugars, amino acids, and fatty acids in the medium of axenically-derived *L. mexicana* promastigotes and amastigotes using a variety of analytical approaches, including enzyme assays, gas chromatography, and radioassays (Hart and Coombs 1982). Their results suggested that promastigotes have a highly glycolytic metabolism with significant amino acid uptake, while Ama^{axenic} exhibit a shift toward greater reliance on fatty acid utilization via β-oxidation (with a concordant decrease in glucose uptake). These conclusions were supported by the increased sensitivity of Ama^{axenic} to the β-oxidation inhibitors, 4-pentenoate, and 2-mercaptoacetate (Hart and Coombs 1982). The researchers also measured several metabolic end products—succinate, acetate, alanine, and CO_2—with the amount secreted varying depending on the developmental stage assayed. In a complementary approach, Rainey and MacKenzie (1991) used ^{13}C nuclear magnetic resonance (NMR) spectroscopy to analyze the end products of *Leishmania pifanio* metabolism. Analysis of the positional isotopologs provided direct evidence for operation of the glycosomal succinate fermentation (GSF) pathway in *Leishmania* in which glycolytic phosphoenolpyruvate is imported back into the glycosomes and reduced to succinate via a series of reactions that replicate those in the TCA cycle. These studies supported a model for promastigote metabolism in which glucose is catabolized via glycolysis to generate phosphoenolpyruvate (PEP) and pyruvate, which are either converted to succinate (via the GSF) or alanine (via alanine transaminase) and secreted or further catabolized in the TCA cycle to generate CO_2, and reducing equivalents for oxidative phosphorylation (Fig. 2). While amastigotes also utilize glucose, they do so at a decreased rate, which coincided with increased use of fatty acids, although these studies did not quantitate the relative contribution of these different carbon sources. Notably, *Leishmania* lack a functional glyoxylate cycle and are therefore unable to generate C_4 precursors for sugar hexose synthesis, suggesting that any increase in fatty acid β-oxidation would primarily be used to sustain ATP synthesis or to top-up (anaplerosis) TCA cycle intermediates.

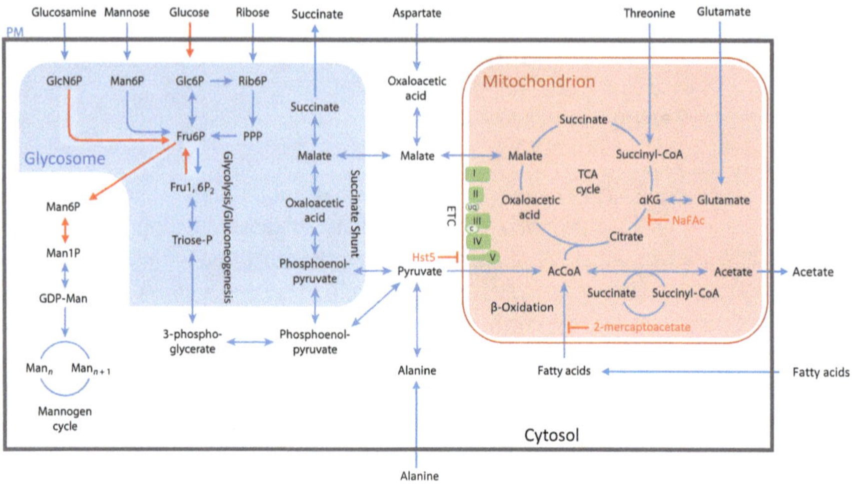

Fig. 2 Central carbon metabolism in *Leishmania*. The schematic shows the acquisition of carbon sources and key metabolic pathways. Glycolysis and gluconeogenesis as well as the GSF pathway occur in specialized peroxisomes termed glycosomes, while the TCA cycle and β-oxidation occur in mitochondria (enzymes for β-oxidation also localize to the glycosomes but mitochondria are likely the main site of β-oxidation). The major carbohydrate reserve material of *Leishmania* is comprised of short oligosaccharides (4–40 mannose residues long), termed mannogen which accumulates in the cytosol and may play a major role in regulating substrate availability for gluconeogenesis and glycolysis. The role of several enzymes/pathways in *Leishmania* metabolism was assessed by generating gene knock-outs or using metabolic inhibitors. Essential pathways are indicated by *red arrows*. In many cases, the effects of gene depletions or metabolic inhibitors on *Leishmania* viability differ between the promastigote and amastigote stage (see text), highlighting the distinct metabolism of the two life-cycle stages. For example parasites lacking the gluconeogenic enzyme FBPase grow normally as promastigotes in rich media but display attenuated virulence in macrophages and mice. Other pathways appear to be essential in both life-cycle stages—e.g., Luque-Ortega et al. (2008) showed that inhibition of the ATP synthase in *Leishmania* through histatin 5 is lethal for promastigotes and amastigotes. The salvage of some essential metabolites such as purine and vitamins are discussed elsewhere (e.g. McConville and Naderer 2011). *AcCoA* Acetylcoenzyme A; *C* Cytochrome C; *ETC* Electron transport chain; *Fru6P* Fructose-6-phosphate; *Fru1,6P₂*, Fructose-1,6-bisphosphate; *GDP-Man* Guanosine diphosphate mannose; *Glc6P* Glucose-6-phosphate; *GlcN6P* Glucosamine-6-phosphate; *Hst5* Histatin 5; *Man6P* Mannose-6-phosphate; *Man1P* Mannose-1-phosphate; *NaFAc* Sodium fluoroacetate; *KG* Ketoglutarate; *Man$_n$* Mannogen oligosaccharides; *NAD* Nicotinamide dinucleotide; *PM* Plasma membrane; *PPP* Pentose phosphate pathway; *Rib5P* Ribose-5-phosphate; *TCA* Tricarboxylic acid; *Triose-P* Triose phosphate; *UQ* Ubiquinone; *I* Complex I (NADH dehydrogenase); *II* Complex II (fumarate reductase); *III* Complex III (cytochrome bc1 complex); *IV* Complex IV (cytochrome c oxidase); *V* Complex V (ATP synthase)

10.2 Intracellular Metabolite Levels

While metabolic footprinting has provided important insights into the distinct metabolism of the *Leishmania* life-cycle stages, the information gained from analyzing extracellular metabolites is limited. An alternative approach is to measure changes in intracellular metabolite pools. Several studies on *Leishmania* promastigotes and amastigotes using NMR spectroscopy (Rainey et al. 1991; Gupta et al. 1999; Arjmand et al. 2016) identified a range of amino acids (Ala, Arg, Glu, Gln, Gly, Ser, Val, Iso/Leu), sugars (mannose), organic acids (lactate, succinate), and other metabolites (acetate, creatinine, β-hydroxybutyrate, glycerol 3-phosphate, glycerol, a-glycerophosphoryl choline, acetoacetate) that differed between the two developmental stages, reinforcing the importance of glucose catabolism in the promastigote stage and a shift to fatty acid/lipid catabolism in the amasitogote stage. In fact, Berg et al. (2015) proposed that the switch from a promastigote-like metabolism (highly glycolytic) to the amastigote-like metabolism (increased reliance on the TCA cycle and β-oxidation for energy generation) occurs gradually, as Pro^{stat} exhibit decreased levels of amino acids and sugar phosphates, which more closely resembles the metabolite profile of Ama^{axenic} than Pro^{log}. Gupta et al. (2001) compared the profiles of Ama^{axenic} with that of Ama^{lesion} and found some striking differences suggesting that Ama^{axenic} may represent an intermediate stage between Pro^{stat} and Ama^{lesion}. While many studies have suggested that Ama^{axenic} closely resemble Ama^{lesion} with regards to morphology, protein expression, infectivity, etc., differences between Ama^{axenic} and Ama^{lesion} potentially reflect inconsistencies in the nutritional conditions of the culture medium and lesion environments. These findings underscore the importance of undertaking future metabolomics work (and proteomics/transcriptomics work) in vivo where possible (see below), alternatively using purified extracted Ama^{lesion} or improving culture conditions to more closely resemble the in vivo milieu.

While these approaches point to the operations of specific pathways, they provide little information on overall flux or relative partitioning between different arms of central carbon metabolism. In particular, they do not allow distinction between the possibilities that changes in metabolite levels can be driven by changes in synthesis/degradation, and uptake/secretion. Even when uptake data are considered, it can be unclear whether a substrate is used for energy generation or metabolite biosynthesis. Given the dynamic nature of metabolite pools, stable isotope (e.g. ^{13}C, ^{2}H, ^{15}N) labeling approaches, provide a powerful means to measure metabolic flux.

10.3 Flux Analyses

A detailed analysis of metabolic networks of *L. mexicana* promastigotes and amastigotes was recently undertaken using comprehensive ^{13}C-stable isotope labeling (Saunders et al. 2014). All major developmental stages (Pro^{log}, Pro^{stat}, Ama^{axenic} isolated Ama^{lesion}) were labeled in defined medium with ^{13}C-labeled glucose, ^{13}C-amino acids, and ^{13}C-fatty acids and the incorporation of tracer carbons into a targeted list of 30 intracellular metabolites was quantitated by GC-MS. Both Ama^{axenic} and Ama^{lesion} were found to exhibit a distinct stringent metabolic response that was characterized by decreased glucose and amino acid uptake and more efficient utilization of these carbon sources (Saunders et al. 2014) (Fig. 3).

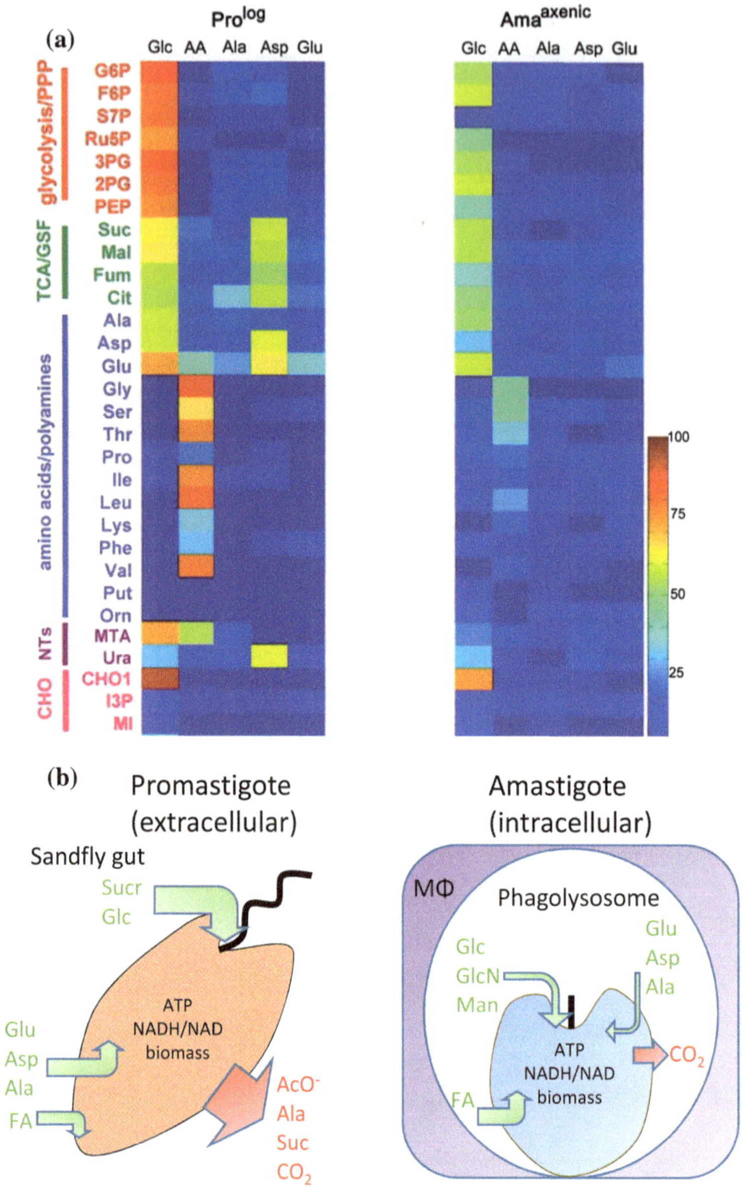

This glucose-sparing metabolism of amastigotes stands in stark contrast to the glucose-wasting metabolism observed in rapidly dividing Pro^{log} as well as nondividing Pro^{stat}, which were characterized by high rates of glucose uptake and secretion of partly oxidized end products (succinate, alanine, acetate) (Saunders et al. 2014). Interestingly, both nondividing (Pro^{stat}) as well as replicating promastigotes (Pro^{log}) exhibited a glucose-wasting metabolism, indicating that the

◀ **Fig. 3** Differences in the central carbon metabolism of *Leishmania* life-cycle stages. **a** ^{13}C-labeling experiments were carried out to determine differences in the central carbon metabolism and carbon source utilization of the *Leishmania* insect stage (Pro^{log}) and the disease causing amastigote stage (Ama^{axenic}) (Saunders et al. 2014). Following incubation of Pro^{log} and Ama^{axenic} in media containing U-^{13}C-glucose (Glc), U-^{13}C-amino acid mix (AA), U-^{13}C-alanine (Ala), U-^{13}C-asparate (Asp), or U-^{13}C glutamate (Glu), the ^{13}C-enrichment was quantified in numerous metabolites using GC-MS. **b** The study revealed marked differences between the two life-cycle stages—promastigotes exhibit a glucose-wasting metabolism which is characterized by rapid uptake and utilization of glucose as well as secretion of organic acids and alanine (*green arrows* indicate uptake of metabolites, *red arrows* indicate secretion of end products). In contrast, amastigotes display a glucose sparing metabolism and show drastically reduced rates of glucose as well as amino acid uptake and utilization together. Both stages scavenge fatty acids, but promastigotes primarily use these for membrane synthesis while amastigotes oxidize acquired fatty acids. *G6P* Glucose 6-phosphate; *F6P* Fructose 6-phosphate; *S7P* Seduheptulose 7-phosphate; *Ru5P* Ribulose 5-phosphate; *3PG* 3-phosphoglycerate; *2PG* 2-phosphoglycerate; *PEP* Phosphoenolpyruvate; *Suc* Succinate; *Mal* Malate; *Fum* Fumarate; *Cit* Citrate; *Ala* Alanine; *Asp* Aspartate; *Glu* Glutamate; *Gly* Glycine; *Ser* Serine; *Thr* Threonine; *Pro* Proline; *Ile* Isoleucine; *Leu* Leucine; *Lys* Lysine; *Phe* Phenylalanine; *Val* Valine; *Put* Putrescine; *Orn* Ornithine; *MTA* 5-methylthioadenosine; *Ura* Uracil; *CHO1* Mannogen; *I3P* Inositol 3-phosphate; *MI* *Myo*-inositol; *G3P* Glycerol 3-phosphate; *Sucr* Sucrose; *Glc* Glucose; *FA* Fatty acids; *AcO*$^{-}$ Acetate; *GlcN* Glucosamine; *Man* Mannose; *ATP* Adenosine triphosphate; *NAD* Nicotinamide adenine dinucleotide. The figures are adapted from Saunders et al. (2014) and McConville et al. (2015)

stringent response was not just a consequence of reduced replication but rather a hard-wired response that is induced during amastigote differentiation. Amastigotes also exhibited increased fatty acid β-oxidation and increased reliance on the TCA cycle metabolism (Saunders et al. 2014), consistent with results from earlier proteomic analyses and metabolomics analysis (see above).

These labeling studies were used to predict pathways essential for amastigote survival in vivo. In particular, the finding that amastigotes stages exhibited negligible rates of glutamate uptake (compared to promastigotes) while the carbon backbones derived from ^{13}C-glucose and ^{13}C-fatty acids were incorporated into glutamate/glutamine, suggested that the TCA cycle may have an important role in generating precursors, such as α-ketoglutarate for glutamate/glutamine synthesis. These amino acids serve as essential amino donors for a number of other essential pathways including aminosugar synthesis, pyrimidine synthesis, and glutathione/trypanothione synthesis. Consistent with the proposal, treatment of infected macrophages with either sodium fluoroacetate (NaFAc, an inhibitor of TCA cycle aconitase enzyme) or methionine sulfoxime (MSO, an inhibitors of glutamine synthetase) resulted in death of intracellular *L. mexicana* amastigotes (Saunders et al. 2014). While all developmental stages exhibited some growth sensitivity to these inhibitors in axenic cultures, particularly when exogenous glutamate or glutamine was absent, Ama^{axenic} were much more sensitive to these inhibitors than Pro^{log}. While Pro^{log} ceased growth during treatment with 5 mM NaFAc, Ama^{axenic} lost viability at 50–100-times lower concentrations. The effect of NaFAc on Ama^{axenic} was only partially rescued by addition of very high concentrations of glutamate (5 mM). Similarly, MSO treatment led to a growth arrest in Pro^{log} which was rescued by supplementation with glutamine. Interestingly, growth of intracellular Ama$^{M\Phi}$ could not be rescued by addition of exogenous glutamine to cultures suggesting that uptake of this amino acid into the phagolysosome by bulk vesicle flow is inefficient and/or that expression of glutamate amino acid transporters on intracellular amastigote stages is highly repressed. Taken together, these results suggest that amastigotes rely heavily on a functional TCA cycle to produce glutamate and glutamine and appear to downregulate the uptake of these amino acids. This is particularly surprising, as the phagolysosome is the major site of protein degradation and is expected to be rich in amino acids (Saunders et al. 2014).

These ^{13}C-labeling studies have also been used to define the objective function of a constraint-based model of *L. infantum* energy metabolism (i.e. not genome-wide) (Subramanian et al. 2015). The resulting model, consisting of over 230 reactions and 5 cellular compartments, was able to accurately predict in silico the growth phenotypes of previously experimentally generated knock-out mutants (for example, the *L. major* ΔFBPase and *L. mexicana* Δphosphomannomutase mutants, Naderer et al. 2006; Garami et al. 2001). Using the model, 61 single reaction combinations and 10,884 double reaction combinations were predicted that, when knocked-out, are anticipated to be lethal in *L. infantum* and therefore may prove useful drug targets. The model was also used to investigate changes to metabolism under different nutritional environments, for example the researchers observed that when optimum oxygen uptake is achieved, glucose is completely catabolized with succinate from the GSF entering the TCA cycle rather than being secreted. As oxygen uptake decreases, partially oxidized end products (acetate and succinate) are secreted from the cell. As described in the profiling and labeling experiments described above, glucose catabolism is critical even when other alternative carbon sources are abundant (the model predicts co-utilization of nonessential amino acids alters fluxes through glycolysis, glutamate biosynthesis and glycine/serine biosynthesis). In support of this, when glucose availability is restricted, the model predicts no parasite proliferation despite the availability of amino acids. Nonetheless, nonessential amino acids were still important when catabolized in conjunction with glucose, increasing biomass 1–1.5-fold over glucose alone (indeed Subramanian et al. highlight the importance of glutamate in ATP generation). Finally, the researchers created promastigote and amastigote scenarios in their model and were able to predict the large reduction in glycolysis, TCA cycle, ATP synthesis, and amino acid metabolism in the amastigote stage (reduced, but essential, glucose uptake, reduced glutamate uptake and reduced overflow metabolite excretion).

11 Studying *Leishmania* Metabolism in Vivo: Amastigote Proliferation and Macromolecule Turnover

All of the studies described above rely on the analysis of axenically cultured amastigotes or isolated Ama^{lesion} that have been incubated in rich media which may not reflect the nutrient levels in vivo (Saunders et al. 2014). Indeed, the nutrients present in the phagolysosome as well as their concentrations are largely undefined (Lorenz and Fink 2002; McConville and Naderer 2011), making it difficult to establish culture media which replicates the conditions in vivo. An alternative approach is to use stable isotope labeling in vivo. A number of studies have infused animals with ^{13}C-labeled precursors (glucose or amino acids) in order to measure metabolic dynamics in vivo utilizing NMR spectroscopy or MS (Neurohr et al. 1983; Stromski et al. 1986; Shalwitz et al. 1989; Magkos and Mittendorfer 2009; Maher et al. 2012; Kowalski et al. 2015). However, these experiments are technically complex and limited by several factors. First, ^{13}C-labeled tracers are costly and in vivo analyses require high quantities of the tracer to achieve detectable levels in the analyzed tissue. While short-term labeling with these tracers can be easily achieved through an oral bolus (gavage) of the tracer, the long-term administration of ^{13}C-tracers requires continuous infusion, which is costly and stressful to the animal. Maintaining constant levels of a tracer within the analyzed tissue can also be challenging due to rapid clearance, catabolism, or zonation effects. Additionally, interpreting in vivo ^{13}C tracer data is expected to be particularly challenging when analyzing intracellular parasites, as the tracer is rapidly metabolized by the host with the pathogens potentially taking up a variety of labeled tracer-derived metabolites and intermediates. Hence, long-term ^{13}C-labeling approaches are largely restricted to in vitro applications and are not well suited for the study of intracellular pathogens.

An alternative approach to using ^{13}C-labeled tracers is the use of ^2H_2O (also known as heavy water, deuterium oxide and D_2O) as the tracer (Hellerstein 2004). ^2H_2O differs from natural abundance water (H_2O) in that it contains two deuterium atoms (^2H) instead of hydrogen (typically available as >99.9 % ^2H_2O, v/v). ^2H_2O is utilized by many enzymes that include water in their catalytic mechanisms, resulting in the incorporation of ^2H into stable C–H/^2H bonds in a wide variety of metabolites. This enzymatic ^2H_2O-labeling differs from ^2H-exchange that occurs across labile N–H or O–H bonds which is commonly used to identify exposed residues in protein folding studies (Englander et al. 1997). While high concentrations of ^2H_2O (>20 %, v/v) can be toxic to some organisms because of a solvent isotope effect (where ^2H_2O/^2H can interfere with the catalytic efficiency of some enzymes) (Reuter et al. 1985; Takeda et al. 1998), few or no adverse effects are observed when cells/animals are exposed to low (≤15 %, v/v) concentrations (Thomson and Klipfel 1960; Lester et al. 1960; Kushner et al. 1999; Busch et al. 2007; Berry et al. 2015). The incorporation of ^2H into different metabolite pools (sugars, amino acids and fatty acids) and downstream macromolecules (DNA, RNA, proteins, lipids and polysaccharides) can be readily quantitated using MS or ^2H-NMR spectroscopy directly or after depolymerization of macromolecules of interest (Dufner and Previs 2003). Importantly, ^2H_2O rapidly equilibrates across all tissues and cells and can be administered to cell cultures or animals safely and easily for weeks or months, making it particularly suitable for measuring processes that have turnover times on the scale of days or longer (Dufner and Previs 2003; Busch et al. 2007). Administration of ^2H_2O also results in no perturbation to external or intracellular metabolite levels in biological systems, such as occurs when a bolus of ^{13}C-labeled metabolites is introduced into cultures or animals (Berry et al. 2015). As a result of these features, $^{2}H_2O$-labeling is increasingly being deployed to measure multiple cellular processes in physiology and nutrition including human studies (Landau et al. 1995; Busch et al. 2006, 2007; Murphy 2006). $^{2}H_2O$ cannot replace but may complement ^{13}C tracers as the processes which can be measured using $^{2}H_2O$ are limited. However, several approaches have been developed which allow the quantification of the following processes in vivo using $^{2}H_2O$: Replication/DNA turnover (Neese et al. 2002; Hsieh et al. 2004; Busch et al. 2007; Pouteau et al. 2009), protein synthesis (Busch et al. 2006; Gasier et al. 2010), lipogenesis and cholesterol synthesis (Murphy 2006; Pouteau et al. 2009; Previs et al. 2011), as well as gluconeogenesis (Landau et al. 1995; Antoniewicz et al. 2011).

Kloehn et al. (2015) have recently shown that heavy water ($^{2}H_2O$–) labeling approaches can be employed to measure multiple physiological and metabolic processes in both cultured and tissue stages of *Leishmania* (Fig. 4). The labeling of the major life-cycle stages of *L. mexicana* with 5 % (v/v) $^{2}H_2O$ in culture can be used to accurately determine replication rates by measuring the incorporation of deuterium into the deoxyribose moiety of DNA using GC-MS. This approach was subsequently used to measure the growth rate of *L. mexicana* amastigotes in inflammatory lesions in infected BALB/c mice following enrichment of the host's body water with $^{2}H_2O$. Previously, the in vivo growth rate of pathogens was typically inferred from changes in the microbial burden in the relevant tissues as determined by direct enumeration of pathogen levels in tissue biopsies following microtitration and limiting cell dilution assays (Titus et al. 1985; Cotterell et al. 2000) or from measurements of pathogens that have been genetically manipulated to express bioluminescent or fluorescent proteins (Lang et al. 2005; Thalhofer et al. 2010; Millington et al. 2010; Michel et al. 2011). However, these methods only provide a measure of the net changes in microbial burden that reflect multiple parameters in addition to replication rate, such as death rate and pathogen dissemination to other tissues. In contrast, $^{2}H_2O$-labelling allows the measurement of cell turnover in vivo (Busch et al. 2007). Lesion amastigotes were found to have a doubling time of nearly 12 days, consistent with the ^{13}C-labeling studies showing that these stages enter a metabolically quiescent state. Although slow, this rate of doubling can still account for the observed increase in parasite burden in lesion granulomas, assuming parasite death is minimal (Kloehn et al. 2015). Taken together, these analyses indicate that activation of the stringent response may allow *Leishmania* amastigotes to sustain a very slow rate of replication and persist within long-lived macrophages. An intriguing implication of the finding that both parasite and macrophage populations are both long lived is that expansion of lesions and parasite numbers may occur via the slow replication of macrophages and partitioning of amastigote-containing phagolysosomes to each of the daughter cells. Such a mechanism would be consistent with microscopy observations that rarely detect extracellular parasites in granuloma tissues. Collectively, these findings indicate that murine inflammatory lesions constrain *Leishmania* growth but

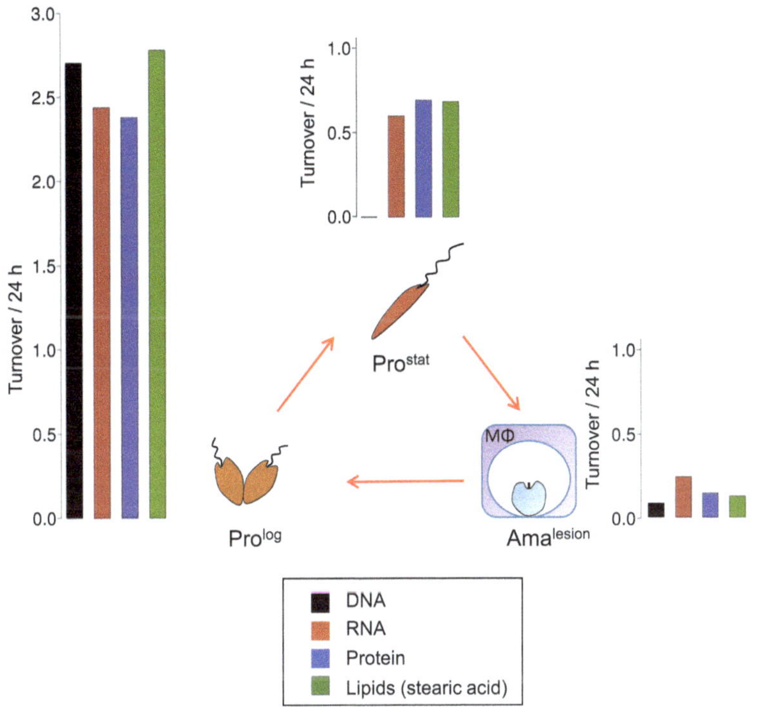

Fig. 4 The physiological state of *Leishmania* life-cycle stages. Differences in the replication rate as well as protein, RNA, and lipid turnover were measured using ^2H_2O-labeling in vitro and in vivo (Kloehn et al. 2015). *Leishmania* Pro^{log} show the fastest turnover rate for all macromolecules indicating rapid replication, protein and RNA, and lipid synthesis. Strikingly, the protein, RNA, and lipid turnover rates of Ama^{lesion} are reduced about 15–30-fold compared to Pro^{log} and are markedly reduced even compared to nondividing Pro^{stat}. These results indicate that *Leishmania* enter a semiquiescent state in the lesion environment which is characterized by slow growth and low activity of other energy intensive cellular processes such as protein synthesis

otherwise provide a highly permissive niche, which allows continuous expansion of the parasite population (Kloehn et al. 2015). This strategy differs from that of other granuloma-inducing pathogens such as *M. tuberculosis* (Munoz-Elias et al. 2005; Gill et al. 2009).

The ^2H_2O-labeling approach was further extended to measure global changes in RNA and protein turnover by measuring ^2H-incorporation into RNA-ribose and protein derived amino acids (Kloehn et al. 2015). As transcription and protein translation represent the most energy consuming processes in cells, these parameters are excellent indicators of the bioenergetic state of a cell. Significant differences were observed in RNA and protein turnover between major life-cycle stages. In particular, Ama^{lesion} were found to have much lower rates of RNA/protein turnover than other stages, including nondividing promastigotes (Kloehn et al. 2015), providing further support for the notion that amastigotes enter into a distinct state of slow growth and metabolic quiescence.

Several studies on Ama^{axenic} have suggested that global rates of protein synthesis are downregulated, based on measurements of rates of ^{35}S-methionine incorporation and polysome profile analysis (Lahav et al. 2011; Cloutier et al. 2012). Another interesting approach for estimating protein turnover/metabolic activity in Ama^{lesion} in vivo has recently been developed using *L. major* that express a photoconvertible fluorescent protein (Muller et al. 2013). The authors suggest that protein synthesis in situ is reduced due to the suppression of metabolism and cell division by sublethal levels of NO. However, a broader application of this approach is limited given that the fluorescence readout is difficult to calibrate to overall protein turnover rates and also requires the generation of transgenic parasites lines with possible associated virulence reduction (da Silva and Sacks 1987; Moreira et al. 2012; Ali et al. 2013).

Additionally, ^2H_2O-labeling was shown to delineate specific metabolic pathways in culture and in vivo. Analysis of ^{2}H-incorporation into total cellular fatty acid pools in Ama^{lesion} showed that this stage largely relies on the scavenging of fatty acids, in contrast to the situation in promastigotes. However, Ama^{lesion} were still dependent on the synthesis of linoleic acid (C18:2), as shown by the labeling of this fatty acid in parasite extracts but not host cell serum and tissue (Kloehn et al. 2015). Linoleic acid is the major polyunsaturated fatty acid of these stages and is synthesized by desaturation of oleic acid. The enzyme oleate desaturase is absent in mammals, and may therefore be a promising drug target.

12 Conclusion and Outlook

Metabolomic approaches have provided important new insights into the biology of *Leishmania* and will continue to be an essential tool for investigating the metabolism of these parasites given their dependence on posttranslational regulatory processes. In particular, metabolomics has been used to map metabolic networks in different developmental stages and parasite mutant lines, as well as to define the mode of action of drugs and understand resistance mechanisms. Importantly, various $^{13}C/^2H$ labeling strategies have now been used to map parasite metabolism and physiology in vitro and in infected tissues. We suggest that metabolomic measurements using noninvasive stable isotope tracers will be increasingly important to understand *Leishmania* metabolism in the mammalian host. Combining metabolomic approaches with other '-omics' approaches as well as molecular biology (e.g. gene depletion), biochemistry (e.g. metabolic inhibitors), and biological data (e.g. virulence) are expected to advance our understanding of *Leishmania* metabolism. Last, phenotypically heterogeneous cells and the mechanism underlying cell-to-cell variability have been identified and studied in a number of bacterial infections (Lewis 2010; Helaine and Holden 2013; Helaine et al. 2014; Bumann 2015; Kopf et al. 2016) but are poorly understood in protozoan infections, despite their crucial role in development (e.g. *Plasmodium* gametocytes and hypnozoites; *Toxoplasma* bradyzoites, *Leishmania* metacyclics), persistence, and the emergence of drug resistance (Seco-Hidalgo et al. 2015). However, the majority of studies described in this chapter rely on bulk measurement, which provides an average of an entire population of typically $>10^7$ cells masking any variability within the population. Hence, novel single cell metabolomics approaches will be invaluable to identify cell-to-cell heterogeneity in protozoan parasite populations (Zenobi 2013; Seco-Hidalgo et al. 2015).

References

Alawieh A et al (2014) Revisiting leishmaniasis in the time of war: the Syrian conflict and the Lebanese outbreak. Int J Infect Dis 29:115–119

Ali KS, Rees RC, Terrell-Nield C, Ali SA (2013) Virulence loss and amastigote transformation failure determine host cell responses to *Leishmania mexicana*. Parasite Immunol 35(12):441–456

Alvar J et al (2012) Leishmaniasis worldwide and global estimates of its incidence. PLoS One 7 (5):e35671

Amaral V, Pirmez C, Goncalves A, Ferreira V, Grimaldi G Jr (2000) Cell populations in lesions of cutaneous leishmaniasis of *Leishmania* (*L.*) amazonensis—infected rhesus macaques, Macaca mulatta. Mem Inst Oswaldo Cruz 95(2):209–216

Andrews KT, Fisher G, Skinner-Adams TS (2014) Drug repurposing and human parasitic protozoan diseases. Int J Parasitol Drugs Drug Resist 4(2):95–111

Anjili C et al (2006) Estimation of the minimum number of *Leishmania major* amastigotes required for infecting *Phlebotomus duboscqi* (Diptera: Psychodidae). East Afr Med J 83(2):68–71

Antoniewicz MR, Kelleher JK, Stephanopoulos G (2011) Measuring deuterium enrichment of glucose hydrogen atoms by gas chromatography/mass spectrometry. Anal Chem 83(8):3211 3216

Arjmand M et al (2016) Metabolomics-based study of logarithmic and stationary phases of promastigotes in *Leishmania major* by 1H NMR spectroscopy. Iran Biomed J 20(2):77–83

Ashutosh Sundar S, Goyal N (2007) Molecular mechanisms of antimony resistance in *Leishmania*. J Med Microbiol 56(Pt 2):143–153

Bakker BM et al (2000) Compartmentation protects *Trypanosomes* from the dangerous design of glycolysis. Proc Natl Acad Sci USA 97(5):2087–2092

Barhoumi M, Tanner NK, Banroques J, Linder P, Guizani I (2006) *Leishmania infantum* LeIF protein is an ATP-dependent RNA helicase and an eIF4A-like factor that inhibits translation in yeast. FEBS J 273(22):5086–5100

Bates PA (2007) Transmission of *Leishmania* metacyclic promastigotes by phlebotomine sand flies. Int J Parasitol 37(10):1097–1106

Beach R, Kiilu G, Hendricks L, Oster C, Leeuwenburg J (1984) Cutaneous leishmaniasis in Kenya: transmission of *Leishmania major* to man by the bite of a naturally infected *Phlebotomus duboscqi*. Trans R Soc Trop Med Hyg 78(6):747–751

Beach R, Kiilu G, Leeuwenburg J (1985) Modification of sand fly biting behavior by *Leishmania* leads to increased parasite transmission. Am J Trop Med Hyg 34(2):278–282

Bente M et al (2003) Developmentally induced changes of the proteome in the protozoan parasite *Leishmania donovani*. Proteomics 3(9):1811–1829

Berg M et al (2015) Experimental resistance to drug combinations in *Leishmania donovani*: metabolic and phenotypic adaptations. Antimicrob Agents Chemother 59(4):2242–2255

Berg M et al (2013) Metabolic adaptations of *Leishmania donovani* in relation to differentiation, drug resistance, and drug pressure. Mol Microbiol 90(2):428–442

Berry D et al (2015) Tracking heavy water (D_2O) incorporation for identifying and sorting active microbial cells. Proc Natl Acad Sci USA 112(2):E194–E203

Bhattacharya A, Biswas A, Das PK (2008) Role of intracellular cAMP in differentiation-coupled induction of resistance against oxidative damage in *Leishmania donovani*. Free Radic Biol Med 44(5):779–794

Blattner J, Helfert S, Michels P, Clayton C (1998) Compartmentation of phosphoglycerate kinase in *Trypanosoma brucei* plays a critical role in parasite energy metabolism. Proc Natl Acad Sci USA 95(20):11596–11600

Bouchard P et al (1982) Diabetes mellitus following pentamidine-induced hypoglycemia in humans. Diabetes 31(1):40–45

Bumann D (2015) Heterogeneous host-pathogen encounters: act locally, think globally. Cell Host Microbe 17(1):13–19

Burchmore RJ et al (2003) Genetic characterization of glucose transporter function in *Leishmania mexicana*. Proc Natl Acad Sci USA 100(7):3901–3906

Busch R et al (2006) Measurement of protein turnover rates by heavy water labeling of nonessential amino acids. Biochim Biophys Acta 1760(5):730–744

Busch R, Neese RA, Awada M, Hayes GM, Hellerstein MK (2007) Measurement of cell proliferation by heavy water labeling. Nat Protoc 2(12):3045–3057

Canuto GA et al (2012) CE-ESI-MS metabolic fingerprinting of *Leishmania* resistance to antimony treatment. Electrophoresis 33(12):1901–1910

Canuto GA et al (2014) Multi-analytical platform metabolomic approach to study miltefosine mechanism of action and resistance in *Leishmania*. Anal Bioanal Chem 406(14):3459–3476

Carvalho FA et al (2002) Diagnosis of American visceral leishmaniasis in humans and dogs using the recombinant *Leishmania donovani* A2 antigen. Diagn Microbiol Infect Dis 43(4):289–295

Castilho-Martins EA et al (2015) Capillary electrophoresis reveals polyamine metabolism modulation in *Leishmania (Leishmania) amazonensis* wild type and arginase knockout mutants under arginine starvation. Electrophoresis 36:2314–2323

Castilla JJ, Sanchez-Moreno M, Mesa C, Osuna A (1995) *Leishmania donovani*: in vitro culture and [1H] NMR characterization of amastigote-like forms. Mol Cell Biochem 142(2):89–97

Cayla M et al (2014) Transgenic analysis of the *Leishmania* MAP kinase MPK10 reveals an auto-inhibitory mechanism crucial for stage-regulated activity and parasite viability. PLoS Pathog 10(9):e1004347

Chakravarty J, Sundar S (2010) Drug resistance in leishmaniasis. J Glob Infect Dis 2(2):167–176

Chang KP, Fong D (1983) Cell biology of host-parasite membrane interactions in leishmaniasis. Ciba Found Symp 99:113–137

Chavali AK et al (2012) Metabolic network analysis predicts efficacy of FDA-approved drugs targeting the causative agent of a neglected tropical disease. BMC Syst Biol 6:27

Chavali AK, Whittemore JD, Eddy JA, Williams KT, Papin JA (2008) Systems analysis of metabolism in the pathogenic trypanosomatid *Leishmania major*. Mol Syst Biol 4:177

Cloutier S et al (2012) Translational control through eIF2alpha phosphorylation during the *Leishmania* differentiation process. PLoS One 7(5):e35085

Collingridge PW, Brown RW, Ginger ML (2010) Moonlighting enzymes in parasitic protozoa. Parasitology 137(9):1467–1475

Cotterell SE, Engwerda CR, Kaye PM (2000) Enhanced hematopoietic activity accompanies parasite expansion in the spleen and bone marrow of mice infected with *Leishmania donovani*. Infect Immun 68(4):1840–1848

Creek DJ, Barrett MP (2014) Determination of antiprotozoal drug mechanisms by metabolomics approaches. Parasitology 141(1):83–92

Croft SL, Coombs GH (2003) Leishmaniasis—current chemotherapy and recent advances in the search for novel drugs. Trends Parasitol 19(11):502–508

Croft SL, Olliaro P (2011) Leishmaniasis chemotherapy—challenges and opportunities. Clin Microbiol Infect 17(10):1478–1483

Croft SL, Sundar S, Fairlamb AH (2006) Drug resistance in leishmaniasis. Clin Microbiol Rev 19 (1):111–126

Cull B et al (2014) Glycosome turnover in *Leishmania major* is mediated by autophagy. Autophagy 10(12):2143–2157

da Silva R, Sacks DL (1987) Metacyclogenesis is a major determinant of *Leishmania* promastigote virulence and attenuation. Infect Immun 55(11):2802–2806

Davidson RN, den Boer M, Ritmeijer K (2009) Paromomycin. Trans R Soc Trop Med Hyg 103 (7):653–660

De Muylder G et al (2011) A screen against *Leishmania* intracellular amastigotes: comparison to a promastigote screen and identification of a host cell-specific hit. PLoS Negl Trop Dis 5(7): e1253

de Oliveira CI, Nascimento IP, Barral A, Soto M, Barral-Netto M (2009) Challenges and perspectives in vaccination against leishmaniasis. Parasitol Int 58(4):319–324

De Rycker M et al (2013) Comparison of a high-throughput high-content intracellular *Leishmania donovani* assay with an axenic amastigote assay. Antimicrob Agents Chemother 57(7):2913–2922

De Souza DP, Saunders EC, McConville MJ, Likic VA (2006) Progressive peak clustering in GC-MS Metabolomic experiments applied to *Leishmania* parasites. Bioinformatics 22 (11):1391–1396

Dereure J et al (2003) Visceral leishmaniasis. Persistence of parasites in lymph nodes after clinical cure. J Infect 47(1):77–81

Dermine JF, Desjardins M (1999) Survival of intracellular pathogens within macrophages. Protoplasma 210(1–2):11–24

Dermine JF, Goyette G, Houde M, Turco SJ, Desjardins M (2005) *Leishmania donovani* lipophosphoglycan disrupts phagosome microdomains in J774 macrophages. Cell Microbiol 7 (9):1263–1270

Dermine JF, Scianimanico S, Prive C, Descoteaux A, Desjardins M (2000) *Leishmania promastigotes* require lipophosphoglycan to actively modulate the fusion properties of phagosomes at an early step of phagocytosis. Cell Microbiol 2(2):115–126

Desjardins M, Descoteaux A (1997) Inhibition of phagolysosomal biogenesis by the *Leishmania* lipophosphoglycan. J Exp Med 185(12):2061–2068

Desjeux P (2004) Leishmaniasis: current situation and new perspectives. Comp Immunol Microbiol Infect Dis 27(5):305–318

Dostalova A, Volf P (2012) *Leishmania* development in sand flies: parasite-vector interactions overview. Parasit Vectors 5:276

Downing T et al (2011) Whole genome sequencing of multiple *Leishmania donovani* clinical isolates provides insights into population structure and mechanisms of drug resistance. Genome Res 21(12):2143–2156

Doyle MA et al (2009) LeishCyc: a biochemical pathways database for *Leishmania major*. BMC Syst Biol 3:57

Dufner D, Previs SF (2003) Measuring in vivo metabolism using heavy water. Curr Opin Clin Nutr Metab Care 6(5):511–517

Eibl H, Unger C (1990) Hexadecylphosphocholine: a new and selective antitumor drug. Cancer Treat Rev 17(2–3):233–242

El Fakhry Y, Ouellette M, Papadopoulou B (2002) A proteomic approach to identify developmentally regulated proteins in *Leishmania infantum*. Proteomics 2(8):1007–1017

El-Sayed NM et al (2005) The genome sequence of *Trypanosoma cruzi*, etiologic agent of Chagas disease. Science 309(5733):409–415

Englander SW, Mayne L, Bai Y, Sosnick TR (1997) Hydrogen exchange: the modern legacy of Linderstrom-Lang. Protein Sci 6(5):1101–1109

Evans KJ, Kedzierski L (2012) Development of Vaccines against Visceral Leishmaniasis. J Trop Med 2012:892817

Freitas-Junior LH, Chatelain E, Kim HA, Siqueira-Neto JL (2012) Visceral leishmaniasis treatment: what do we have, what do we need and how to deliver it? Int J Parasitol Drugs Drug Resist 2:11–19

Fukushima A et al (2014) Metabolomic characterization of knockout mutants in *Arabidopsis*: development of a metabolite profiling database for knockout mutants in *Arabidopsis*. Plant Physiol 165(3):948–961

Gangneux JP et al (2007) Recurrent American cutaneous leishmaniasis. Emerg Infect Dis 13 (9):1436–1438

Gannavaram S et al (2012) Deletion of mitochondrial associated ubiquitin fold modifier protein Ufm1 in *Leishmania donovani* results in loss of beta-oxidation of fatty acids and blocks cell division in the amastigote stage. Mol Microbiol 86(1):187–198

Gannavaram S, Sharma P, Duncan RC, Salotra P, Nakhasi HL (2011) Mitochondrial associated ubiquitin fold modifier-1 mediated protein conjugation in *Leishmania donovani*. PLoS One 6 (1):e16156

Garami A, Mehlert A, Ilg T (2001) Glycosylation defects and virulence phenotypes of *Leishmania mexicana* phosphomannomutase and dolicholphosphate-mannose synthase gene deletion mutants. Mol Cell Biol 21(23):8168–8183

Gasier HG, Fluckey JD, Previs SF (2010) The application of 2H2O to measure skeletal muscle protein synthesis. Nutr Metab (Lond) 7:31

Gill WP et al (2009) A replication clock for *Mycobacterium tuberculosis*. Nat Med 15(2):211–214

Glaser TA, Baatz JE, Kreishman GP, Mukkada AJ (1988) pH homeostasis in *Leishmania donovani* amastigotes and promastigotes. Proc Natl Acad Sci U S A 85(20):7602–7606

Goldman-Pinkovich A et al (2016) An arginine deprivation response pathway is induced in *Leishmania* during macrophage invasion. PLoS Pathog 12(4):e1005494

Gossage SM, Rogers ME, Bates PA (2003) Two separate growth phases during the development of *Leishmania* in sand flies: implications for understanding the life cycle. Int J Parasitol 33 (10):1027–1034

Grigore D, Meade JC (2006) A COOH-terminal domain regulates the activity of *Leishmania* proton pumps LDH1A and LDH1B. Int J Parasitol 36(4):381–393

Gupta N, Goyal N, Rastogi AK (2001) In vitro cultivation and characterization of axenic amastigotes of *Leishmania*. Trends Parasitol 17(3):150–153

Gupta N et al (1999) Characterization of intracellular metabolites of axenic amastigotes of *Leishmania donovani* by 1H NMR spectroscopy. Acta Trop 73(2):121–133

Haanstra JR, Gonzalez-Marcano EB, Gualdron-Lopez M, Michels PA (2016) Biogenesis, maintenance and dynamics of glycosomes in Trypanosomatid parasites. Biochim Biophys Acta 1863(5):1038–1048

Haile S, Papadopoulou B (2007) Developmental regulation of gene expression in Trypanosomatid parasitic protozoa. Curr Opin Microbiol 10(6):569–577

Haldar AK, Sen P, Roy S (2011) Use of antimony in the treatment of leishmaniasis: current status and future directions. Mol Biol Int 2011:571242

Handman E (2001) Leishmaniasis: current status of vaccine development. Clin Microbiol Rev 14 (2):229–243

Handman E, Bullen DV (2002) Interaction of *Leishmania* with the host macrophage. Trends Parasitol 18(8):332–334

Handman E, Osborn AH, Symons F, van Driel R, Cappai R (1995) The *Leishmania promastigote* surface antigen 2 complex is differentially expressed during the parasite life cycle. Mol Biochem Parasitol 74(2):189–200

Hart DT, Coombs GH (1982) *Leishmania mexicana*: energy metabolism of amastigotes and promastigotes. Exp Parasitol 54(3):397–409

Helaine S et al (2014) Internalization of *Salmonella* by macrophages induces formation of nonreplicating persisters. Science 343(6167):204–208

Helaine S, Holden DW (2013) Heterogeneity of intracellular replication of bacterial pathogens. Curr Opin Microbiol 16(2):184–191

Helfert S, Estevez AM, Bakker B, Michels P, Clayton C (2001) Roles of triosephosphate isomerase and aerobic metabolism in *Trypanosoma brucei*. Biochem J 357(Pt 1):117–125

Hellerstein MK (2003) In vivo measurement of fluxes through metabolic pathways: the missing link in functional genomics and pharmaceutical research. Annu Rev Nutr 23:379–402

Hellerstein MK (2004) New stable isotope-mass spectrometric techniques for measuring fluxes through intact metabolic pathways in mammalian systems: introduction of moving pictures into functional genomics and biochemical phenotyping. Metab Eng 6(1):85–100

Holmes E (2010) The evolution of metabolic profiling in parasitology. Parasitology 137(9):1437–1449

Hoppe A (2012) What mRNA abundances can tell us about metabolism. Metabolites 2(3):614–631

Hsieh EA, Chai CM, de Lumen BO, Neese RA, Hellerstein MK (2004) Dynamics of keratinocytes in vivo using HO labeling: a sensitive marker of epidermal proliferation state. J Invest Dermatol 123(3):530–536

Huna-Baron R et al (2000) Mucosal leishmaniasis presenting as sinusitis and optic neuropathy. Arch Ophthalmol 118(6):852–854

Ivens AC et al (2005) The genome of the kinetoplastid parasite, *Leishmania major*. Science 309 (5733):436–442

Jamdhade MD et al (2015) Comprehensive proteomics analysis of glycosomes from *Leishmania donovani*. OMICS 19(3):157–170

Joshi S et al (2014) Visceral Leishmaniasis: advancements in vaccine development via classical and molecular approaches. Front Immunol 5:380

Kassi M, Kassi M, Afghan AK, Rehman R, Kasi PM (2008) Marring leishmaniasis: the stigmatization and the impact of cutaneous leishmaniasis in Pakistan and Afghanistan. PLoS Negl Trop Dis 2(10):e259

Kaye P, Scott P (2011) Leishmaniasis: complexity at the host-pathogen interface. Nat Rev Microbiol 9(8):604–615

Kedzierski L (2010) Leishmaniasis vaccine: where are we today? J Glob Infect Dis 2(2):177–185

Kell DB et al (2005) Metabolic footprinting and systems biology: the medium is the message. Nat Rev Microbiol 3(7):557–565

Killick-Kendrick R, Molyneux DH, Hommel M, Leaney AJ, Robertson ES (1977) Leishmania in phlebotomid sandflies. V. The nature and significance of infections of the pylorus and ileum of the sandfly by *Leishmaniae* of the *Braziliensis* complex. Proc R Soc Lond B Biol Sci 198 (1131):191–199

Kim DH, Creek DJ (2015) What role can metabolomics play in the discovery and development of new medicines for infectious diseases? Bioanalysis 7(6):629–631

Kimblin N et al (2008) Quantification of the infectious dose of *Leishmania major* transmitted to the skin by single sand flies. Proc Natl Acad Sci USA 105(29):10125–10130

Kloehn J et al (2016) Using metabolomics to dissect host-parasite interactions. Curr Opin Microbiol 32:59–65

Kloehn J, Saunders EC, O'Callaghan S, Dagley MJ, McConville MJ (2015) Characterization of metabolically quiescent *Leishmania* parasites in murine lesions using heavy water labeling. PLoS Pathog 11(2):e1004683

Kopf SH et al (2016) Trace incorporation of heavy water reveals slow and heterogeneous pathogen growth rates in cystic fibrosis sputum. Proc Natl Acad Sci USA 113(2):E110–E116

Kowalski GM et al (2015) In vivo cardiac glucose metabolism in the high-fat fed mouse: Comparison of euglycemic-hyperinsulinemic clamp derived measures of glucose uptake with a dynamic metabolomic flux profiling approach. Biochem Biophys Res Commun 463(4):818–824

Krishnamurthy G et al (2005) Hemoglobin receptor in *Leishmania* is a hexokinase located in the flagellar pocket. J Biol Chem 280(7):5884–5891

Kumar R, Engwerda C (2014) Vaccines to prevent leishmaniasis. Clin Transl Immunology 3(3): e13

Kushner DJ, Baker A, Dunstall TG (1999) Pharmacological uses and perspectives of heavy water and deuterated compounds. Can J Physiol Pharmacol 77(2):79–88

Lahav T et al (2011) Multiple levels of gene regulation mediate differentiation of the intracellular pathogen *Leishmania*. FASEB J 25(2):515–525

Lamour SD, Choi BS, Keun HC, Muller I, Saric J (2012) Metabolic characterization of *Leishmania major* infection in activated and nonactivated macrophages. J Proteome Res 11 (8):4211–4222

Landau BR et al (1995) Use of 2H2O for estimating rates of gluconeogenesis. Application to the fasted state. J Clin Invest 95(1):172–178

Lang T, Goyard S, Lebastard M, Milon G (2005) Bioluminescent *Leishmania* expressing luciferase for rapid and high throughput screening of drugs acting on amastigote-harbouring macrophages and for quantitative real-time monitoring of parasitism features in living mice. Cell Microbiol 7(3):383–392

Laskay T, van Zandbergen G, Solbach W (2003) Neutrophil granulocytes–Trojan horses for *Leishmania major* and other intracellular microbes? Trends Microbiol 11(5):210–214

Lau SK et al (2015) Metabolomic profiling of *Burkholderia pseudomallei* using UHPLC-ESI-Q-TOF-MS reveals specific biomarkers including 4-methyl-5-thiazoleethanol and unique thiamine degradation pathway. Cell Biosci 5:26

Leifso K, Cohen-Freue G, Dogra N, Murray A, McMaster WR (2007) Genomic and proteomic expression analysis of *Leishmania* promastigote and amastigote life stages: the *Leishmania* genome is constitutively expressed. Mol Biochem Parasitol 152(1):35–46

Lester W Jr, Sun SH, Seber A (1960) Observations on the influence of deuterium on bacterial growth. Ann NY Acad Sci 84:667–677

Lewis K (2010) Persister cells. Annu Rev Microbiol 64:357–372

Lorenz MC, Fink GR (2002) Life and death in a macrophage: role of the glyoxylate cycle in virulence. Eukaryot Cell 1(5):657–662

Luque-Ortega JR, van't Hof W, Veerman EC, Saugar JM, Rivas L (2008) Human antimicrobial peptide histatin 5 is a cell-penetrating peptide targeting mitochondrial ATP synthesis in *Leishmania*. FASEB J 22(6):1817–1828

Magkos F, Mittendorfer B (2009) Stable isotope-labeled tracers for the investigation of fatty acid and triglyceride metabolism in humans in vivo. Clin Lipidol 4(2):215–230

Maher EA et al (2012) Metabolism of [U-13 C]glucose in human brain tumors in vivo. NMR Biomed 25(11):1234–1244

Maier T, Guell M, Serrano L (2009) Correlation of mRNA and protein in complex biological samples. FEBS Lett 583(24):3966–3973

Marovich MA et al (2001) Leishmaniasis recidivans recurrence after 43 years: a clinical and immunologic report after successful treatment. Clin Infect Dis 33(7):1076–1079

McCall LI, Zhang WW, Ranasinghe S, Matlashewski G (2013) Leishmanization revisited: immunization with a naturally attenuated cutaneous *Leishmania donovani* isolate from Sri Lanka protects against visceral leishmaniasis. Vaccine 31(10):1420–1425

McConville M (2014) Open questions: microbes, metabolism and host-pathogen interactions. BMC Biol 12:18

McConville MJ, de Souza D, Saunders E, Likic VA, Naderer T (2007) Living in a phagolysosome; metabolism of *Leishmania* amastigotes. Trends Parasitol 23(8):368–375

McConville MJ, Naderer T (2011) Metabolic pathways required for the intracellular survival of *Leishmania*. Annu Rev Microbiol 65:543–561

McConville MJ, Saunders EC, Kloehn J, Dagley MJ (1988) *Leishmania* carbon metabolism in the macrophage phagolysosome—feast or famine? F1000Research 4(F1000 Faculty Rev):938

McElrath MJ, Murray HW, Cohn ZA (1988) The dynamics of granuloma formation in experimental visceral leishmaniasis. J Exp Med 167(6):1927–1937

Michel G et al (2011) Luciferase-expressing *Leishmania infantum* allows the monitoring of amastigote population size, in vivo, ex vivo and in vitro. PLoS Negl Trop Dis 5(9):e1323

Michels PA, Bringaud F, Herman M, Hannaert V (2006) Metabolic functions of glycosomes in Trypanosomatids. Biochim Biophys Acta 1763(12):1463–1477

Millington OR, Myburgh E, Mottram JC, Alexander J (2010) Imaging of the host/parasite interplay in cutaneous leishmaniasis. Exp Parasitol 126(3):310–317

Mittra B et al (2013) Iron uptake controls the generation of *Leishmania* infective forms through regulation of ROS levels. J Exp Med 210(2):401–416
Miyamoto S et al (2001) Discrepancies between the gene expression, protein expression, and enzymatic activity of thymidylate synthase and dihydropyrimidine dehydrogenase in human gastrointestinal cancers and adjacent normal mucosa. Int J Oncol 18(4):705–713
Moore EM, Lockwood DN (2010) Treatment of visceral leishmaniasis. J Glob Infect Dis 2 (2):151–158
Moradin N, Descoteaux A (2012) *Leishmania* promastigotes: building a safe niche within macrophages. Front Cell Infect Microbiol 2:121
Morales MA et al (2010a) Phosphoproteome dynamics reveal heat-shock protein complexes specific to the *Leishmania donovani* infectious stage. Proc Natl Acad Sci USA 107(18):8381–8386
Morales MA, Pescher P, Spath GF (2010b) Leishmania major MPK7 protein kinase activity inhibits intracellular growth of the pathogenic amastigote stage. Eukaryot Cell 9(1):22–30
Moreira D et al (2012) Impact of continuous axenic cultivation in *Leishmania infantum* virulence. PLoS Negl Trop Dis 6(1):e1469
Moreira D et al (2015) *Leishmania infantum* modulates host macrophage mitochondrial metabolism by hijacking the SIRT1-AMPK axis. PLoS Pathog 11(3):e1004684
Muller AJ et al (2013) Photoconvertible pathogen labeling reveals nitric oxide control of *Leishmania major* infection in vivo via dampening of parasite metabolism. Cell Host Microbe 14(4):460–467
Munoz-Elias EJ et al (2005) Replication dynamics of *Mycobacterium tuberculosis* in chronically infected mice. Infect Immun 73(1):546–551
Murphy EJ (2006) Stable isotope methods for the in vivo measurement of lipogenesis and triglyceride metabolism. J Anim Sci 84(Suppl):E94–E104
Myler PJ et al (2000) Genomic organization and gene function in *Leishmania*. Biochem Soc Trans 28(5):527–531
Naderer T et al (2006) Virulence of *Leishmania major* in macrophages and mice requires the gluconeogenic enzyme fructose 1,6 bisphosphatase. Proc Natl Acad Sci USA 103(14):5502–5507
Naderer T, Heng J, McConville MJ (2010) Evidence that intracellular stages of *Leishmania major* utilize amino sugars as a major carbon source. PLoS Pathog 6(12):e1001245
Naderer T et al (2015) Intracellular survival of *Leishmania major* depends on uptake and degradation of extracellular matrix glycosaminoglycans by macrophages. PLoS Pathog 11(9): e1005136
Nasereddin A, Schweynoch C, Schonian G, Jaffe CL (2010) Characterization of *Leishmania (Leishmania) tropica* axenic amastigotes. Acta Trop 113(1):72–79
Neese RA et al (2002) Measurement in vivo of proliferation rates of slow turnover cells by 2H2O labeling of the deoxyribose moiety of DNA. Proc Natl Acad Sci USA 99(24):15345–15350
Neurohr KJ, Barrett EJ, Shulman RG (1983) In vivo carbon-13 nuclear magnetic resonance studies of heart metabolism. Proc Natl Acad Sci USA 80(6):1603–1607
Noronha FS, Cruz JS, Beirao PS, Horta MF (2000) Macrophage damage by *Leishmania amazonensis* cytolysin: evidence of pore formation on cell membrane. Infect Immun 68 (8):4578–4584
Nugent PG, Karsani SA, Wait R, Tempero J, Smith DF (2004) Proteomic analysis of *Leishmania mexicana* differentiation. Mol Biochem Parasitol 136(1):51–62
Olliaro PL et al (2005) Treatment options for visceral leishmaniasis: a systematic review of clinical studies done in India, 1980–2004. Lancet Infect Dis 5(12):763–774
Oppenheim RD et al (2014) BCKDH: the missing link in apicomplexan mitochondrial metabolism is required for full virulence of *Toxoplasma gondii* and *Plasmodium berghei*. PLoS Pathog 10 (7):e1004263
Ouellette M, Drummelsmith J, Papadopoulou B (2004) Leishmaniasis: drugs in the clinic, resistance and new developments. Drug Resist Updat 7(4–5):257–266

Pawar H, Kulkarni A, Dixit T, Chaphekar D, Patole MS (2014) A bioinformatics approach to reanalyze the genome annotation of kinetoplastid protozoan parasite *Leishmania donovani*. Genomics 104(6 Pt B):554–561

Peacock CS et al (2007) Comparative genomic analysis of three *Leishmania* species that cause diverse human disease. Nat Genet 39(7):839–847

Pearson RD, Sousa AQ (1996) Clinical spectrum of Leishmaniasis. Clin Infect Dis 22(1):1–13

Pimenta PF, Modi GB, Pereira ST, Shahabuddin M, Sacks DL (1997) A novel role for the peritrophic matrix in protecting *Leishmania* from the hydrolytic activities of the sand fly midgut. Parasitology 115(Pt 4):359–369

Pouteau E, Beysen C, Saad N, Turner S (2009) Dynamics of adipose tissue development by 2H2O labeling. Methods Mol Biol 579:337–358

Previs SF et al (2011) Quantifying cholesterol synthesis in vivo using ^2H_2O: enabling back-to-back studies in the same subject. J Lipid Res 52(7):1420–1428

Prina E, Antoine JC, Wiederanders B, Kirschke H (1990) Localization and activity of various lysosomal proteases in *Leishmania amazonensis*-infected macrophages. Infect Immun 58 (6):1730–1737

Quinones MP et al (2007) CCL2-independent role of CCR2 in immune responses against *Leishmania major*. Parasite Immunol 29(4):211–217

Raamsdonk LM et al (2001) Co-consumption of sugars or ethanol and glucose in a *Saccharomyces cerevisiae* strain deleted in the HXK2 gene. Yeast 18(11):1023–1033

Rabinowitz JD, Purdy JG, Vastag L, Shenk T, Koyuncu E (2011) Metabolomics in drug target discovery. Cold Spring Harb Symp Quant Biol 76:235–246

Rainey PM, MacKenzie NE (1991) A carbon-13 nuclear magnetic resonance analysis of the products of glucose metabolism in *Leishmania pifanoi* amastigotes and promastigotes. Mol Biochem Parasitol 45(2):307–315

Rainey PM, Spithill TW, McMahon-Pratt D, Pan AA (1991) Biochemical and molecular characterization of *Leishmania pifanoi* amastigotes in continuous axenic culture. Mol Biochem Parasitol 49(1):111–118

Rakotomanga M, Blanc S, Gaudin K, Chaminade P, Loiseau PM (2007) Miltefosine affects lipid metabolism in *Leishmania donovani* promastigotes. Antimicrob Agents Chemother 51 (4):1425–1430

Rakotomanga M, Loiseau PM, Saint-Pierre-Chazalet M (2004) Hexadecylphosphocholine interaction with lipid monolayers. Biochim Biophys Acta 1661(2):212–218

Rangel H, Dagger F, Hernandez A, Liendo A, Urbina JA (1996) Naturally azole-resistant *Leishmania braziliensis* promastigotes are rendered susceptible in the presence of terbinafine: comparative study with azole-susceptible *Leishmania mexicana* promastigotes. Antimicrob Agents Chemother 40(12):2785–2791

Rastrojo A et al (2013) The transcriptome of *Leishmania major* in the axenic promastigote stage: transcript annotation and relative expression levels by RNA-seq. BMC Genomics 14:223

Real F et al (2013) The genome sequence of *Leishmania* (*Leishmania*) *amazonensis*: functional annotation and extended analysis of gene models. DNA Res 20(6):567–581

Reithinger R (2008) Leishmaniases' burden of disease: ways forward for getting from speculation to reality. PLoS Negl Trop Dis 2(10):e285

Reuter HD, Fischer JH, Thiele S (1985) Investigations on the effects of heavy water (D_2O) on the functional activity of human platelets. Haemostasis 15(3):157–163

Ridley DS, De Magalhaes AV, Marsden PD (1989) Histological analysis and the pathogenesis of mucocutaneous leishmaniasis. J Pathol 159(4):293–299

Ritter U, Frischknecht F, van Zandbergen G (2009) Are neutrophils important host cells for *Leishmania* parasites? Trends Parasitol 25(11):505–510

Rogers MB et al (2011) Chromosome and gene copy number variation allow major structural change between species and strains of *Leishmania*. Genome Res 21(12):2129–2142

Rogers ME, Bates PA (2007) *Leishmania* manipulation of sand fly feeding behavior results in enhanced transmission. PLoS Pathog 3(6):e91

Rojo D et al (2015) A multiplatform metabolomic approach to the basis of antimonial action and resistance in *Leishmania infantum*. PLoS One 10(7):e0130675
Rosenzweig D et al (2008a) Retooling *Leishmania* metabolism: from sand fly gut to human macrophage. FASEB J 22(2):590–602
Rosenzweig D, Smith D, Myler PJ, Olafson RW, Zilberstein D (2008b) Post-translational modification of cellular proteins during *Leishmania donovani* differentiation. Proteomics 8 (9):1843–1850
Russell DG, Vanderven BC, Glennie S, Mwandumba H, Heyderman RS (2009) The macrophage marches on its phagosome: dynamic assays of phagosome function. Nat Rev Immunol 9 (8):594–600
Saar Y et al (1998) Characterization of developmentally-regulated activities in axenic amastigotes of *Leishmania donovani*. Mol Biochem Parasitol 95(1):9–20
Sacks DL (1989) Metacyclogenesis in *Leishmania* promastigotes. Exp Parasitol 69(1):100–103
Sacks DL, da Silva RP (1987) The generation of infective stage *Leishmania major* promastigotes is associated with the cell-surface expression and release of a developmentally regulated glycolipid. J Immunol 139(9):3099–3106
Sacks DL, Perkins PV (1984) Identification of an infective stage of *Leishmania* promastigotes. Science 223(4643):1417–1419
Sacks DL, Perkins PV (1985) Development of infective stage *Leishmania* promastigotes within phlebotomine sand flies. Am J Trop Med Hyg 34(3):456–459
Saunders EC et al (2015) Use of (13)C stable isotope labelling for pathway and metabolic flux analysis in *Leishmania* parasites. Methods Mol Biol 1201:281–296
Saunders EC et al (2012) LeishCyc: a guide to building a metabolic pathway database and visualization of metabolomic data. Methods Mol Biol 881:505–529
Saunders EC et al (2014) Induction of a stringent metabolic response in intracellular stages of *Leishmania mexicana* leads to increased dependence on mitochondrial metabolism. PLoS Pathog 10(1):e1003888
Saxena A et al (2007) Analysis of the *Leishmania donovani* transcriptome reveals an ordered progression of transient and permanent changes in gene expression during differentiation. Mol Biochem Parasitol 152(1):53–65
Schlein Y, Jacobson RL, Messer G (1992) *Leishmania* infections damage the feeding mechanism of the sandfly vector and implement parasite transmission by bite. Proc Natl Acad Sci USA 89 (20):9944–9948
Schriefer A, Wilson ME, Carvalho EM (2008) Recent developments leading toward a paradigm switch in the diagnostic and therapeutic approach to human leishmaniasis. Curr Opin Infect Dis 21(5):483–488
Seco-Hidalgo V, Osuna A, Pablos LM (2015) To bet or not to bet: deciphering cell to cell variation in protozoan infections. Trends Parasitol 31(8):350–356
Seifert K (2011) Structures, targets and recent approaches in anti-leishmanial drug discovery and development. Open Med Chem J 5:31–39
Seifert K et al (2003) Characterisation of *Leishmania donovani* promastigotes resistant to hexadecylphosphocholine (miltefosine). Int J Antimicrob Agents 22(4):380–387
Seifert K et al (2007) Inactivation of the miltefosine transporter, LdMT, causes miltefosine resistance that is conferred to the amastigote stage of *Leishmania donovani* and persists in vivo. Int J Antimicrob Agents 30(3):229–235
Shalwitz RA et al (1989) Hepatic glycogen synthesis from duodenal glucose and alanine. An in situ 13C NMR study. J Biol Chem 264(7):3930–3934
Silva LA, Vinaud MC, Castro AM, Cravo PV, Bezerra JC (2015) In silico search of energy metabolism inhibitors for alternative leishmaniasis treatments. Biomed Res Int 2015:965725
Sindermann H, Engel J (2006) Development of miltefosine as an oral treatment for leishmaniasis. Trans R Soc Trop Med Hyg 100(Suppl 1):S17–20
Sollelis L et al (2015) First efficient CRISPR-Cas9-mediated genome editing in *Leishmania* parasites. Cell Microbiol 17(10):1405–1412

Souza-Lemos C, de Campos SN, Teva A, Porrozzi R, Grimaldi G Jr (2011) In situ characterization of the granulomatous immune response with time in nonhealing lesional skin of *Leishmania braziliensis*-infected rhesus macaques (Macaca mulatta). Vet Immunol Immunopathol 142(3–4):147–155

Stromski ME, Arias-Mendoza F, Alger JR, Shulman RG (1986) Hepatic gluconeogenesis from alanine: 13C nuclear magnetic resonance methodology for in vivo studies. Magn Reson Med 3 (1):24–32

Stuart K et al (2008) Kinetoplastids: related protozoan pathogens, different diseases. J Clin Invest 118(4):1301–1310

Subramanian A, Jhawar J, Sarkar RR (2015) Dissecting *Leishmania infantum* energy metabolism —a systems perspective. PLoS One 10(9):e0137976

Sundar S, Goyal AK, More DK, Singh MK, Murray HW (1998) Treatment of antimony-unresponsive Indian visceral leishmaniasis with ultra-short courses of amphotericin-B-lipid complex. Ann Trop Med Parasitol 92(7):755–764

Sundar S, Jha TK, Thakur CP, Sinha PK, Bhattacharya SK (2007) Injectable paromomycin for Visceral leishmaniasis in India. N Engl J Med 356(25):2571–2581

Takeda H et al (1998) Mechanisms of cytotoxic effects of heavy water (deuterium oxide: D_2O) on cancer cells. Anticancer Drugs 9(8):715–725

Thalhofer CJ, et al (2010) In vivo imaging of transgenic *Leishmania* parasites in a live host. J Vis Exp (41)

Thi EP, Lambertz U, Reiner NE (2012) Sleeping with the enemy: how intracellular pathogens cope with a macrophage lifestyle. PLoS Pathog 8(3):e1002551

Thiel M, Bruchhaus I (2001) Comparative proteome analysis of *Leishmania donovani* at different stages of transformation from promastigotes to amastigotes. Med Microbiol Immunol 190(1–2):33–36

Thomson JF, Klipfel FJ (1960) Some effects of deuterium oxide in vivo and in vitro on certain enzymes of rat tissues. Biochem Pharmacol 3:283–288

Titus RG, Marchand M, Boon T, Louis JA (1985) A limiting dilution assay for quantifying *Leishmania major* in tissues of infected mice. Parasite Immunol 7(5):545–555

Titus RG, Ribeiro JM (1988) Salivary gland lysates from the sand fly *Lutzomyia longipalpis* enhance *Leishmania* infectivity. Science 239(4845):1306–1308

t'Kindt R et al (2010) Metabolomics to unveil and understand phenotypic diversity between pathogen populations. PLoS Negl Trop Dis 4(11):e904

Tonkin ML et al (2015) Structural and functional divergence of the aldolase fold in *Toxoplasma gondii.* J Mol Biol 427(4):840–852

Trindade S et al (2016) *Trypanosoma brucei* parasites occupy and functionally adapt to the adipose tissue in mice. Cell Host Microbe 19(6):837–848

Tsigankov P et al (2014) Regulation dynamics of *Leishmania* differentiation: deconvoluting signals and identifying phosphorylation trends. Mol Cell Proteomics 13(7):1787–1799

Tsigankov P, Gherardini PF, Helmer-Citterich M, Spath GF, Zilberstein D (2013) Phosphoproteomic analysis of differentiating *Leishmania* parasites reveals a unique stage-specific phosphorylation motif. J Proteome Res 12(7):3405–3412

van Griensven J et al (2010) Combination therapy for visceral leishmaniasis. Lancet Infect Dis 10 (3):184–194

van Zandbergen G et al (2004) Cutting edge: neutrophil granulocyte serves as a vector for *Leishmania* entry into macrophages. J Immunol 173(11):6521–6525

van Zandbergen G, Solbach W, Laskay T (2007) Apoptosis driven infection. Autoimmunity 40 (4):349–352

Vanegas G et al (2007) Enolase as a plasminogen binding protein in *Leishmania mexicana.* Parasitol Res 101(6):1511–1516

Vermeersch M et al (2009) In vitro susceptibilities of *Leishmania donovani* promastigote and amastigote stages to antileishmanial reference drugs: practical relevance of stage-specific differences. Antimicrob Agents Chemother 53(9):3855–3859

Vince JE, Tull D, Landfear S, McConville MJ (2011) Lysosomal degradation of *Leishmania* hexose and inositol transporters is regulated in a stage-, nutrient- and ubiquitin-dependent manner. Int J Parasitol 41(7):791–800

Vincent IM, Barrett MP (2015) Metabolomic-based strategies for anti-parasite drug discovery. J Biomol Screen 20(1):44–55

Vincent IM et al (2014) Untargeted metabolomic analysis of miltefosine action in *Leishmania* infantum reveals changes to the internal lipid metabolism. Int J Parasitol Drugs Drug Resist 4 (1):20–27

Vogel C, Marcotte EM (2012) Insights into the regulation of protein abundance from proteomic and transcriptomic analyses. Nat Rev Genet 13(4):227–232

Volf P, Hajmova M, Sadlova J, Votypka J (2004) Blocked stomodeal valve of the insect vector: similar mechanism of transmission in two trypanosomatid models. Int J Parasitol 34(11):1221–1227

Walker J et al (2006) Identification of developmentally-regulated proteins in *Leishmania panamensis* by proteome profiling of promastigotes and axenic amastigotes. Mol Biochem Parasitol 147(1):64–73

Westrop GD et al (2015) Metabolomic analyses of *Leishmania* reveal multiple species differences and large differences in amino acid metabolism. PLoS One 10(9):e0136891

Worthey EA et al (2003) *Leishmania major* chromosome 3 contains two long convergent polycistronic gene clusters separated by a tRNA gene. Nucleic Acids Res 31(14):4201–4210

Zenobi R (2013) Single-cell metabolomics: analytical and biological perspectives. Science 342 (6163):1243259

Zhang WW, Matlashewski G (2015) CRISPR-Cas9-mediated genome editing in *Leishmania donovani*. MBio 6(4):e00861

Zilberstein D, Philosoph H, Gepstein A (1989) Maintenance of cytoplasmic pH and proton motive force in promastigotes of *Leishmania donovani*. Mol Biochem Parasitol 36(2):109–117

Online Reference

WHO.int. URL: http://www.who.int/, July 2016

Chapter 5
Exometabolomics for Linking Soil Carbon Dynamics to Microbial Communities

Andrea Lubbe and Trent Northen

1 Introduction

Microbial metabolism has helped shape the world into what it is today, and continues to drive biogeochemical cycles (Falkowski et al. 2008) including the carbon cycle. Soil microorganisms play a central role in the global carbon cycle, with an estimated soil carbon pool of 2500 Gt, over three times the size of the atmospheric carbon pool (Lal 2004). Inputs of organic carbon into soil is largely plant and microbial biomass-derived, and carbon is released from soil into the atmosphere mainly as CO_2, the product of plant root, and microbial respiration (Johnston et al. 2004). While we are able to measure emergent properties such as the total release of CO_2 from soil and total organic carbon in soil at a particular time, the underlying processes that occur between input and release are not well defined. This limits our ability to understand how human activities are altering the balance of the global carbon cycle, and how this will affect soil carbon dynamics (Lal 2004) that are mediated by soil microbial community metabolism.

The bulk of microbial community studies have been based on metagenomics approaches, where total genomic DNA from soil is sequenced (Delmont et al. 2011; Fierer et al. 2012; Roesch et al. 2007). This culture-independent approach yields insights into the microbial community structure (phylogenetic makeup), and has become a very active field of research due to advances in sequencing technologies (Franzosa et al. 2015). Besides community structure, metagenomes also reveal the complement of genes present in soil microbial communities, reflecting their potential metabolic functions which are inferences based on often poorly annotated

A. Lubbe · T. Northen (✉)
Lawrence Berkeley National Laboratory, 1 Cyclotron Road, Berkeley, CA, USA
e-mail: trnorthen@lbl.gov

A. Lubbe
e-mail: alubbe@lbl.gov

D.J. Beale et al. (eds.), *Microbial Metabolomics*,
DOI 10.1007/978-3-319-46326-1_5

genomes. These inferences of in situ metabolic processes can be strengthened through metatranscriptomics and metaproteomic studies of environmental samples, since these analyses enable correlation of genes actively transcribed and translated (respectively) in the community with environmental variables and stresses (Morales and Holben 2011).

Metabolomics is emerging as a very promising complement to soil metagenomics approaches as it can provide direct insights into the functioning of soil microbial communities in their environment. Exometabolomics, the study of how cells transform their extracellular small molecule environment (Silva and Northen 2015), is particularly relevant for studying soil metabolic processes and provides an experimental approach to link organic carbon in the soil to the metabolism of particular microorganisms or taxonomic group. Soils are complex mixtures of organic and inorganic components, and are estimated to contain more than two-thirds of the carbon in the terrestrial biosphere (Lal 2004). The organic components, known as soil organic matter (SOM), make up most of this pool, and thereby comprise the largest reservoir of carbon on Earth. While total pools of organic carbon in soil can be estimated, the form it takes has been contested (Lehmann and Kleber 2015). Most organic matter inputs to soil decompose within a year (Jenkinson and Rayner 1977). Initial degradation is performed by exoenzymes released by fungi and bacteria that break down organic matter into pieces small enough to be assimilated by microbial cells (Baldock and Nelson 2000; Weiss et al. 1991). A longstanding view was that some of the degraded organic carbon was assimilated into microbial biomass, and the rest was converted to large stable polymeric compounds called humic substances (Stevenson 1994). Their stability was thought to account for the large belowground pool of organic carbon. However, advances in analytical techniques in the last few decades revealed a lack of evidence for polymeric humic substances in soil (Piccolo 2002). Recent evidence suggests that soil organic matter is rather a continuum of progressively decomposing organic compounds (Fig. 1) (Lehmann and Kleber 2015). The new view suggests that much of the SOM exists in the form of lower molecular weight molecules (below 600 Da). Their persistence in soils is not due to any inherent recalcitrance of these molecules, but rather to factors related to the environment, such as absence of degraders or consumers in the immediate environment, sorption onto mineral surfaces, formation of noncovalently bonded aggregates, water availability, pH, and redox state (Schmidt et al. 2011).

2 Exometabolomics for Analysis of Soil Organic Matter

Metabolomics involves the study of the metabolome, defined as the low molecular-weight metabolites (typically less than 2000 Da) present in a cell or living organism under a given set of physiological conditions (Harrigan and Goodacre 2003; Oliver et al. 1998). By contrast, exometabolomics aims to characterize extracellular small metabolites (Silva and Northen 2015). By studying

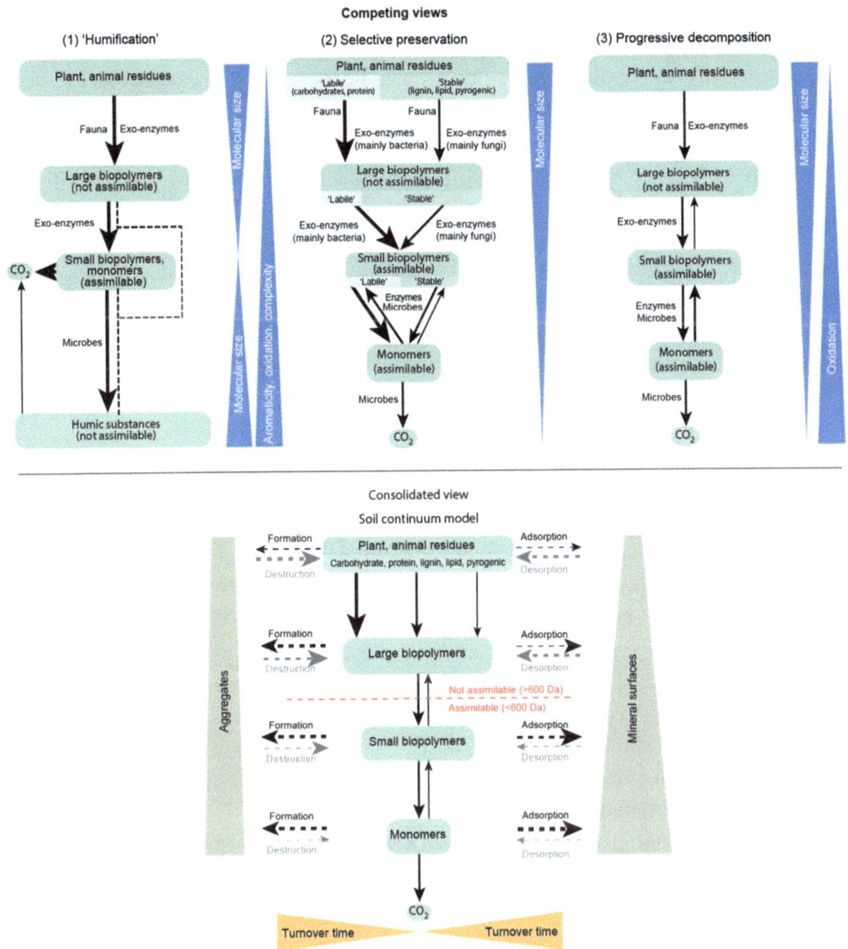

Fig. 1 Schematic representation of three competing models for the fate of organic inputs to soil (*top*), and the recently proposed soil continuum model (*below*). Selective preservation assumes that some organic materials are preferentially mineralized, leaving intrinsically 'stable' decomposition products behind. Progressive decomposition reflects the concept of microbial processing of large plant biopolymers to smaller molecules. In the proposed SCM, a continuum of organic fragments is continuously processed by the decomposer community from large plant and animal residues toward smaller molecular size. At the same time, greater oxidation of the organic materials increases solubility in water as well as the opportunity for protection against further decomposition through greater reactivity toward mineral surfaces and incorporation into aggregates. *Dashed arrow lines* denote mainly abiotic transfer; *solid lines* denote mainly biotic transfer; *thicker lines* indicate more rapid rates; *larger boxes* and *ends of wedges* illustrate greater pool sizes; all differences are illustrative. All *arrows* represent processes that are a function of temperature, moisture, and the biota present. Reprinted from Lehmann and Kleber (2015), with permission

metabolites consumed from or secreted into the extracellular environment, insights can be gained into the metabolic activity of the cell (Kell et al. 2005). This approach (also known as metabolic footprinting) has been applied to characterize yeast mutant metabolism and phenotypes (Allen et al. 2003; Castrillo et al. 2007; Mas et al. 2007). Exometabolomics has also been applied in industrial settings, where analysis of extracellular fermentation media are part of the process of optimizing yeast fermentation conditions (Devantier et al. 2005; Fu et al. 2014), for monitoring various industrially important bacterial and yeast strains in bioreactor cultures (Paczia et al. 2012), and to study the breakdown of polysaccharides by anaerobic bacterial strains (Villas-Boas et al. 2006). Apart from some recent studies (Swenson et al. 2015b; Warren 2014), few examples of the application of exometabolomics to characterize soil microbial communities have been reported, though dissolved organic matter has been characterized in sea and river water with an exometabolomics approach (Kido Soule et al. 2015; Morales-Cid et al. 2009).

A major reason for the paucity of soil exometabolomics studies is the complexity of soil, and the associated challenges of extraction and sample preparation. The extraction method used in any metabolomics experiment is critical to the quality of the data obtained. The choice of extraction method should allow effective extraction of metabolites from the system under study, without artifact formation or compound degradation. In our current understanding of the nature of SOM (Lehmann and Kleber 2015), the organic compounds that make up SOM exist in different compartments with different degrees of biological accessibility. The soluble component of SOM is the most accessible to processing by soil microbes, and is referred to as dissolved organic matter (DOM). DOM is often defined as dissolved metabolites able to pass through a 0.45 μm filter (Gregorich et al. 2000), to differentiate it from particulate organic matter. In order to characterize DOM in traditional soil science, various methodological approaches involving extraction from soils have been developed (Zsolnay 2003). These often involve extraction of soil under relatively gentle conditions (e.g. aqueous salt solutions) to yield a fraction referred to as water extractable organic matter (WEOM). This fraction conceptually consists of the mobile and available portion of the total DOM pool (Corvasce et al. 2006). An example of such an extraction procedure involves extraction of soil with concentrated salt solutions (e.g. up to 500 mM K_2SO_4) for a few hours, followed by filtration or centrifugation, and analysis for total organic carbon (Jones and Willett 2006). The high salt concentration in the extraction buffer helps extract mineral sorbed metabolites, but can cause issues with downstream sample preparation and metabolite analysis in metabolomics methods used to characterize individual components of DOM (e.g. formation of salt crystals in samples, ion suppression in mass spectrometry, and decrease in sensitivity in NMR) (Annesley 2003; Kelly et al. 2002). Therefore, in recent metabolomics studies, water-based extraction methods were developed to extract organic matter from soils. Warren (2013a, b) extracted field-moist soils in water by shaking for 10 min, followed by centrifugation and filtration. The relatively short extraction time addressed concerns over continued metabolism during extraction, which could give rise to altered metabolite profiles (Rousk and Jones 2010). Swenson et al. (2015b) followed a similar process but performed aqueous extraction of soils for one h at 4 °C to slow any metabolic activity present. Another concern with aqueous extraction is enrichment of the metabolite profile by intracellular metabolites. Osmotic shock can potentially lyze microbial cells and cause leakage of metabolites (Gregorich et al. 2000). Swenson et al. (2015b) compared aqueous soil extracts to samples extracted in 10 mM K_2SO_4 and 10 mM NH_4HCO_3, and found no significant qualitative differences in metabolites detected. It was concluded that water is a suitable extractant for soil exometabolomics of DOM and that these extracts would be most representative of the types of resources available for soil microbes.

In some cases, an experiment may require analysis of soil intracellular and extracellular metabolites. In this case, cell lysis is an important and desirable step in sample preparation. To access intracellular metabolites, a traditional approach used in soil science involves chloroform fumigation of soil to lyze microbial cells, followed by extraction (Brookes et al. 1985; Vance et al. 1987). Swenson et al. (2015b) compared metabolite profiles of water extracts of fumigated and unfumigated soil samples (Fig. 2). A significant increase in the number and abundance of metabolites was observed, however, the fumigation technique requires long times of exposure to chloroform vapors, which raise concerns about continued metabolic activity or increased enzymatic degradation of metabolites (Warren 2013a, b). The use of organic solvents which are able to lyze microbial cells is another way to obtain soil extracts containing intracellular and extracellular metabolites. This was demonstrated by Swenson et al. (2015b) who used hierarchical cluster analysis to show similarity in metabolite patterns between fumigated soil extracts and organic solvent extracts of unfumigated soil. Soil was also directly extracted with chloroform and K_2SO_4 solution (1:4, v/v) to obtain extracts containing intracellular metabolites (Kakumanu et al. 2013). Rochfort et al. (2015) extracted freeze-dried, finely ground soil with an 8:2 methanol-water solution by sonication for 10 min. Since this method breaks up soil aggregates in addition to using organic solvents, it is not surprising that it provided extracts with a wide coverage of metabolite classes, derived from the intracellular and extracellular soil metabolite pools.

The major analytical methods used in the field of metabolomics are based on Mass Spectrometry (MS) and Nuclear Magnetic Resonance (NMR) (Dettmer et al. 2007; Dunn et al. 2005; Forseth and Schroeder 2011). Each method has advantages and disadvantages related to sensitivity, structural information, ease of quantitation, breadth of metabolite coverage, and availability of structural databases for identification (Lenz and Wilson 2007). Although studies on the soil microbial exometabolome are limited, a few recent examples demonstrate the utility of these methods to this field.

Gas Chromatography coupled with Mass Spectrometry (GC-MS) has previously been applied to targeted analyses of particular chemical classes of small soil metabolites such as sugars or amino acids (Kakumanu et al. 2013). GC-MS is also well suited for measuring a broad range of small metabolite classes, and has been widely used in untargeted metabolomics studies in plants (Jenkins et al. 2004), human biofluids (Garcia and Barbas 2011), and microbial biomass (Koek et al. 2006). Analysis of hydrophilic/polar metabolites by GC-MS requires derivatization

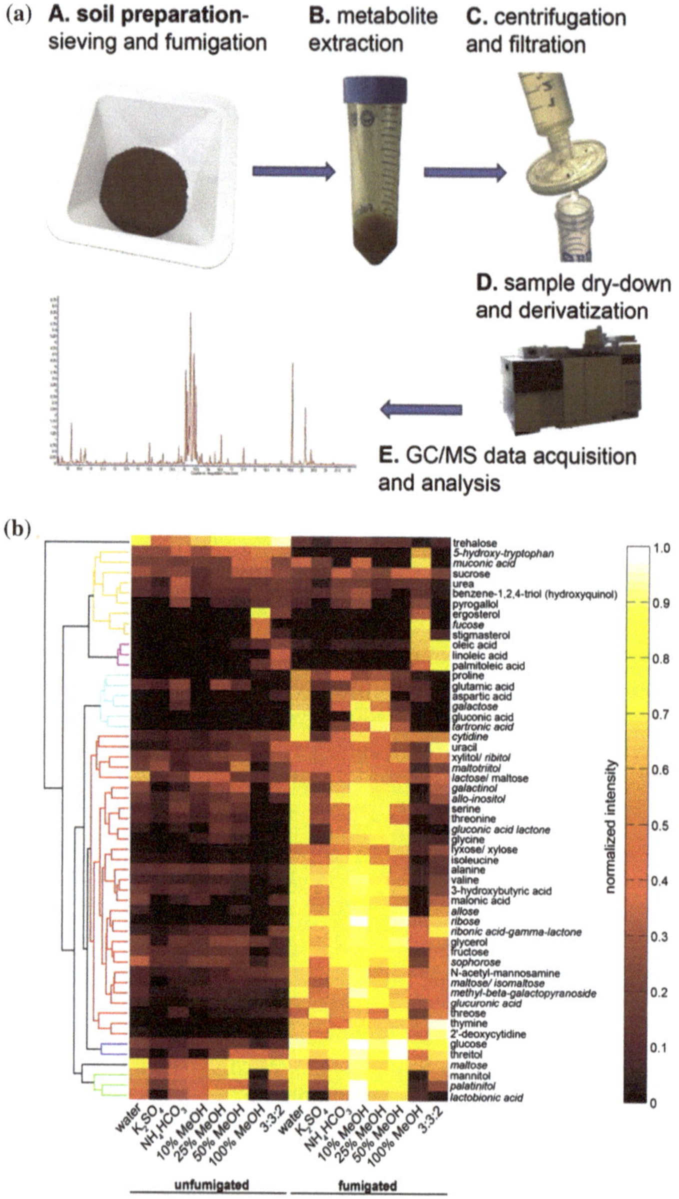

Fig. 2 **a** Workflow for soil DOM extraction: *A* Soil is sieved through 2 mm and fumigated with chloroform for 24 h to access intracellular metabolites or left unfumigated for extracellular metabolites. *B* Soil is extracted with the appropriate solution in a 1:4 ratio (2 g soil: 8 mL extractant) on an orbital shaker for 1 h at 4 °C. *C* Extract is centrifuged to pellet soil and the supernatant filtered through 0.45 μm filter discs. *D* Dried down and derivatized for GC-MS. *E* Data are analyzed for metabolite identification. **b** Relative intensity of metabolites in extracts of unfumigated and fumigated soil prepared with different extractants, and analyzed by GC-MS. Reprinted from (Swenson et al. 2015b), with permission

to increase the volatility of compounds. Swenson et al. (2015) characterized soil extracts using GC-MS. After extraction with different solvents, samples were derivatized by methoxyamination and trimethylsilylation. Hundreds of unique features were detected, of these 55 were confidently annotated using the Fiehn spectral metabolite database (Kind et al. 2009) and comparison with authentic standards. Metabolites detected in all samples included sugars, sugar alcohols, amino acids and amino acid metabolites, nucleobases, carboxylic acids, and sterols (Swenson et al. 2015b).

Liquid Chromatography coupled with Mass Spectrometry (LC-MS) has become an important analytical tool in metabolomics, and has also been applied in studies on many biological systems (Theodoridis et al. 2008; Zhou et al. 2012). Separation of metabolites is achieved by LC using various stationary phases depending on the polarity of the target metabolites. There are various options available for ion sources and mass analyzers in LC-MS systems (reviewed by Zhou et al. 2012). Due to the high structural diversity of metabolites, a particular sample typically needs to be analyzed in positive and negative ionization mode to obtain a good coverage of the metabolome. DOM is by definition composed of small metabolites dissolved in water in situ (Leenheer and Croué 2003). This fraction of SOM is therefore amenable to separation by hydrophilic interaction liquid chromatography (HILIC), a variant of normal phase chromatography (Alpert 1990). Baran et al. (2015) analyzed extracellular soil water, as well as intracellular metabolites of isolates from biological soil crust, with LC-MS using zwitterionic HILIC chromatography and electrospray ionization (ESI). Out of nearly 500 molecular features detected in this study, 79 metabolites were identified based on MS/MS data and comparison with authentic reference standards. A similar method was used by Swenson et al. (2015a) to study sorption of microbially derived metabolites onto mineral surfaces.

Capillary electrophoresis mass spectrometry (CE-MS) is suitable for the analysis of charged metabolites and has found applications in metabolomics studies summarized in a series of reviews (Ramautar et al. 2009, 2011, 2013). In a study focusing on the pool of nitrogen metabolites in soil (dissolved organic nitrogen: DON), Warren (2013a) employed a CE-MS procedure. This method allowed detection of small metabolites ionizable by electrospray that are cationic at low pH. Approximately 100 nitrogen-containing metabolites with a wide range of polarities were detected, of which 57 were identified (Warren 2013a).

Fourier transform-ion cyclotron resonance-mass spectrometry (FT-ICR-MS) is an established method for analyzing natural organic matter, and has been widely used to characterize complex organic mixtures in environmental samples (Kujawinski 2002). This method allows the detection of ions with excellent mass accuracy and resolving power, so that unique empirical formulas can be assigned to most of the thousands of signals detected (Hockaday et al. 2006). Based on atomic ratios (e.g. H:C, O:C) these formulas can be assigned to chemical classes such as carbohydrates, lipids, lignins, tannins, and proteins (Ohno et al. 2014). While individual features are not unambiguously identified using this technique, FT-ICR-MS is very useful for obtaining overviews of patterns in soil DOM dynamics (Hockaday et al. 2006, Kujawinski et al. 2004; Ohno et al. 2014).

NMR is an established analytical platform in the field of metabolomics. It has been applied in analysis of human biofluids (Nicholson and Lindon 2008), plants (Kim et al. 2010), and microbiological samples (Grivet et al. 2003). NMR exometabolomics has been extensively applied to study microbial cell culture systems (Behrends et al. 2014; Resmer and White 2011; Szeto et al. 2010). While solid-state NMR techniques have been employed to analyze macromolecules and structural aspects in soils (Baldock et al. 1992; Kögel-Knabner 1997), there are few examples of NMR used for the characterization of the small metabolite complement (microbial- or plant-derived) of SOM. Jones et al. (2014) analyzed extracts of soils from former mine sites by NMR. The aim was to obtain a survey of the naturally occurring products of soil community metabolism (including intracellular metabolites). NMR spectra were dominated by sugars, and a range of other metabolites such as amino acids and nucleosides were detected. A recent study also characterized soil extracts by NMR for a comparison of native versus agricultural soils (Rochfort et al. 2015). Complex spectra were obtained that were dominated by sugar resonances. Lipophilic compounds (terpenes, lipids) were also detected due to the extraction solvents having a higher organic solvent composition than that used by Jones et al. (2014).

3 Exometabolomics for Analysis of Whole Microbial Communities

The exometabolome of a complex soil microbial community comprises the sum of small metabolites being produced or released, and consumed by all the metabolic activity in the soil. The exometabolome is thus a reflection of the net metabolic state of the community. Studying differences in the microbial community exometabolome under different conditions can lead to insights into the response of communities as a whole. In one of the first exometabolomics studies on complex microbial communities, Henriques et al. (2007) applied an LC-MS based approach to analyze soluble metabolites in wastewater treatment plant communities. Activated sludge cultures from four different wastewater treatment plants were exposed to four different chemical stressors known to affect the processing ability of such communities. Comparisons of metabolite profiles between untreated and treated samples using multivariate statistical methods revealed clear patterns between the different toxin-stressed cultures. A limited number of variables were able to discriminate samples based on chemical treatment, which was community-independent. It was concluded that the discriminant metabolites may be universal biomarkers for these stress conditions, and that these may be used in developing early warning tools for toxins in these systems (Henriques et al. 2007). Exometabolomics has also been applied to analyze uptake and release of extracellular metabolites from microbial biofilm consortia occurring in water pipes (Beale et al. 2010). Small metabolite profiles, obtained by GC-MS, of water flowing through copper pipe systems differentiated samples exposed to copper corroding microbial biofilms from those that were not (Beale et al. 2012). In a pilot study this approach was applied to a water supply network, where it provided information on biofilm activity in the system. This approach showed potential for elucidating the relationship between specific metabolites in water supply networks and issues related to water quality, caused by microbial biofilms (Beale et al. 2013).

The effect of human activities on soil systems have been the topic of metabolomics field studies. In a report on soils from former mine sites in the UK, Jones et al. (2014) employed NMR and principal component analysis (PCA) to compare metabolite profiles of soil extracts. Soil sites under study were geographically dispersed and had a range of physicochemical properties. The PCA grouped some sites together based on similarity of their overall profiles. The authors concluded that the observed patterns are likely due to the similar pollution patterns at the sites, but did not do further in-depth analysis of the factors potentially underlying the observed patterns. Another NMR-based study compared soil extracts (intra- and extracellular metabolites) of geochemically matched remnant (native) and agricultural (managed) soils (Rochfort et al. 2015). When subjected to multivariate data analysis, samples were grouped together based on land use. NMR resonances responsible for the observed groupings were assigned to characteristic terpene and aromatic compound signals in the remnant soils, and sugar and lipid signals in the agricultural soils. Soil samples were analyzed in parallel by mid-infrared (MIR) spectroscopy, a technique that employs absorption and transmission of photons in the infrared energy range (about 2500–25,000 nm in the electromagnetic spectrum), to characterize molecules based on their constituent bonds. (Bellon-Maurel and McBratney 2011). Multivariate data analysis of these data resulted in samples clustering together based on location, irrespective of land use. Soil extracts were also tested for antibacterial activity, and the most active extracts were from native soil samples that clustered together by PCA analysis. This study established that the two analytical methods captured different aspects of the soil, namely soil biochemistry (NMR) and soil physicochemistry (MIR). It also demonstrated how biochemical characteristics as measured in this metabolomics study can be related to functional aspects of soil communities as a whole.

The above studies followed an untargeted metabolomics approach, where metabolite profiles were measured and compared between samples without identifying the compounds responsible for discriminating groups. Rochfort et al. (2015) were able to assign important discriminating NMR signals to compound classes (e.g. terpenes and aromatics), but noted that further characterization would be needed to confidently identify individual metabolites. This untargeted approach is widely used in other fields employing metabolomics (Sévin et al. 2015), but also points to a larger issue in soil exometabolomics, i.e., very little data on the composition of soil metabolites. The studies mentioned in the analytical methods section comprise the few that have contributed to the broad qualitative profiling of multiple compound classes in soil (as opposed to targeted methods for one compound class at a time). Even fewer have attempted to characterize soil metabolites in a quantitative manner. One exception is Warren (2013b) who performed a broad analysis of small nitrogen-containing metabolites in different soil types dominated by different vegetation. The relative proportions of the different compound classes in this pool of small metabolites were determined. The relative abundance of the top ten small nitrogen-containing metabolites in each soil type was also analyzed. Even though the study of Warren (2013b) focused on a particular subset of metabolites, such a detailed quantitative analysis lays an important foundation for understanding what is in the soil exometabolome. Similar characterizations are needed that include a broader range of metabolite classes, and relate these to differences in factors such as vegetation type and physicochemical factors.

Warren (2013b) pointed out that soil water extracts may not accurately represent what is biologically available, since differential adsorption to the soil stationary phase may occur. The mineral content and surface area of soils are known to affect the solid-state partitioning of and thereby the accessibility of DOM components to microorganisms (Kalbitz et al. 2000). However, these processes are not understood down to a metabolite-specific level. Recently, Swenson et al. (2015a) investigated the sorption of small metabolites from a soil bacterial lysate on an iron oxide mineral, ferrihydrite. Different metabolite classes were adsorbed to different degrees, with phosphate-containing metabolites, for example, showing the highest sorption (Fig. 3), while other metabolites were not adsorbed, suggesting their higher degree of bioavailability in iron-rich soils. Since high-sorbing metabolites were able to displace sorbed phosphate from the ferrihydrite, the authors hypothesized that the release of such metabolites by soil microbes may be a strategy to access phosphate in soils where it is limiting. More studies of the effects of minerals on the bioavailability of small metabolites will help elucidate the role abiotic factors play in SOM dynamics of different environments.

4 Who Does What in Soil Community: Characterizing Metabolism of Individual Members

One approach to understanding the dynamics of a microbial community is to characterize the individual members of the community in isolation. Studying the uptake and release of metabolites through a particular microbial isolate in the laboratory, insights can be gained into its metabolic interactions with the environment. Baran et al. (2011) used an untargeted metabolite footprinting approach to characterize the marine cyanobacterium *Synechococcus* sp. PCC 7002 cultured in different growth media (Fig. 4). A wide variety of metabolites were found to be taken up by this strain, and analysis of intracellular metabolites also provided insights into which metabolites were actively turned over and which were maintained in cells in their native states. A study on acid mine drainage used an exometabolomics approach to study the role of the primary producer *Euglena mutabilis* in these oligotrophic environments (Halter et al. 2012). The exo- and endo-metabolome of *E. mutabilis* was profiled in situ and also for laboratory grown cultures. A number of metabolites

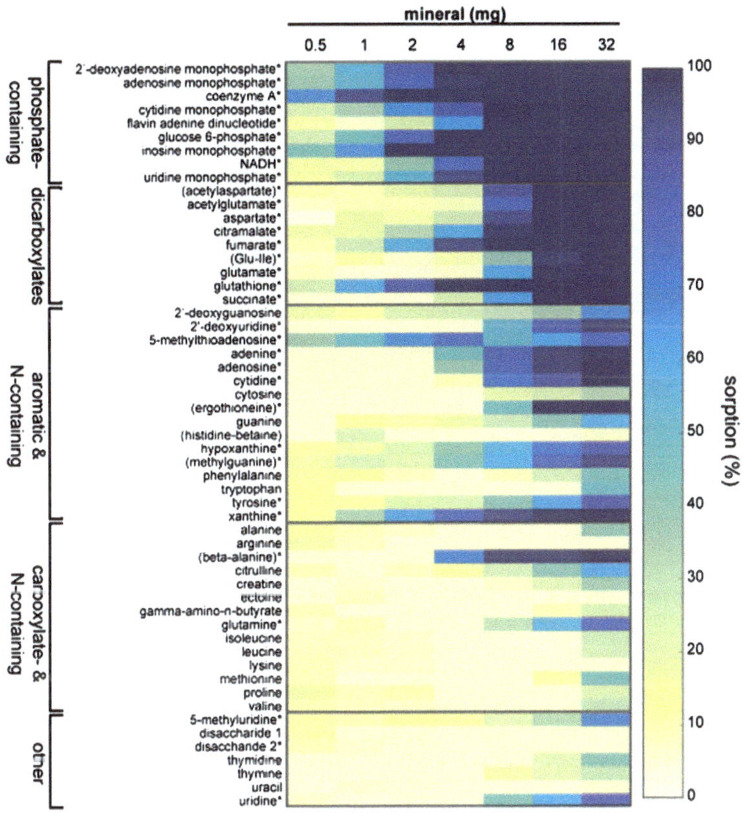

Fig. 3 Sorption of small metabolites from a soil bacterial lysate on an iron oxide mineral, ferrihydrite. For each metabolite, the percent sorption (relative to the non-mineral control) is displayed as mineral concentration increased from 0.5–32 mg. Metabolites were analyzed by LC-MS. Putatively identified metabolites are indicated by parentheses. Reprinted from Swenson et al. (2015a), with permission

in the acid mine drainage exometabolome were found to be secreted by the cells in laboratory cultures. This suggested an important role in organic matter production by *E. mutabilis* for consumption by other microbial strains in this ecosystem.

Baran et al. (2015) extended this approach by characterizing multiple isolates from a biological soil crust (biocrust) community. The primary producer in this community, the filamentous cyanobacterium *Microcoleus vaginatus*, was cultured in the laboratory. Exo- and endo-metabolite profiling revealed that many metabolites were released into the culture medium by this strain. Seven bacterial isolates, representing diverse phyla from the biocrust environment, were cultured individually in different-rich media to characterize their substrate preferences. Only a small proportion of metabolites detected in the media were taken up by any given strain, and there was little overlap between the strains' preferred substrates. Metabolite

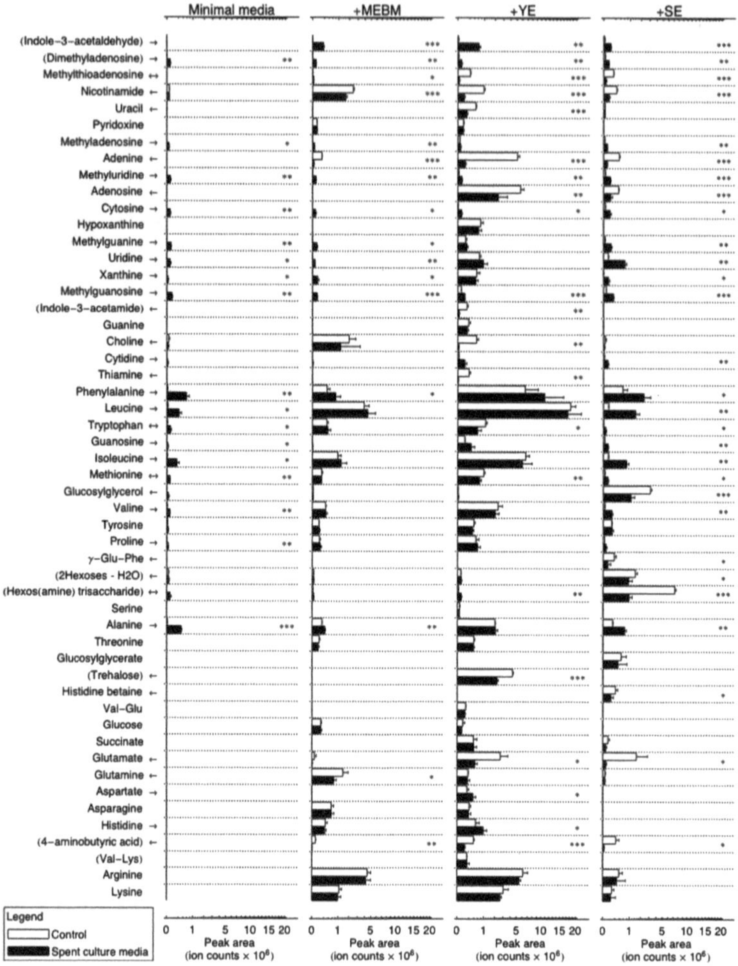

Fig. 4 Comparison of levels of selected metabolites in the growth media following growth of *Synechococcus* (*full bars*) against their levels in control media (*open bars*, $n = 4$), as determined by LC-MS. The peak areas axis was scaled with a square root to improve the visualization of smaller peaks. Statistically significant differences are indicated as "*" ($p < 0.05$), "**" ($p < 0.01$), or "***" ($p < 0.001$). An *arrow* is shown next to the name of a metabolite if it was found to be significantly consumed (←), released (→), or both consumed and released (↔). Reprinted from Baran et al. (2011), with permission

profiling of the biocrust soil water was also performed to link the observed patterns from the isolates to the intact microbial community. Changes in metabolite profiles at different times following wet up of desiccated biocrust showed patterns similar to those observed in the individual isolate experiments. This study revealed the particular substrate preferences of sympatric isolates from a soil community, which suggest that exometabolite niche partitioning may be an important driver in maintaining soil microbial diversity. Conversely, if different microbial phyla have different roles in processing soil organic matter, it follows that changes in soil microbial diversity may affect carbon cycling in soils.

Integrating exometabolomics data from various soil isolates would be a useful way to form hypotheses about the relationships between different strains in a particular environment. An online data repository, Web of Microbes, has been developed for such exometabolomics data (webofmicrobes.org). This tool allows rapid visualizations of large exometabolomics datasets of individual isolates that enables predictions to be made about how they behave in a community. This includes interactions such as potential resource competition and cross-feeding between strains, and how these relationships would be affected by changes in the chemical environment. Characterizing individual isolates from a soil community can shed light on how they behave in relation to other soil community members. However, this approach is limited to strains that can be cultured outside of their native habitat, thereby excluding the vast majority of the soil microbial diversity (Schloss and Handelsman 2003). Hence, there is a need for methods that enable the study of soil microbial communities in situ, to link specific functions to particular community members, and to elucidate the metabolic interactions between them.

5 Stable Isotope Probing: Tracking Flow of Substrates Through Communities

Stable isotope probing (SIP) techniques involves addition of a stable isotope enriched substrate, and tracking its fate as it is transformed by the metabolism of community members into labeled molecules/biomarkers (Dumont and Murrell 2005). Variations of SIP target different biomarkers that become labeled as a result of growth on the labeled substrate. One approach targets microbial phospholipid fatty acids (PLFAs). Since different microbial classes possess characteristic fatty acids as part of their cell membranes, selective extraction and analysis of PLFA patterns is an established approach for determining the composition of microbial communities (Zelles 1999). In PLFA-SIP, tracking which is the characteristic phospholipid fatty acids become labeled with the stable isotope yields information about which groups of microbes were responsible for metabolizing the labeled substrate. This approach has been used to identify groups of microorganisms performing particular functions in soils based on labeling with substrates such as $^{13}CH_4$ (Bull et al. 2000), ^{13}C-acetate (Boschker et al. 1998), ^{13}C- glucose, and -ribose (Apostel et al. 2015). Uniformly labeled ^{13}C-cellulose and ^{13}C-lignin were substrates in a PLFA-SIP study on the role of diverse microbial groups in plant polymer degradation (Torres et al. 2014). Treonis et al. (2004) combined a $^{13}CO_2$-labeling experiment with PLFA analysis to identify microbes assimilating plant root exudates. The disadvantage of this approach is based on PLFA biomarkers determined from cultivated microbes limiting application to uncultivable microorganisms.

Another approach relies on combining SIP with nucleic acid analysis. This relies on the incorporation of the isotopic label into DNA or RNA, so that subsequent separation and sequencing of the labeled fraction identifies community members actively incorporating the labeled substrate (Dumont and Murrell 2005). DNA-SIP has mostly been used with ^{13}C-labeled substrates, such as $^{13}CH_3OH$ to study soil methylotrophs (Radajewski et al. 2000), and ^{13}C-naphthalene and other organic compounds to characterize pollutant biodegraders (Padmanabhan et al. 2003). Using ^{13}C-cellulose as a substrate, El Zahar Haicher et al. (2007) identified cellulose degraders in a soil community using a DNA-SIP approach. These included bacteria not previously known to have this ability, as well as a number of uncultured strains. A disadvantage of DNA-SIP is that a relatively large amount of labeled substrate is needed, together with long incubation times to allow production of highly labeled ^{13}C genomic DNA. Artificially high substrate concentrations are thus often applied to soils, which can cause biases in how the community behaves (Dumont and Murrell 2005). The incorporation of ^{13}C into RNA occurs earlier than DNA (Manefield et al. 2002), therefore targeting RNA as the labeled biomarker molecules in RNA-SIP allows the use of shorter incubation times. RNA-SIP also allows detection of cells which are metabolically active, even though they are not dividing or growing (El Zahar Haichar et al. 2007). A combination of DNA- and RNA-SIP can potentially be a very powerful approach. Recently, $H_2^{18}O$ was applied to a soil bacterial community as a universal substrate, and was found to be an effective label for DNA- and RNA-based SIP approaches for studying active members of the community (Rettedal and Brözel 2015).

SIP approaches monitoring the changes in labeling patterns over time can yield valuable information on how nutrients flow through a microbial community. Labeled substrates will become incorporated into cells (e.g. as part of PLFA, DNA, RNA) but a proportion will be transformed and transported out of the cell, where it may be consumed by other members of the community food web (DeRito et al. 2005). Extending the analysis of labeled biomolecules to the community exometabolite pool is thus a potentially powerful approach for elucidating how these trophic interactions occur. Date et al. (2010) combined DNA-SIP with NMR exometabolomics to study microbial variability and metabolic dynamics in a mouse gut microbial community. Using labeled glucose as a sole carbon source, metabolic-microbial correlation analysis was performed, allowing identification of glucose-utilizing gut microbes and their extracellular metabolites. Microbial strains consuming the metabolites produced by the glucose utilizers were also identified, together with their extracellular metabolites. The study demonstrated that the feasibility of this approach for tracking carbon flux within a microbial community by identifying members of the community involved at different steps, as well as the metabolites that mediate their interactions. This approach clearly has great potential to address questions about carbon flux in the context of soil microbial communities.

Labeling experiments with stable isotopes can also aid analysis of the highly complex soil exometabolome. For example, in NMR-based studies, the low natural abundance of the magnetic isotope of carbon (^{13}C) results in low sensitivity of detection in unlabeled systems. Signals are dramatically enhanced as metabolites become labeled with ^{13}C isotopes, thereby facilitating identification of metabolites downstream of the labeled substrate (Schneider et al. 2003). In mass spectrometry based methods coupled to chromatographic separations, labeled metabolites can be detected at the same retention time as their unlabeled counterparts, with characteristic shifts in mass spectral features corresponding to the number of incorporated labeled isotopes (Rodgers et al. 2000). Computational methods have been developed for the quantitative detection of features derived from a particular labeled compound, even when not all metabolites are identified (Hiller et al. 2010). However, stable isotope labeling can greatly facilitate unambiguous assignment of chemical formulas, and thereby identification of unknown features (Baran et al. 2010). Thus, SIP methods show great promise for reducing noise by highlighting relevant metabolites and pathways, and identifying unknown metabolites in complex datasets such as those generated in soil exometabolomics experiments.

6 Metabolite Imaging: Microbial Interactions Through Space

Soil is a very heterogeneous matrix, in which biotic and abiotic factors combine to create diverse microclimates. Studies on soil microbial communities are often conducted on homogenized bulk soil samples, in which the spatial structure of soil and soil microorganisms are disrupted (Becker et al. 2006). Yet, observations of microbial communities at the micron scale have revealed defined spatial organization. For example, in dental plaque, a nine taxon microbial consortium was observed to be radially arranged around cells of filamentous bacteria (Welch et al. 2016). Obligate aerobes were arranged around the periphery of the consortium, anaerobes were found in the interior, and others were localized in ways suggestive of their functional roles in the consortium. Such structured assemblages have been observed in biofilms and consortia occurring in aquatic systems, on leaf and root surfaces, and in pathogenic or commensal associations with humans (Almstrand et al. 2013; Wessel et al. 2013). It is believed that the spatial arrangement of soil microbial communities is a very important driver of microbial diversity in soil, thereby also of microbial community functioning (Ettema and Wardle 2002). Therefore, the next level of detail required to understand microbial soil communities is characterizing their functioning in space. To achieve this, experimental methods are needed to observe specific community members and their activities in relation to other community members in their native spatial arrangement.

Experimental approaches utilizing labeled substrates have been successfully used to visualize and identify labeled microbial cells taking up a specific compound. FISH-microautoradiography or fluorescence are approaches for detecting bacterial cells that have consumed and metabolized a specific radioactive substrate, and identification of these cells using an oligonucleotide probe (Adamczyk et al. 2003; Lee et al. 1999). The use of radioactive labels are less desirable, and recent technological advances have yielded promising alternative approaches for analyzing interactions between microorganisms and their chemical environment (Wessel et al. 2013). Orphan et al. (2001) used FISH in combination with Secondary Ion Mass Spectrometry (SIMS) to detect and visualize ^{13}C profiles in microbial consortia composed of archaea and sulfate-reducing bacteria. Lower $^{13}C/^{12}C$ ratios in both the archaea and associated bacteria provided evidence that methane was consumed by the former, and that methane-derived carbon was transferred to the latter consortium members. SIMS is ideally suited to detect isotopes at very fine resolution, for example, nanoSIMS can image with some 50 nm resolution. Therefore, the combination of SIMS with stable isotope probing (SIP) shows great promise for spatially resolved analysis of single microbial cells and their utilization of particular substrates (Behrens et al. 2008; Chandra et al. 2008; Cliff et al. 2002).

The above-mentioned methods allowed tracking of substrates into identifiable microbial cells. Ideally, the metabolites released into the environment and exchanged between community members should also be characterized. Besides SIMS, other Mass Spectrometry Imaging techniques are potentially well suited to do this, since a broad range of metabolites can be detected without the need for labeling with a radioactive or stable isotope (reviewed by Watrous and Dorrestein 2011). In Mass Spectrometry Imaging techniques, ionization probes generate ions on the sample surface, and the sample stage is moved in the *x*–*y* plane so that this is done across a defined sample area. Mass analyzers detect the generated ions, resulting in a grid of data points each with its own mass spectrum. An ion map can be made from these data showing the location and intensity of detected ions across the measured sample surface.

In specific Mass Spectrometry Imaging techniques (e.g. nanoDESI-IMS), samples are detected at atmospheric pressure, and samples can be wet (e.g. fresh samples of bacteria on an agar plate can be analyzed directly), while for others (e.g. MALDI-IMS), dry samples are covered in matrix and ionization and detection occurs under high vacuum (Wessel et al. 2013). In nanostructure-initiator mass spectrometry (NIMS), microbial agar cultures cannot be analyzed directly since desorption/ionization occurs at the bottom of the sample (Woo et al. 2008). For this technique, "replica extraction transfer" is used to transfer metabolites from the culture surface to the NIMS surface, allowing spatial arrangement of metabolites to be retained (Louie et al. 2013). These approaches have been used to characterize metabolites produced and released by microorganisms on solid culture media (Fig. 5). For example, Traxler et al. (2013) detected a suite of secondary metabolites released by *Streptomyces coelicolor* in response to interactions with various Actinomycetes. Watrous et al. (2013) used nanoDESI to profile metabolites in single colonies of *Schewanella oneidensis* MR-1 and *Bacillus subtilis* 3610, as well

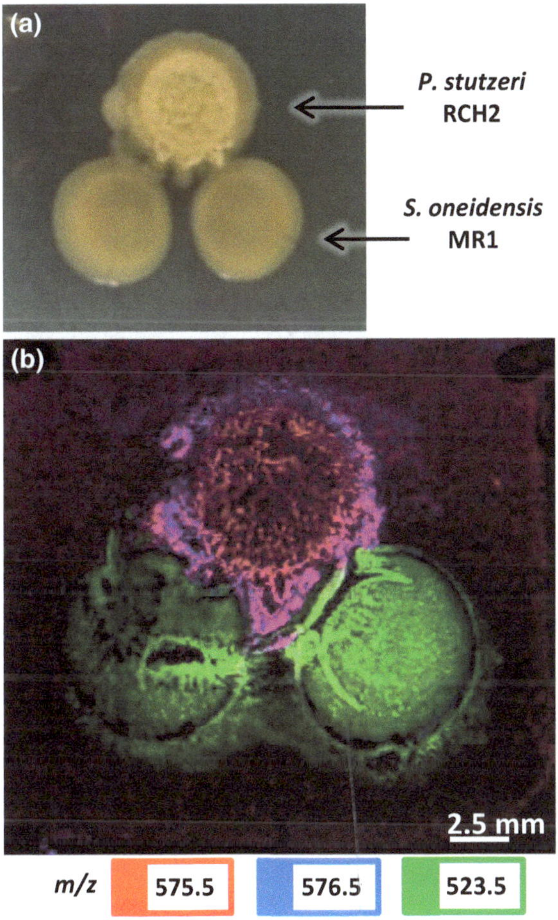

Fig. 5 Mass spectrometry imaging of *P. stutzeri* RCH2 and *S. oneidensis* MR1 coculture. **a** Optical image of coculture on solid medium. **b** Tricolor mass spectrometry image of coculture with *m*/*z* corresponding to species-specific lipids (Katherine Louie, unpublished)

as a mixed biofilm of these strains. A range of peptides, lipids, and small molecules were detected (Watrous et al. 2013). A REX-NIMS approach was used by Louie et al. (2013) to identify ions localized to regions within and between bacterial colonies cultured individually and in coculture on agar media. In a study on methanotrophic microbial mats from sea shelf methane seeps, Time-of-Flight Secondary Ion Mass Spectrometry (TOF-SIMS) was used to characterize microbial lipid biomarkers (Thiel et al. 2007). Characteristic lipid classes were detected in distinguished areas, which matched the presence of different microbial colonies as determined by conventional microscopic techniques.

Mass Spectrometry Imaging techniques show great promise for characterization of microbial communities in space, however, technical challenges limit their application to complex microbial communities in soil environments. Benefits and challenges of different imaging mass spectrometry techniques were comprehensively reviewed by Watrous and Dorrestein (2011). While able to detect metabolites with a wide mass range, the resolution of methods such as MALDI-IMS and nanoDESI currently does not allow imaging down to the single-cell level. Dynamic SIMS offers the best spatial resolution (sub-μm scale) but does not provide molecular information beyond elements and small atomic clusters. Another challenge of imaging experiments targeting small metabolites lies in the sample preparation. Methods for preparing thin soil sections as used in soil sciences typically involve fixing in formaldehyde, washing and impregnation in resin (Nunan et al. 2001), which would not be suitable for IMS. Tissue samples (e.g. mammalian, plant) are usually embedded in a substance such as OTC polymer, gelatin, or agarose gel for stabilization during cryosectioning (Cornett et al. 2007; Lee et al. 2012). Even if such treatment would not cause delocalization of metabolites in soil samples, the heterogeneous physical structure of soil may hamper cutting thin sections for imaging. Most IMS experiments to date involve laboratory cultured agar samples which can be analyzed fresh or mounted and dehydrated in preparation for imaging experiments (Traxler et al. 2013; Yang et al. 2012). Studies where microbial consortia from a natural environment were used either involved smears of soil on the sample target (Orphan et al. 2001), or cryosectioning of well-structured communities such as microbial mats (Thiel et al. 2007) or symbionts associated with other organisms (Lechene et al. 2007; Schoenian et al. 2011). Given the great potential of these approaches, there is an urgent need for improved sample preparation methods that will enable small metabolite imaging of soil microbial communities in their natural spatial arrangement.

7 Conclusions and Future Outlook

Recent technological and methodological advances have led to great progress in understanding the linkages between microbial diversity and ecosystem functioning (Bardgett et al. 2008). Metagenomics approaches have enabled characterization of the members of the soil community, including uncultivable microorganisms. Other molecular methods and SIP approaches have improved understanding of the particular members of the soil community's metabolic capabilities. What members of a particular soil microbial community actually do, will depend on the substrates that are available in their environment. Therefore, exometabolomics is a very promising approach in that it provides direct evidence of the soil metabolites available to soil microorganisms and how the available substrates are transformed by microbial community metabolism.

This chapter reviewed the handful of reports where an exometabolomics approach was applied to the study of intact soil microbial communities, or to laboratory experiments focusing on a particular aspect of such complex systems. Soil microbial communities are very complex, and soils are extremely heterogeneous matrices, so it is not surprising that there are many technical challenges that remain to be resolved in this field. Care should be taken to use sample preparation methods appropriate for the specific question being asked and analytical method being used. There is no single analytical method that can detect the massive diversity of metabolites across large dynamic ranges in an unbiased way. The choice of analytical method will depend on the focus and needs of the study, and combinations of complementary techniques may offer a more comprehensive coverage of diverse chemical classes (Simpson et al. 2004; Werf et al. 2007). It is important to keep in mind that the mineral composition and other factors may confound analysis by preferentially sorbing certain metabolites (e.g. ferrihydrite sorbing phosphate containing metabolites) making it challenging to compare soil types.

As with any metabolomics workflow, soil community exometabolomics experiments generate large datasets. Untargeted metabolomics results usually include many detected features that remain unidentified. There are several well-established mass spectrometry and NMR databases that can aid in identification of such unknowns (Kind et al. 2009; Smith et al. 2005; Wishart 2007). Many of these target intracellular metabolism of organisms such as yeast or plants (Bais et al. 2010; Hummel et al. 2007; Jewison et al. 2011). Since much soil organic matter is derived from plant and microbial biomass, these are useful to soil organic matter characterizations. There is currently great interest in secondary metabolites from soil microorganisms, and increasing the number of entries in databases of such compounds will also be very helpful in the context of soil exometabolomics (Hadjithomas et al. 2015). Many workflows have been developed for the analysis of large metabolomics datasets, which are also applicable to exometabolomics data analysis (Bowen and Northen 2010; Rübel et al. 2013; Tautenhahn et al. 2012; Xia et al. 2012). Data analysis tools for interpreting data in SIP experiments will be of particular value (Hiller et al. 2010; Huang et al. 2014). Any experimental setup and data reported should meet the quality and reporting standards as set by the larger metabolomics community (Goodacre et al. 2007).

The full potential of soil microbial community exometabolomics will be realized when it can be integrated with other approaches such as metagenomics, metatranscriptomics, and metaproteomics. A recent review describes examples where such multiomics approaches were applied to understanding microbial communities

(Franzosa et al. 2015). Careful planning of experimental design and data integration strategies are needed to derive the most value out of such combined approaches (Muller et al. 2013). Such data integration should result in improved mechanistic models of the structure and functioning of soil microbial communities that can be tested in combinations of laboratory and field experiments (Franzosa et al. 2015). This will enable better predictions of the effects of environmental perturbations on soil carbon cycling by soil microorganisms.

Acknowledgments T.R.N. gratefully acknowledges support from ENIGMA—Ecosystems and Networks Integrated with Genes and Molecular Assemblies (http://enigma.lbl.gov), a Scientific Focus Area Program at Lawrence Berkeley National Laboratory is based upon work supported by the US Department of Energy, Office of Science, Office of Biological & Environmental Research under contract number DE-AC02-05CH11231. A.L. was supported by the Office of Science, Office of Biological and Environmental Research, of the US Department of Energy, Award No. DE-SC0012627. We thank Katherine Louie for providing the mass spectrometry image of a microbial coculture.

References

Adamczyk J, Hesselsoe M, Iversen N, Horn M, Lehner A, Nielsen PH, Schloter M, Roslev P, Wagner M (2003) The isotope array, a new tool that employs substrate-mediated labeling of rrna for determination of microbial community structure and function. Appl Environ Microbiol 69:6875–6887. doi:10.1128/AEM.69.11.6875-6887.2003

Allen J, Davey HM, Broadhurst D, Heald JK, Rowland JJ, Oliver SG, Kell DB (2003) High-throughput classification of yeast mutants for functional genomics using metabolic foot-printing. Nat Biotechnol 21:692–696. doi:10.1038/nbt823

Almstrand R, Daims H, Persson F, Sörensson F, Hermansson M (2013) New methods for analysis of spatial distribution and co-aggregation of microbial populations in complex biofilms. Appl Environ Microbiol 79:5978–5987. doi:10.1128/AEM.01727-13

Alpert AJ (1990) Hydrophilic-interaction chromatography for the separation of peptides, nucleic acids and other polar compounds. J Chromatogr 499:177–196

Annesley TM (2003) Ion suppression in mass spectrometry. Clin Chem 49:1041–1044. doi:10.1373/49.7.1041

Apostel C, Dippold M, Kuzyakov Y (2015) Biochemistry of hexose and pentose transformations in soil analyzed by position-specific labeling and 13C-PLFA. Soil Biol Biochem 80:199–208. doi:10.1016/j.soilbio.2014.09.005

Bais P, Moon SM, He K, Leitao R, Dreher K, Walk T, Sucaet Y, Barkan L, Wohlgemuth G, Roth MR, Wurtele ES, Dixon P, Fiehn O, Lange BM, Shulaev V, Sumner LW, Welti R, Nikolau BJ, Rhee SY, Dickerson JA (2010) Plantmetabolomics.org: a web portal for plant metabolomics experiments. Plant Physiol 152:1807–1816. doi:10.1104/pp

Baldock JA, Nelson PN (2000) Soil organic matter. In: Sumner ME (ed) Handbook of soil science. CRC Press, Boca Raton, pp B25–B84

Baldock JA, Oades JM, Waters AG, Peng X, Vassallo AM, Wil-son MA (1992) Aspects of the chemical structure of soil organic materials as revealed by solid-state13C NMR spectroscopy. Biogeochemistry 16:1–42. doi:10.1007/BF00024251

Baran R, Brodie EL, Mayberry-Lewis J, Hummel E, Da Rocha UN, Chakraborty R, Bowen BP, Karaoz U, Cadillo-Quiroz H, Garcia-Pichel F, Northen TR (2015) Exometabolite niche partitioning among sympatric soil bacteria. Nat Commun 6:8289. doi:10.1038/ncomms9289

Baran R, Bowen BP, Northen TR (2011) Untargeted metabolic footprinting reveals a surprising breadth of metabolite uptake and release by Synechococcus sp. PCC 7002. Mol BioSyst 7:3200. doi:10.1039/c1mb05196b

Baran R, Bowen BP, Bouskill NJ, Brodie EL, Yannone SM, Northen TR (2010) Metabolite identification in synechococcus sp. PCC 7002 using untargeted stable isotope assisted metabolite profiling. Anal Chem 82:9034–9042. doi:10.1021/ac1020112

Bardgett RD, Freeman C, Ostle NJ (2008) Microbial contributions to climate change through carbon cycle feedbacks. ISME J 2:805–814. doi:10.1038/ismej.2008.58

Beale DJ, Barratt R, Marlow DR, Dunn MS, Palombo EA, Mor-rison PD, Key C (2013) Application of metabolomics to understanding biofilms in water distribution systems: a pilot study. Biofouling 29:283–294. doi:10.1080/08927014.2013.772140

Beale DJ, Dunn MS, Marney D (2010) Application of GC–MS metabolic profiling to "blue-green water" from microbial influenced corrosion in copper pipes. Corros Sci 52:3140–3145. doi:10.1016/j.corsci.2010.04.039

Beale DJ, Dunn MS, Morrison PD, Porter NA, Marlow DR (2012) Characterisation of bulk water samples from copper pipes undergoing microbially influenced corrosion by diagnostic metabolomic pro-filing. Corros Sci 55:272–279. doi:10.1016/j.corsci.2011.10.026

Becker JM, Parkin T, Nakatsu CH, Wilbur JD, Konopka A (2006) Bacterial activity, community structure, and centimeter-scale spatial heterogeneity in contaminated soil. Microb Ecol 51:220–231. doi:10.1007/s00248-005-0002-9

Behrends V, Williams H, Bundy J (2014) Metabolic footprinting: extracellular metabolomic analysis. In: Filloux A, Ramos J-L (eds) Pseudomonas methods and protocols. Methods in molecular biology. Springer, New York, pp 281–292

Behrens S, Lösekann T, Pett-Ridge J, Weber PK, Ng WO, Stevenson BS, Hutcheon ID, Relman DA, Spormann AM (2008) Linking microbial phylogeny to metabolic activity at the single-cell level by using enhanced element labeling-catalyzed reporter deposition fluorescence in situ hybridization (El-Fish) and nanosims. Appl Environ Microbiol 74:3143–3150. doi:10.1128/AEM.00191-08

Bellon-Maurel V, McBratney A (2011) Near-infrared (NIR) and mid-infrared (MIR) spectroscopic techniques for assessing the amount of carbon stock in soils–critical review and research perspectives. Soil Biol Biochem 43:1398–1410. doi:10.1016/j.soilbio.2011.02.019

Boschker HTS, Nold SC, Wellsbury P, Bos D, de Graaf W, Pel R, Parkes RJ, Cappenberg TE (1998) Direct linking of microbial populations to specific biogeochemical processes by 13C-labelling of biomarkers. Nature 392:801–805. doi:10.1038/33900

Bowen BP, Northen TR (2010) Dealing with the unknown: metabolomics and metabolite atlases. J Am Soc Mass Spectrom 21:1471–1476. doi:10.1016/j.jasms.2010.04.003

Brookes PC, Landman A, Pruden G, Jenkinson DS (1985) Chloroform fumigation and the release of soil nitrogen: a rapid direct extraction method to measure microbial biomass nitrogen in soil. Soil Biol Biochem 17:837–842. doi:10.1016/0038-0717(85)90144-0

Bull ID, Parekh NR, Hall GH, Ineson P, Evershed RP (2000) Detection and classification of atmospheric methane oxidizing bacteria in soil. Nature 405:175–178. doi:10.1038/35012061

Castrillo JI, Zeef LA, Hoyle DC, Zhang N, Hayes A, Gardner DC, Cornell MJ, Petty J, Hakes L, Wardleworth L, Rash B, Brown M, Dunn WB, Broadhurst D, O'Donoghue K, Hester SS, Dunkley TP, Hart SR, Swainston N, Li P, Gaskell SJ, Pa-ton NW, Lilley KS, Kell DB, Oliver SG (2007) Growth control of the eukaryote cell: a systems biology study in yeast. J Biol 6:4. doi:10.1186/jbiol54

Chandra S, Pumphrey G, Abraham JM, Madsen EL (2008) Dynamic SIMS ion microscopy imaging of individual bacterial cells for studies of isotopically labeled molecules. Appl Surf Sci 255:847–851. doi:10.1016/j.apsusc.2008.05.129 (Proceedings of the Sixteenth International Conference on Secondary Ion Mass Spectrometry, SIMS XVI)

Cliff JB, Gaspar DJ, Bottomley PJ, Myrold DD (2002) Exploration of inorganic C and N assimilation by soil microbes with time-of-flight secondary ion mass spectrometry. Appl Environ Microbiol 68:4067–4073. doi:10.1128/AEM.68.8.4067-4073.2002

Cornett DS, Reyzer ML, Chaurand P, Caprioli RM (2007) MALDI imaging mass spectrometry: molecular snapshots of biochemical systems. Nat Methods 4:828–833. doi:10.1038/nmeth1094

Corvasce M, Zsolnay A, D'Orazio V, Lopez R, Miano TM (2006) Characterization of water extractable organic matter in a deep soil pro-file. Chemosphere 62:1583–1590. doi:10.1016/j.chemosphere.2005.07.065

Date Y, Nakanishi Y, Fukuda S, Kato T, Tsuneda S, Ohno H, Kikuchi J (2010) New monitoring approach for metabolic dynamics in microbial ecosystems using stable-isotope-labeling technologies. J Biosci Bioeng 110:87–93. doi:10.1016/j.jbiosc.2010.01.004

Delmont TO, Robe P, Cecillon S, Clark IM, Constancias F, Simonet P, Hirsch PR, Vogel TM (2011) Accessing the soil metagenome for studies of microbial diversity. Appl Environ Microbiol 77:1315–1324. doi:10.1128/AEM.01526-10

DeRito CM, Pumphrey GM, Madsen EL (2005) Use of field-based stable isotope probing to identify adapted populations and track carbon flow through a phenol-degrading soil microbial community. Appl Environ Microbiol 71:7858–7865. doi:10.1128/AEM.71.12.7858-7865.2005

Dettmer K, Aronov PA, Hammock BD (2007) Mass spectrometry-based metabolomics. Mass Spectrom Rev 26:51–78. doi:10.1002/mas.20108

Devantier R, Scheithauer B, Villas-Bôas SG, Pedersen S, Olsson L (2005) Metabolite profiling for analysis of yeast stress response during very high gravity ethanol fermentations. Biotechnol Bioeng 90:703–714. doi:10.1002/bit.20457

Dumont MG, Murrell JC (2005) Stable isotope probing—linking microbial identity to function. Nat Rev Microbiol 3:499–504. doi:10.1038/nrmicro1162

Dunn WB, Bailey NJC, Johnson HE (2005) Measuring the metabolome: current analytical technologies. Analyst 130:606–625. doi:10.1039/B418288J

El Zahar Haichar F, Achouak W, Christen R, Heulin T, Marol C, Marais MF, Mougel C, Ranjard L, Balesdent J, Berge O (2007) Identification of cellulolytic bacteria in soil by stable isotope probing. Environ Microbiol 9:625–634. doi:10.1111/j.1462-2920.2006.01182.x

Ettema CH, Wardle DA (2002) Spatial soil ecology. Trends Ecol Evol 17:177–183. doi:10.1016/S0169-5347(02)02496-5

Falkowski PG, Fenchel T, Delong EF (2008) The microbial engines that drive earth's biogeochemical cycles. Science 320:1034–1039. doi:10.1126/science.1153213

Fierer N, Leff JW, Adams BJ, Nielsen UN, Bates ST, Lauber CL, Owens S, Gilbert JA, Wall DH, Caporaso JG (2012) Cross-biome metagenomic analyses of soil microbial communities and their functional attributes. Proc Natl Acad Sci USA 109:21390–21395. doi:10.1073/pnas.1215210110

Forseth RR, Schroeder FC (2011) NMR-spectroscopic analysis of mixtures: from structure to function. Curr Opin Chem Biol 15:38–47. doi:10.1016/j.cbpa.2010.10.010

Franzosa EA, Hsu T, Sirota-Madi A, Shafquat A, Abu-Ali G, Morgan XC, Huttenhower C (2015) Sequencing and beyond: integrating molecular "omics" for microbial community profiling. Nat Rev Microbiol 13:360–372. doi:10.1038/nrmicro3451

Fu Z, Verderame TD, Leighton JM, Sampey BP, Appelbaum ER, Patel PS, Aon JC (2014) Exometabolome analysis reveals hypoxia at the up-scaling of a *Saccharomyces cerevisiae* high-cell density fed-batch biopharmaceutical process. Microb Cell Factories 13:32. doi:10.1186/1475-2859-13-32

Garcia A, Barbas C (2011) Gas chromatography-mass spectrometry (GC-MS)-based metabolomics. In: Metz TO (ed) Metabolic profiling, methods in molecular biology. Humana Press, New York, pp 191–204

Goodacre R, Broadhurst D, Smilde AK, Kristal BS, Baker JD, Beger R, Bessant C, Connor S, Capuani G, Craig A, Ebbels T, Kell DB, Manetti C, Newton J, Paternostro G, Somorjai R, Sjöström M, Trygg J, Wulfert F (2007) Proposed minimum reporting standards for data analysis in metabolomics. Metabolomics 3:231–241. doi:10.1007/s11306-007-0081-3

Gregorich EG, Liang BC, Drury CF, Mackenzie AF, McGill WB (2000) Elucidation of the source and turnover of water soluble and microbial biomass carbon in agricultural soils. Soil Biol Biochem 32:581–587. doi:10.1016/S0038-0717(99)00146-7

Grivet JP, Delort AM, Portais JC (2003) NMR and microbiology: from physiology to metabolomics. Biochimie Metabolic NMR 85:823–840. doi:10.1016/j.biochi.2003.08.004

Hadjithomas M, Chen IMA, Chu K, Ratner A, Palaniappan K, Szeto E, Huang J, Reddy TBK, Cimermančič P, Fischbach MA, Ivanova NN, Markowitz VM, Kyrpides NC, Pati A (2015) IMG-ABC: a knowledge base to fuel discovery of biosynthetic gene clusters and novel secondary metabolites. mBio 6:e00932-15. doi:10.1128/mBio.00932-15

Halter D, Goulhen-Chollet F, Gallien S, Casiot C, Hamelin J, Gi-lard F, Heintz D, Schaeffer C, Carapito C, Van Dorsselaer A, Tcherkez G, Arsène-Ploetze F, Bertin PN (2012) In situ proteo-metabolomics reveals metabolite secretion by the acid mine drainage bio-indicator. Euglena mutabilis. ISME J. 6:1391–1402. doi:10.1038/ismej.2011.198

Harrigan GG, Goodacre R (eds) (2003) Metabolic profiling: its role in biomarker discovery and gene function analysis. Springer, Boston

Henriques IDS, Aga DS, Mendes P, O'Connor SK, Love NG (2007) Metabolic footprinting: a new approach to identify physiological changes in complex microbial communities upon exposure to toxic chemicals. Environ Sci Technol 41:3945–3951. doi:10.1021/es062796t

Hiller K, Metallo CM, Kelleher JK, Stephanopoulos G (2010) Non-targeted elucidation of metabolic pathways using stable-isotope tracers and mass spectrometry. Anal Chem 82:6621–6628. doi:10.1021/ac1011574

Hockaday WC, Grannas AM, Kim S, Hatcher PG (2006) Direct molecular evidence for the degradation and mobility of black carbon in soils from ultrahigh-resolution mass spectral analysis of dissolved organic matter from a fire-impacted forest soil. Org Geochem 37:501–510. doi:10.1016/j.orggeochem.2005.11.003

Huang X, Chen Y-J, Cho K, Nikolskiy I, Crawford PA, Patti GJ (2014) X13CMS: Global tracking of isotopic labels in untargeted metabolomics. Anal Chem 86:1632–1639. doi:10.1021/ac403384n

Hummel J, Selbig J, Walther D, Kopka J (2007) The golm metabolome database: a database for GC-MS based metabolite profiling. In: Nielsen J, Jewett MC (eds) Metabolomics, topics in current genetics. Springer, Berlin, pp 75–95

Jenkins H, Hardy N, Beckmann M, Draper J, Smith AR, Taylor J, Fiehn O, Goodacre R, Bino RJ, Hall R, Kopka J, Lane GA, Lange BM, Liu JR, Mendes P, Nikolau BJ, Oliver SG, Paton NW, Rhee S, Roessner-Tunali U, Saito K, Smedsgaard J, Sumner LW, Wang T, Walsh S, Wurtele ES, Kell DB (2004) A proposed framework for the description of plant metabolomics experiments and their results. Nat Biotechnol 22:1601–1606. doi:10.1038/nbt1041

Jenkinson DS, Rayner JH (1977) The turnover of soil organic matter in some of the Rothamsted classical experiments. Soil Sci 123:298–305

Jewison T, Knox C, Neveu V, Djoumbou Y, Guo AC, Lee J, Liu P, Mandal R, Krishnamurthy R, Sinelnikov I, Wilson M, Wishart DS (2011) YMDB: the Yeast Metabolome Database. Nucleic Acids Res. doi:10.1093/nar/gkr916

Johnston CA, Groffman P, Breshears DD, Cardon ZG, Currie W, Emanuel W, Gaudinski J, Jackson RB, Lajtha K, Nadelhoffer K, Nelson D, Post WM, Retallack G, Wielopolski L (2004) Carbon cycling in soil. Front Ecol Environ 2:522–528. doi:10.2307/3868382

Jones OAH, Sdepanian S, Lofts S, Svendsen C, Spurgeon DJ, Maguire ML, Griffin JL (2014) Metabolomic analysis of soil communities can be used for pollution assessment. Environ Toxicol Chem 33:61–64. doi:10.1002/etc.2418

Jones DL, Willett VB (2006) Experimental evaluation of methods to quantify dissolved organic nitrogen (DON) and dissolved organic carbon (DOC) in soil. Soil Biol Biochem 38:991–999. doi:10.1016/j.soilbio.2005.08.012

Kakumanu ML, Cantrell CL, Williams MA (2013) Microbial community response to varying magnitudes of desiccation in soil: a test of the osmolyte accumulation hypothesis. Soil Biol Biochem 57:644–653. doi:10.1016/j.soilbio.2012.08.014

Kalbitz K, Solinger S, Park JH, Michalzik B, Matzner E (2000) Controls on the dynamics of dissolved organic matter in soils: a review. Soil Sci 165:277–304

Kell DB, Brown M, Davey HM, Dunn WB, Spasic I, Oliver SG (2005) Metabolic footprinting and systems biology: the medium is the message. Nat Rev Microbiol 3:557–565. doi:10.1038/nrmicro1177

Kelly AE, Ou HD, Withers R, Dötsch V (2002) Low-conductivity buffers for high-sensitivity NMR measurements. J Am Chem Soc 124:12013–12019. doi:10.1021/ja026121b

Kido Soule MC, Longnecker K, Johnson WM, Kujawinski EB (2015) Environmental metabolomics: analytical strategies. Mar Chem 177:374–387. doi:10.1016/j.marchem.2015.06.029

Kim HK, Choi YH, Verpoorte R (2010) NMR-based metabolomic analysis of plants. Nat Protoc 5:536–549. doi:10.1038/nprot.2009.237

Kind T, Wohlgemuth G, Lee DY, Lu Y, Palazoglu M, Shahbaz S, Fiehn O (2009) FiehnLib: mass spectral and retention index li-braries for metabolomics based on quadrupole and time-of-flight gas chromatography/mass spectrometry. Anal Chem 81:10038–10048. doi:10.1021/ac9019522

Koek MM, Muilwijk B, van der Werf MJ, Hankemeier T (2006) Microbial metabolomics with gas chromatography/mass spectrometry. Anal Chem 78:1272–1281. doi:10.1021/ac051683

Kögel-Knabner I (1997) ^{13}C and ^{15}N NMR spectroscopy as a tool in soil organic matter studies. Geoderma 80:243–270. doi:10.1016/S0016-7061(97)00055-4

Kujawinski EB (2002) Electrospray ionization fourier transform ion cyclotron resonance mass spectrometry (ESI FT-ICR MS): characterization of complex environmental mixtures. Environ Forensics 3:207–216. doi:10.1080/713848382

Kujawinski EB, Del Vecchio R, Blough NV, Klein GC, Marshall AG (2004) Probing molecular-level transformations of dissolved organic matter: insights on photochemical degradation and protozoan modification of DOM from electrospray ionization Fourier transform ion cyclotron resonance mass spectrometry. Mar Chem 92:23–37. doi:10.1016/j.marchem.2004.06.038 (New Approaches in Marine Organic Biogeochemistry: A Tribute to the Life and Science of John I Hedges)

Lal R (2004) Soil carbon sequestration impacts on global climate change and food security. Science 304:1623–1627. doi:10.1126/science.1097396

Lechene CP, Luyten Y, McMahon G, Distel DL (2007) Quantitative imaging of nitrogen fixation by individual bacteria within animal cells. Science 317:1563–1566. doi:10.1126/science.1145557

Leenheer JA, Croué JP (2003) Characterizing aquatic dissolved organic matter. Environ Sci Technol 37:18A–26A

Lee N, Nielsen PH, Andreasen KH, Juretschko S, Nielsen JL, Schleifer K-H, Wagner M (1999) Combination of fluorescent in situ hybridization and microautoradiography—a new tool for structure-function analyses in microbial ecology. Appl Environ Micro-biol 65:1289–1297

Lee YJ, Perdian DC, Song Z, Yeung ES, Nikolau BJ (2012) Use of mass spectrometry for imaging metabolites in plants. Plant J 70:81–95. doi:10.1111/j.1365-313X.2012.04899.x

Lehmann J, Kleber M (2015) The contentious nature of soil organic matter. Nature 528:60–68. doi:10.1038/nature16069

Lenz EM, Wilson ID (2007) Analytical strategies in metabonomics. J Proteome Res 6:443–458. doi:10.1021/pr0605217

Louie KB, Bowen BP, Cheng X, Berleman JE, Chakraborty R, Deutschbauer A, Arkin A, Northen TR (2013) "Replica-extraction-transfer" nanostructure-initiator mass spectrometry imaging of acoustically printed bacteria. Anal Chem 85:10856–10862. doi:10.1021/ac402240q

Manefield M, Whiteley AS, Griffiths RI, Bailey MJ (2002) RNA stable isotope probing, a novel means of linking microbial community function to phylogeny. Appl Environ Microbiol 68:5367–5373. doi:10.1128/AEM.68.11.5367-5373.2002

Mas S, Villas-Bôas SG, Edberg Hansen M, Åkesson M, Nielsen J (2007) A comparison of direct infusion MS and GC-MS for metabolic footprinting of yeast mutants. Biotechnol Bioeng 96:1014–1022. doi:10.1002/bit.21194

Morales-Cid G, Gebefugi I, Kanawati B, Harir M, Hertkorn N, Rosselló-Mora R, Schmitt-Kopplin P (2009) Automated microextraction sample preparation coupled on-line to FT-ICR-MS:

application to desalting and concentration of river and marine dissolved organic matter. Anal Bioanal Chem 395:797–807. doi:10.1007/s00216-009-3025-0
Morales SE, Holben WE (2011) Linking bacterial identities and eco-system processes: can "omic" analyses be more than the sum of their parts? FEMS Microbiol Ecol 75:2–16. doi:10.1111/j.1574-6941.2010.00938.x
Muller EEL, Glaab E, May P, Vlassis N, Wilmes P (2013) Con-densing the omics fog of microbial communities. Trends Microbiol 21:325–333. doi:10.1016/j.tim.2013.04.009
Nicholson JK, Lindon JC (2008) Systems biology: metabonomics. Nature 455:1054–1056. doi:10.1038/4551054a
Nunan N, Ritz K, Crabb D, Harris K, Wu K, Crawford JW, Young IM (2001) Quantification of the in situ distribution of soil bacteria by large-scale imaging of thin sections of undisturbed soil. FEMS Microbiol Ecol 37:67–77. doi:10.1111/j.1574-6941.2001.tb00854.x
Ohno T, Parr TB, Gruselle MI, Fernandez IJ, Sleighter RL, Hatcher PG (2014) Molecular composition and biodegradability of soil organic matter: a case study comparing two new england forest types. Environ Sci Technol 48:7229–7236. doi:10.1021/es405570c
Oliver SG, Winson MK, Kell DB, Baganz F (1998) Systematic functional analysis of the yeast genome. Trends Biotechnol 16:373–378. doi:10.1016/S0167-7799(98)01214-1
Orphan VJ, House CH, Hinrichs KU, McKeegan KD, DeLong EF (2001) Methane-consuming archaea revealed by directly coupled isotopic and phylogenetic analysis. Science 293:484–487. doi:10.1126/science.1061338
Paczia N, Nilgen A, Lehmann T, Gätgens J, Wiechert W, Noack S (2012) Extensive exometabolome analysis reveals extended overflow metabolism in various microorganisms. Microb Cell Factories 11:122. doi:10.1186/1475-2859-11-122
Padmanabhan P, Padmanabhan S, DeRito C, Gray A, Gannon D, Snape JR, Tsai CS, Park W, Jeon C, Madsen EL (2003) Respiration of 13C-labeled substrates added to soil in the field and subsequent 16S rRNA gene analysis of 13C-labeled soil DNA. Appl Environ Microbiol 69:1614–1622
Piccolo A (2002) The supramolecular structure of humic substances: a novel understanding of humus chemistry and implications in soil science. Adv Agronomy 75:57–134
Radajewski S, Ineson P, Parekh NR, Murrell JC (2000) Stable-isotope probing as a tool in microbial ecology. Nature 403:646–649. doi:10.1038/35001054
Ramautar R, Mayboroda OA, Somsen GW, de Jong GJ (2011) CE-MS for metabolomics: developments and applications in the period 2008–2010. Electrophoresis 32:52–65. doi:10.1002/elps.201000378
Ramautar R, Somsen GW, de Jong GJ (2013) CE-MS for metabolomics: developments and applications in the period 2010-2012. Electrophoresis 34:86–98. doi:10.1002/elps.201200390
Ramautar R, Somsen GW, de Jong GJ (2009) CE-MS in metabolomics. Electrophoresis 30:276–291. doi:10.1002/elps.200800512
Resmer KL, White RL (2011) Metabolic footprinting of the anaerobic bacterium Fusobacterium varium using 1H NMR spectroscopy. Mol Bio Syst 7:2220. doi:10.1039/c1mb05105a
Rettedal EA, Brözel VS (2015) Characterizing the diversity of active bacteria in soil by comprehensive stable isotope probing of DNA and RNA with H218O. Microbiology Open 4:208–219. doi:10.1002/mbo3.230
Rochfort S, Ezernieks V, Mele P, Kitching M (2015) NMR metabolomics for soil analysis provide complementary, orthogonal data to MIR and traditional soil chemistry approaches—a land use study. Magn Reson Chem 53:719–725. doi:10.1002/mrc.4187
Rodgers RP, Blumer EN, Hendrickson CL, Marshall AG (2000) Stable isotope incorporation triples the upper mass limit for determination of elemental composition by accurate mass measurement. J Am Soc Mass Spectrom 11:835–840. doi:10.1016/S1044-0305(00)00158-6
Roesch LF, Fulthorpe RR, Riva A, Casella G, Hadwin AK, Kent AD, Daroub SH, Camargo FA, Farmerie WG, Triplett EW (2007) Pyrosequencing enumerates and contrasts soil microbial diversity. ISME J 1:283–290. doi:10.1038/ismej.2007.53

Rousk J, Jones DL (2010) Loss of low molecular weight dissolved organic carbon (DOC) and nitrogen (DON) in H2O and 0.5 M K2SO4 soil extracts. Soil Biol Biochem 42:2331–2335. doi:10.1016/j.soilbio.2010.08.017

Rübel O, Greiner A, Cholia S, Louie K, Bethel EW, Northen TR, Bowen BP (2013) OpenMSI: a high-performance web-based platform for mass spectrometry imaging. Anal Chem 85:10354–10361. doi:10.1021/ac402540a

Schloss PD, Handelsman J (2003) Biotechnological prospects from metagenomics. Curr Opin Biotechnol 14:303–310. doi:10.1016/S0958-1669(03)00067-3

Schmidt MWI, Torn MS, Abiven S, Dittmar T, Guggenberger G, Janssens IA, Kleber M, Kögel-Knabner I, Lehmann J, Manning DAC, Nannipieri P, Rasse DP, Weiner S, Trumbore SE (2011) Persistence of soil organic matter as an ecosystem property. Nature 478:49–56. doi:10.1038/nature10386

Schneider B, Gershenzon J, Graser G, Hölscher D, Schmitt B (2003) One-dimensional 13C NMR and HPLC-1H NMR techniques for observing carbon-13 and deuterium labelling in biosynthetic studies. Phytochem Rev 2:31–43. doi:10.1023/B:PHYT.0000004196.73829.4e

Schoenian I, Spiteller M, Ghaste M, Wirth R, Herz H, Spiteller D (2011) Chemical basis of the synergism and antagonism in microbial communities in the nests of leaf-cutting ants. Proc Natl Acad Sci USA 108:1955–1960. doi:10.1073/pnas.1008441108

Sévin DC, Kuehne A, Zamboni N, Sauer U (2015) Biological in-sights through nontargeted metabolomics. Curr Opin Biotechnol 34:1–8. doi:10.1016/j.copbio.2014.10.001

Silva LP, Northen TR (2015) Exometabolomics and MSI: deconstructing how cells interact to transform their small molecule environment. Curr Opin Biotechnol 34:209–216. doi:10.1016/j.copbio.2015.03.015

Simpson AJ, Tseng L-H, Simpson MJ, Spraul M, Braumann U, Kingery WL, Kelleher BP, Hayes MHB (2004) The application of LC-NMR and LC-SPE-NMR to compositional studies of natural organic matter. Analyst 129:1216. doi:10.1039/b408064e

Smith CA, Maille GO, Want EJ, Qin C, Trauger SA, Brandon TR, Custodio DE, Abagyan R, Siuzdak G (2005) METLIN: a metabolite mass spectral database. Ther Drug Monit 27

Stevenson FJ (1994) Humus chemistry: genesis, composition, reactions. Wiley, New York

Swenson TL, Bowen BP, Nico PS, Northen TR (2015a) Competitive sorption of microbial metabolites on an iron oxide mineral. Soil Biol Biochem 90:34–41. doi:10.1016/j.soilbio.2015.07.022

Swenson TL, Jenkins S, Bowen BP, Northen TR (2015b) Untargeted soil metabolomics methods for analysis of extractable organic matter. Soil Biol Biochem 80:189–198. doi:10.1016/j.soilbio.2014.10.007

Szeto SSW, Reinke SN, Sykes BD, Lemire BD (2010) Mutations in the saccharomyces cerevisiae succinate dehydrogenase result in distinct metabolic phenotypes revealed through 1H NMR-based metabolic footprinting. J Proteome Res 9:6729–6739. doi:10.1021/pr100880y

Tautenhahn R, Patti GJ, Rinehart D, Siuzdak G (2012) XCMS online: a web-based platform to process untargeted metabolomic data. Anal Chem 84:5035–5039. doi:10.1021/ac300698c

Theodoridis G, Gika HG, Wilson ID (2008) LC-MS-based methodology for global metabolite profiling in metabonomics/metabolomics. TrAC Trends Anal Chem 27:251–260. doi:10.1016/j.trac.2008.01.008

Thiel V, Heim C, Arp G, Hahmann U, Sjövall P, Lausmaa J (2007) Biomarkers at the microscopic range: ToF-SIMS molecular imaging of Archaea-derived lipids in a microbial mat. Geobiology 5:413–421. doi:10.1111/j.1472-4669.2007.00119.x

Torres IF, Bastida F, Hernández T, Bombach P, Richnow HH, García C (2014) The role of lignin and cellulose in the carbon-cycling of degraded soils under semiarid climate and their relation to microbial biomass. Soil Biol Biochem 75:152–160. doi:10.1016/j.soilbio.2014.04.007

Traxler MF, Watrous JD, Alexandrov T, Dorrestein PC, Kolter R (2013) Interspecies interactions stimulate diversification of the streptomyces coelicolor secreted metabolome. mBio 4:e00459-13. doi:10.1128/mBio.00459-13

Treonis AM, Ostle NJ, Stott AW, Primrose R, Grayston SJ, Ineson P (2004) Identification of groups of metabolically-active rhizosphere microorganisms by stable isotope probing of PLFAs. Soil Biol Biochem 36:533–537. doi:10.1016/j.soilbio.2003.10.015

Vance ED, Brookes PC, Jenkinson DS (1987) An extraction method for measuring soil microbial biomass C. Soil Biol Biochem 19:703–707. doi:10.1016/0038-0717(87)90052-6

Villas-Bôas SG, Noel S, Lane GA, Attwood G, Cookson A (2006) Extracellular metabolomics: a metabolic footprinting approach to assess fiber degradation in complex media. Anal Biochem 349:297–305. doi:10.1016/j.ab.2005.11.019

Warren CR (2014) Response of osmolytes in soil to drying and re-wetting. Soil Biol Biochem 70:22–32. doi:10.1016/j.soilbio.2013.12.008

Warren CR (2013a) High diversity of small organic N observed in soil water. Soil Biol Biochem 57:444–450. doi:10.1016/j.soilbio.2012.09.025

Warren CR (2013b) Quaternary ammonium compounds can be abundant in some soils and are taken up as intact molecules by plants. New Phytol 198:476–485. doi:10.1111/nph.12171

Watrous JD, Dorrestein PC (2011) Imaging mass spectrometry in microbiology. Nat Rev Microbiol 9:683–694. doi:10.1038/nrmicro2634

Watrous J, Roach P, Heath B, Alexandrov T, Laskin J, Dorrestein PC (2013) Metabolic profiling directly from the Petri dish using nanospray desorption electrospray ionization imaging mass spectrometry. Anal Chem 85:10385–10391. doi:10.1021/ac4023154

Weiss MS, Abele U, Weckesser J, Welte W, Schiltz E, Schulz GE (1991) Molecular architecture and electrostatic properties of a bacterial porin. Science 254:1627–1630. doi:10.1126/science.1721242

Welch JLM, Rossetti BJ, Rieken CW, Dewhirst FE, Borisy GG (2016) Biogeography of a human oral microbiome at the micron scale. Proc Natl Acad Sci USA 113:E791–E800. doi:10.1073/pnas.1522149113

van der Werf MJ, Overkamp KM, Muilwijk B, Coulier L, Hankemeier T (2007) Microbial metabolomics: toward a platform with full metabolome coverage. Anal Biochem 370:17–25. doi:10.1016/j.ab.2007.07.022

Wessel AK, Hmelo L, Parsek MR, Whiteley M (2013) Going local: technologies for exploring bacterial microenvironments. Nat Rev Microbiol 11:337–348. doi:10.1038/nrmicro3010

Wishart DS (2007) Current Progress in computational metabolomics. Brief Bioinform 8:279–293. doi:10.1093/bib/bbm030

Woo H-K, Northen TR, Yanes O, Siuzdak G (2008) Nanostructure-initiator mass spectrometry: a protocol for preparing and applying NIMS surfaces for high-sensitivity mass analysis. Nat Protoc 3:1341–1349. doi:10.1038/nprot.2008.110

Xia J, Mandal R, Sinelnikov IV, Broadhurst D, Wishart DS (2012) MetaboAnalyst 2.0—a comprehensive server for metabolomic data analysis. Nucleic Acids Res 40:W127–W133. doi:10.1093/nar/gks374

Yang JY, Phelan VV, Simkovsky R, Watrous JD, Trial RM, Fleming TC, Wenter R, Moore BS, Golden SS, Pogliano K, Dorrestein PC (2012) Primer on agar-based microbial imaging mass spectrometry. J Bacteriol 194:6023–6028. doi:10.1128/JB.00823-12

Zelles L (1999) Fatty acid patterns of phospholipids and lipopolysaccharides in the characterisation of microbial communities in soil: a review. Biol Fertil Soils 29:111–129. doi:10.1007/s003740050533

Zhou B, Xiao JF, Tuli L, Ressom HW (2012) LC-MS-based metabolomics. Mol Bio Syst 8:470–481. doi:10.1039/c1mb05350g

Zsolnay Á (2003) Dissolved organic matter: artefacts, definitions, and functions. Geoderma, Ecological aspects of dissolved organic matter in soils 113:187–209. doi:10.1016/S0016-7061(02)00361-0

Chapter 6
Soil Microbial Metabolomics

Michael W. Heaven and Devin Benheim

1 Soil Complexity

The world demand for food is currently expected to increase by two- to fivefold by 2030 (Ewert et al. 2005; Food and Agriculture Organization (FAO) 2002). This projection requires food production to increase from 60 to 70 % (Clair and Lynch 2010), although there is some confusion due to the types of food stuffs the global population will be consuming at that time (Alexandratos and Bruinsma 2013) and the exclusion from earlier studies of important food types (e.g. fruit and vegetables). Agricultural practices over the last century have succeeded in significantly increasing crop yields. For instance, global cereal production doubled between 1960 and 2000 (Tilman et al. 2002). However, the yield increases were driven largely by intensification in the use of non-renewable synthetic fertilisers (Lynch 2007). They were seminal in improving western lifestyles but provided limited relief in many regions of the world such as Africa, Asia and South America. Moreover, the technologies were double edged, with gains in agricultural production coinciding with increased soil erosion (Matson et al. 1997), industrial agricultural pollution (Horrigan et al. 2002), declines in water quality (Foley et al. 2005) and, possibly most importantly, loss of biodiversity (including genetic erosion) (Aguilar et al. 2008; Balestrini et al. 2015). Arguably, a sustainable way forward is 'ecological intensification'. This paradigm expands agricultural intensification that promotes the development of food systems with enhanced nutrient uptake and water use efficiency (Cassman 1999), reductions in pest and disease

M.W. Heaven (✉)
Department of Economic Development, Jobs, Transport and Resources,
1301 Hazeldean Road, Ellinbank, Victoria, Australia
e-mail: michael.heaven@ecodev.vic.gov.au

D. Benheim
Department of Physiology, Anatomy and Microbiology,
School of Life Sciences, La Trobe University, Victoria, Australia

D.J. Beale et al. (eds.), *Microbial Metabolomics*,
DOI 10.1007/978-3-319-46326-1_6

control (Bommarco et al. 2013) and a restoration in soil fertility (Balestrini et al. 2015; Matson et al. 1997; Tittonell 2014). Many of these goals are only achievable today due to the advances in analytical technologies and scientific knowledge that underpins methodologies like metabolomics. As useful as the broad brush of traditional physicochemical analyses has been for agricultural systems (Sinsabaugh et al. 2016) (to understand how to manipulate plant growth), it would be advantageous to determine biogeochemical processes that occur in soil (Rockström et al. 2009; Sardans et al. 2011).

The soil matrix is one of the most diverse ecological systems on the planet (Torsvik and Øvreås 2002). Home to a plethora of organisms, from the largest redwood to the smallest microbe, there are numerous chemical (McBride 1994), physical (Marshall and Holmes 1980) and biological (Barea et al. 2005; Lorenz and Wackernagel 1994) interactions occurring in soil. Until recently, quantifying these biogeochemical processes as a metabolome of "soil" was not considered feasible, with only narrow components of the soil matrix, such as plants and animals, being studied (Maron et al. 2011; Mendes et al. 2013; Mosier et al. 2013). However, a holistic approach to understanding the soil community is something of increasing interest with examples in organic nutrient pools [(Warren 2013, 2014), Table 1], pollution assessment (Jones et al. 2014; Tremaroli et al. 2009), effects of climate change [(Pang et al. 2016), Table 2], plant [(Badri et al. 2013a; van Dam and Bouwmeester 2016), Table 3] and/or microbial metabolism (Ponomarova and Patil 2015; Warren 2015). Recent progress in performing untargeted metabolomics to identify soil organic matter (SOM) metabolites (e.g. lipids and organic acids), which can be linked to nutrient uptake [(Swenson et al. 2015b), Table 3], has shown that "soil metabolomics" is now beyond theory. Still, the ability to study soil holistically is at an early stage of development. The potential for using metabolomics to advance study into how the soil matrix operates will increase as the technologies underpinning the methodology (i.e. gas and liquid chromatography–mass spectrometry (GC–MS and LC–MS), and nuclear magnetic resonance (NMR)) improve (Singh 2006). Likewise, metabolomics can provide a holistic understanding of the impact of the increasing resource demands has on soil (Rockström et al. 2009). Through understanding of the important metabolomic pools and fluxes in soil, so as to understand and monitor soil health as it is managed by farmers, foresters and the community, is likely to improve productivity and the environment while reducing costs (Abhilash et al. 2012; Desai et al. 2010).

Metabolomics of the soil would identify the largest component of metabolites coming from the most varied and numerous collection of microbes known (Barea et al. 2005; Huang et al. 2014; Mendes et al. 2013), with estimates of between 1 and 40 million species per gram of soil (Burns et al. 2013; Desai et al. 2010; Řezanka and Sigler 2009). Despite the importance of microbes to the agricultural and environmental communities, knowledge of the composition of the microbial biomass is limited. For instance, estimates of fungi range from 700 000 to 1.5 million (Lumbsch and Leavitt 2011; Ponomarova and Patil 2015; Rastogi and Sani 2011). However, only 100,000 have been detailed. A large part of this struggle to understand the composition of the soil microbial community is that <1 % has been

Table 1 Typical metabolites found in soil-related samples using other techniques

Metabolite class	Technique (Column for GC or LC)	Spectroscopic range (min)	Why?	Publication
Amino acids, N containing compounds, dipeptides	CE–MS (bare fused silica capillary)	10.99–35.32	Identification of organic N molecules in soil water	Warren (2013)
FAME	GC-FAME	8.23–11.71	Influence of nanoparticles on soil microbes	Shah and Belozerova (2009), Sasser (2006)
Essential oils	GC-FID (Supelcowax 10)	10.70–25.20	Rhizobacteria inoculation	Cappellari et al. (2013)
PLFA	GC-IRMS (HP 5-MS fused silica capillary column)	22.30–40.79	Microbial PLFA biomarkers using stable isotope methods	Watzinger (2015)
Carbohydrates, amino acids, phenolics	HPLC	8.07 18.62	Effects of VAM fungi on maize	Azaizeh et al. (1995)
Flavonoids, organic acids, resorcinols	RP-HPLC (DAD) and LC–MS (Nucleodur Sphinx RP)	15.34–69.73	Secondary metabolite profiling to identify bacterial–rice associations	Chamam et al. (2013)

able to be cultured (Kirk et al. 2004; van den Berg et al. 2015). Metabolomics, being a microbe-independent technique, could shed light on this rich network via metabolite fluxes and identifying previously unknown metabolites and biogeochemical processes, with genomic studies identifying numerous other microbial clades that have been identified only by their molecular sequences (Cesco et al. 2012; Cheynier et al. 2013; Swenson et al. 2015b). There is potential to extract useful information using metabolomics (Alivisatos et al. 2015); although it is acknowledged that to date, studies have been limited due to the complexity of the soil matrix (Ponomarova and Patil 2015).

This chapter uses the lens of metabolomics to discuss connections between soil microbes and the complex interactions they undergo with plants, animals and human intervention. Although studies of soil microbes using metabolomic techniques are scant, associated studies that can shed light on the complex biogeochemical interactions have been undertaken and provide direction. Due to the drive to feed the planet and increase agricultural productivity, the rhizosphere of plants and how to manipulate plant–microbe interactions has probably been of the greatest research interest in this area (Bertin et al. 2003; De-la-Pena and Loyola-Vergas

Table 2 Typical metabolites found in soil-related samples using 1H-NMR

Sample; treatments	Metabolite class	Techniques used	Metabolites	Spectroscopic range (ppm)	Why?	Publication
(a) Grape skin; soil, weather, cultivar. (b) Soil; bacterial strain, location. (c) Worms; soil type, organic versus chemical fertiliser	Amino acids	1H-NMR (D_2O, TSP), 1H-NMR (D_2O, TSP or $CDCl_3$)	Isoleucine, leucine, valine	0.90–1.10	Biomarkers	(a) Pereira et al. (2006). (b) Liebeke et al. (2009). (c) Rochfort et al. (2009)
(a) Plant cuts; time treatments. (b) Seedlings and plants; NaCl and $NaHCO_3$. (c) Worms; soil type, organic versus chemical fertiliser	Amino acids	1H-NMR TOCSY, NOESY, HSCQ; (D_2O with TSP), 1H-NMR NOESY (D_2O, Na_2HPO_4, NaH_2PO_4, TSP), 1H-NMR (D_2O, TSP or $CDCl_3$)	Threonine	1.30, 1.45	Biomarkers, Salinity	(a) Bertram et al. (2010). (b) Pang et al. (2016). (c) Rochfort et al. (2009)
Worms; soil type, organic vs. chemical fertiliser	α-Hydroxy acids	1H-NMR (D_2O, TSP or $CDCl_3$)	Lactate	1.30	Biomarkers	Rochfort et al. (2009)
Wine; bacterial strain	Alcohols	1H-NMR NOESYPREAST (D_2O, oxalate, DSS)	2,3-Butanediol	1.37	Vintage	Lee et al. (2009)
(a) Grape skin; Soil, weather, cultivar. (b) Plant cuts; time treatments. (c) Soil; bacterial strain, location.	Amino acids	1H-NMR (D_2O, TSP), 1H-NMR TOCSY, NOESY, HSCQ; (D_2O with TSP), 1H-NMR NOESY (D2O,	Alanine	1.46–1.70	Biomarkers, salinity, vintage	(a) Pereira et al. (2006). (b) Bertram et al. (2010). (c) Liebeke et al. (2009). (d) Pang et al. (2016).

(continued)

Table 2 (continued)

Sample; treatments	Metabolite class	Techniques used	Metabolites	Spectroscopic range (ppm)	Why?	Publication
(d) Seedlings and plants; NaCl and $NaHCO_3$. (e) Worms; soil type, organic versus chemical fertiliser. (f) Wine; bacterial strain		Na2HPO4, NaH2PO4, TSP), 1H-NMR (D_2O, TSP or $CDCl_3$), 1H-NMR NOESYPREAST (D_2O, oxalate, DSS)				(e) Rochfort et al. (2009). (f) Lee et al. (2009)
Grape skin; soil, weather, cultivar	Amino acids	1H-NMR (D_2O, TSP)	Arginine	1.70	Biomarkers	Pereira et al. (2006)
(a) Soil; bacterial strain, location, (b) Earthworms; various	Organic acids	1H-NMR (D_2O, TMSP), 1H-NMR (Various)	Acetic acid	1.92	Biomarkers, ecotoxicity	(a) Liebeke et al. (2009). (b) Simpson and McKelvie (2009)
Grape skin; soil, weather, cultivar	Amino acids	1H-NMR (D_2O, TSP)	GABA + proline	1.94	Biomarkers	Pereira et al. (2006)
Grape skin; soil, weather, cultivar	Amino acids	1H-NMR (D_2O, TSP)	Proline	1.98	Biomarkers	Pereira et al. (2006)
Soil; bacterial strain, location	Amino acids	1H-NMR (D_2O, TMSP)	Glutamic acid	2.06	Biomarkers	Liebeke et al. (2009)
Seedlings and plants; NaCl and $NaHCO_3$	Amino acids	1H-NMR NOESY (D_2O, Na_2HPO_4, NaH_2PO_4, TSP)	Glutamate	2.10	Salinity	Pang et al. (2016)
Seedlings and plants; NaCl and $NaHCO_3$	Amino acids	1H-NMR NOESY (D_2O, Na_2HPO_4, NaH_2PO_4, TSP)	Glutamine	2.20	Salinity	Pang et al. (2016)

(continued)

Table 2 (continued)

Sample; treatments	Metabolite class	Techniques used	Metabolites	Spectroscopic range (ppm)	Why?	Publication
Grape skin; soil, weather, cultivar	Organic acid	1H-NMR (D_2O, TSP)	Shikimic acid	2.22	Biomarkers	Pereira et al. (2006)
Earthworms; various	Organic acids	1H-NMR (Various)	Acetoacetate	2.23	Ecotoxicity	Simpson and McKelvie (2009)
Soil; bacterial strain, location	Organic acids	1H-NMR (D_2O, TMSP)	Pyruvic acid	2.27	Biomarkers	Liebeke et al. (2009)
Wine; bacterial strain	Organic acids	1H-NMR NOESYPREAST (D_2O, oxalate, DSS)	Acetate	2.28	Vintage	Lee et al. (2009)
(a) Seedlings and plants; NaCl and $NaHCO_3$. (b) Soil; bacterial strain, location. (c) Earthworms; various. (d) Wine; bacterial strain	Organic acid	1H-NMR NOESY (D_2O, Na_2HPO_4, NaH_2PO_4, TSP), 1H-NMR (D_2O, TMSP), 1H-NMR NOESYPREAST (D_2O, oxalate, DSS)	Succinate	2.40, 2.42, 2.82	Salinity, biomarkers, ecotoxicity, vintage	(a) Pang et al. (2016), (b) Liebeke et al. (2009). (c) Simpson and McKelvie (2009). (d) Lee et al. (2009)
Grape skin; soil, weather, cultivar	Amino acids	1H-NMR (D_2O, TSP)	GABA + glutamine	2.46	Biomarkers	Pereira et al. (2006)
(a) Earthworms; various. (b) Seedlings and plants; NaCl and $NaHCO_3$	Amine	1H-NMR (Various), 1H-NMR NOESY (D_2O, Na_2HPO_4, NaH_2PO_4, TSP)	Dimethylamine	2.75, 2.80	Ecotoxicity, salinity	(a) Simpson and McKelvie (2009). (b) Pang et al. (2016)
Soil; bacterial strain, location	Amino acids	1H-NMR (D_2O, TMSP)	Aspartic acid	2.79	Biomarkers	Liebeke et al. (2009)

(continued)

Table 2 (continued)

Sample; treatments	Metabolite class	Techniques used	Metabolites	Spectroscopic range (ppm)	Why?	Publication
Seedlings and plants; NaCl and $NaHCO_3$	Organic acid	1H-NMR NOESY (D_2O, Na_2HPO_4, NaH_2PO_4, TSP)	2-Oxoglutarate	3.00	Salinity	Pang et al. (2016)
(a) Grape skin; soil, weather, cultivar. (b) Plant cuts; time treatments. (c) Plant cuts; time treatments	Neurotransmitter (Amino acid)	1H-NMR (D_2O, TSP)	GABA	1.85, 2.17, 2.25, 2.84, 3.02, 3.06, 3.10, 3.28	Biomarkers, vintage	(a) Pereira et al. (2006). (b) Bertram et al. (2010). (c) Lee et al. (2009)
(a) Seedlings and plants; NaCl and $NaHCO_3$. (b) Plant cuts; time treatments	Amino acids	1H-NMR NOESY (D_2O, Na_2HPO_4, NaH_2PO_4, TSP), 1H-NMR TOCSY, NOESY, HSCQ; (D_2O with TSP)	Choline	3.20, 3.27	Salinity, biomarker	(a) Pang et al. (2016). (b) Bertram et al. (2010)
in situ; Cytochrome P450 inhibitor; xenobiotic	Xenobiotics	1H-NMR (D_2O + TSP), ISP-MS	Thiodiglycolic acid	3.25	Degradation of xenobiotics	Delort and Combourieu (2001)
Seedlings and plants; NaCl and $NaHCO_3$	Amino acids	1H-NMR NOESY (D_2O, Na_2HPO_4, NaH_2PO_4, TSP)	Betaine	3.30	Salinity	Pang et al. (2016)
(a) Grape skin; soil, weather, cultivar. (b) Plant cuts; time treatments	Sugars	1H-NMR (D_2O, TSP), 1H-NMR TOCSY, NOESY, HSCQ; (D_2O with TSP)	Glucose, fructose, sucrose	3.42–4.22	Biomarkers	(a) Pereira et al. (2006). (b) Bertram et al. (2010)

(continued)

Table 2 (continued)

Sample; treatments	Metabolite class	Techniques used	Metabolites	Spectroscopic range (ppm)	Why?	Publication
(a) Soil; bacterial strain, location. (b) Earthworms; various. (c) Seedlings and plants; NaCl and $NaHCO_3$	Amino acids	1H-NMR (D_2O, TMSP), 1H-NMR (Various), 1H-NMR NOESY (D_2O, Na_2HPO_4, NaH_2PO_4, TSP)	Glycine	3.56–3.60	Biomarkers, ecotoxicity, salinity	(a) Liebeke et al. (2009). (b) Simpson and McKelvie (2009). (c) Pang et al. (2016)
in situ; Cytochrome P450 inhibitor; xenobiotic	Xenobiotics	1H-NMR (D_2O + TSP), ISP-MS	Glycolate	3.95	Degradation of xenobiotics	Delort and Combourieu (2001)
Seedlings and plants; NaCl and $NaHCO_3$	Amino acids	1H-NMR NOESY (D_2O, Na_2HPO_4, NaH_2PO_4, TSP)	Malate	4.40	Salinity	Pang et al. (2016)
(a) Grape skin; soil, weather, cultivar. (b) Soil; bacterial strain, location. (c) Plant cuts; time treatments. (d) Worms; soil type, organic versus chemical fertiliser	Sugars	1H-NMR (D_2O, TSP), 1H-NMR (D_2O, TMSP), 1H-NMR TOCSY, NOESY, HSCQ; (D_2O with TSP), 1H-NMR (D_2O, TSP or $CDCl_3$)	Glucose, maltose, sucrose	4.62–5.42	Biomarkers	(a) Pereira et al. (2006). (b) Liebeke et al. (2009). (c) Bertram et al. (2010). (d) Rochfort et al. (2009)
Wine; bacterial strain	Organic acids	1H-NMR NOESYPREAST (D_2O, oxalate, DSS)	Tartrate	4.63	Biomarkers	Lee et al. (2009)
Soil; bacterial strain, location	Sugars	1H-NMR (D_2O, TMSP)	Trehalose	5.19	Biomarkers	Liebeke et al. (2009)

(continued)

Table 2 (continued)

Sample; treatments	Metabolite class	Techniques used	Metabolites	Spectroscopic range (ppm)	Why?	Publication
Soil; bacterial strain, location	Organic acids	1H-NMR (D_2O, TMSP)	Malic acid	6.03	Biomarkers	Liebeke et al. (2009)
(a) Worms; soil type, organic versus chemical fertiliser. (b) Seedlings and plants; NaCl and $NaHCO_3$. (c) Soil; bacterial strain, location. (d) Earthworms; various	Organic acids	1H-NMR (D_2O, TSP or $CDCl_3$), 1H-NMR NOESY (D_2O, Na_2HPO_4, NaH_2PO_4, TSP), 1H-NMR (D_2O, TMSP), 1H-NMR (Various)	Fumarate	6.20–6.52	Biomarkers, ecotoxicity	(a) Rochfort et al. (2009). (b) Pang et al. (2016). (c) Liebeke et al. (2009). (d) Simpson and McKelvie (2009)
Earthworms; various	Organic Acids	1H-NMR (Various)	Orotic acid	6.20	Ecotoxicity	Simpson and McKelvie (2009
(a) Soil; bacterial strain, location. (b) Worms; soil type, organic versus chemical fertiliser	Amino acids	1H-NMR (D_2O, TMSP), 1H-NMR (D_2O, TSP or $CDCl_3$)	Tyrosine	6.92–7.00	Biomarkers	(a) Liebeke et al. (2009). (b) Rochfort et al. (2009)
Worms; soil type, organic versus chemical fertiliser	Amino acids, Sulfonate, Lipids, Triacylglycerides (TAG)	1H-NMR (D_2O, TSP or $CDCl_3$)	Phenylalanine	7.40	Biomarkers	Rochfort et al. (2009)
(a) Worms; soil type, organic versus chemical fertiliser. (b) Soil; bacterial strain, location. (c) Earthworms; various	Organic acids	1H-NMR (D_2O, TSP or $CDCl_3$), 1H-NMR (D_2O, TMSP), 1H-NMR (Various)	Formate	8.35	Biomarkers, ecotoxicity	(a) Rochfort et al. (2009). (b) Liebeke et al. (2009). (c) Simpson and McKelvie (2009)

Table 3 Details of selected GC–MS analyses reported in this review

Sample; treatments	Metabolite class	Technique (Column)	Spectroscopic range (min)	Why?	Publication
(a) Soil: time, fumigation, ^{13}C incorporation. (b) Soil columns on farm; time	PLFA	(a) GC–MS (DB1-MS column). (b) GC–MS (HP-1 methylpolysiloxane)	1.50–48.83	(a) Transformations of soil monosaccharides (b) Incorporation of organic compounds into microbes	(a) Apostel et al. (2015). (b) Gunina et al. (2014)
Root exudates; soil; solvent extraction fraction	Sugars, Phenolics, Amino acids, Sugar alcohols	GC–MS (rtx5Sil-MS column)	3.57–17.14	Identifying root exudates that affect soil microbiome	Badri et al. (2013a)
Roots; microbe inoculation, humic acids, time	Nitrogenous compounds, Carbohydrates, Fatty acids, Organic acids, Aromatic and phenol derivatives, Terpenoids and steroids, Alcohols	GC–MS (Rtx-5MS WCOT capillary column)	3.98–45.54	Root exudates after inoculation with microbes and humic acids	Lima et al. (2014)
Plant rosettes; NaCl loading, plant species, water concentration	Amino acids, Organic acids, Alcohols, Carbohydrates, Amines	GC-QQQ-MS (Rtx-5Sil MS capillary column) and GC-FID (J&W DB5)	10.68–47.99	How *Arabidopsis* and *Thellungiella* react to salt and/or water stress	Lugan et al. (2010)
Untargeted soil profiling, fumigation, labelled (^{13}C)	Amino acids, Organic acids, Sugar alcohols, Carbohydrate, Lipids, Nucleosides, Nucleobases, Sugars, Sterols and others	GC–MS (Rtx5Sil-MS column)	8.42–24.32	Development of a simple soil metabolomics workflow and a novel spike recovery approach using ^{13}C bacterial lysates to assess the types of metabolites remaining in the water extractable organic fraction	Swenson et al. (2015b)

2014; van Dam and Bouwmeester 2016; Zahir et al. 2004). Other areas of discussion beyond the plant–microbe interactions include specific foci on how soil microbes are affected by pollution, diseases, pests and potential climate change. A theme of the chapter will be how metabolomics may be used to improve soil management (Rochfort et al. 2015; Zhang et al. 2015), both for increasing productivity of the soil and mitigating environmental effects.

2 Soil Metabolomics Is a Nascent Field

Soil is often a secondary topic for metabolomics analyses. The focus of the research is often on how microbes interact with plants such as grasses (e.g. *Trifolium*: clover) or legumes [(Bertram et al. 2010), Table 2], weeds such as *Arabidopsis* [(Gamir et al. 2014), Table 4], trees such as *Aspen* (Wallenstein et al. 2010) or invertebrates that inhabit the soil such as earthworms *Aporrectodea caliginosa* [(Rochfort et al. 2009), Table 2] or *Eisenia fetida* (Simpson and McKelvie 2009). Recent reviews have discussed soil microbial metabolites within the concept of environmental metabolomics. For instance, a review on environmental metabolomics suggests the field is fragmented, as this new "holistic" methodology was mainly being used to study single species of the researchers' interest, including those involving soil microbes (Bundy et al. 2009). The authors reiterated that this misses one of environmental metabolomics assets in gathering an understanding of the interactions between species and their environment. Another review focused on using metabolomics to assess soil contamination (Hernandez-Soriano and Jimenez-Lopez 2014). Earthworms, usually of the genus *Eisenia*, were found to be typical subjects for metabolomic analyses. Studies related to microbial metabolomics were small, with the only reference to work on the response to tellurite by *Pseudomonas pseudoalcaligenes* (Tremaroli et al. 2009). Earthworms were also considered the main target for soil metabolomics of another review on environmental sciences and metabolomics (Lin et al. 2006). This review detailed the various effects of contaminants on the metabolic profiles of a variety of earthworms, particularly halogenated compounds and metal-contaminated soils. Another review discussed 'eco-metabolomics', the use of metabolomics and how it relates to ecology, with regard to interactions between organisms, including those in the soil (Sardans et al. 2011). While mainly related to how plants and worms react to changes in environment, some discussion was made on how fungi *Magnaporthe grisea* and *Sclerontinia sclerotiorum* release metabolites to suppress plant defences.

Various studies that have examined microbes and/or the metabolites associated with them. However, only in 2015 has there been research that specifically mentions the use of "soil metabolomics" to identify biogeochemical processes occurring in soils [(Swenson et al. 2015a; Swenson et al. 2015b), Table 4]. The research sought to understand the fluxes of microbial metabolites in SOM that might occur due to climate change. Comparing fumigated and unfumigated soil samples using

Table 4 Details of selected LC–MS analyses reported in this review

Sample; treatments	Metabolite class	Technique (Column)	Spectroscopic range (min)	Why?	Publication
Plant roots; arbuscular mycorrhizal fungi (AMF) strains, time	Amino acids, Organic acids, Phosphate compounds, Acetates, Nucleosides, Vitamins, Indoles, Purines	LC-ESI Q-TOF-MS (SunFire C18 analytical column)	0.50–13.40	Mycorrhizal association of AMF with tomato plants	Rivero et al. (2015)
Roots; AMF and/or plant growth-promoting rhizobacteria (PGPR)	Lipids (PE, PC, LPE, Carnitine)	LC-HILIC-Q-TOF-MS (Acuity 1.7 μm BEH HILIC) or GC-TOF-MS (Rtx-5SilMS column)	2.60–8.82	AMF effects on wheat in N limited, P-rich environment	Saia et al. (2015b)
Arabidopsis cells; Nutrients	Phytochelatin (PC)	LC–MS (X-Terra MS C18 column)	9.70–25.00	Plant response to cadmium stress	Sarry et al. (2006)
Ferrihydrite: Organic Carbon, temperature	Phosphate, Dicarboxylate, Aromatic, N-carboxylate, other N organic compounds	LC–MS (ZIC-pHILIC)	1.70–23.6	Microbial metabolites on ferrihydrite	Swenson et al. (2015a)
Leaves; pre- and post-fungal treatments	Amino acids, Organic acids, Aldehydes, Indoles	LC-QQQ-MS (Pack ODS-A reversed-phase C18)	0.53–13.40	Understanding priming of plants against microbes	Gamir et al. (2014)

water extractable SOM, it was possible to identify metabolites associated with microbial species. Initial studies using GC–MS identified 55 metabolites via comparison with accurate standards or carbon-labelled samples (Swenson et al. 2015b). Up to 300 molecular features (after extensive sample preparation using GC-friendly water soluble solvents) were identified in soil samples following optimisation of the extraction media water, aqueous potassium sulphate or ammonium carbonate, isopropanol and methanol. Stable isotope labelling studies, using ^{13}C-acetate as a growth medium, allowed the differentiation of microbial metabolites from other compounds in the soil. This method identified sugars and amino acids as most likely to be co-extracted with other metabolites that contained hydrogen bonding functional groups [e.g. fatty acids (FAs) and sterols]. A followup study using LC–MS identified a further 55 metabolites that interacted with iron oxides in soil, reducing access to these compounds by microbes (Swenson et al. 2015a). Metabolites from this study could be grouped into phosphate containing (e.g. AMP), dicarboxylates (e.g. fumarate), aromatic and nitrogen containing (e.g. phenylalanine), carboxylate and nitrogen containing (e.g. creatinine) and others such as thymine. Future research that draws from the work described below should allow for a more holistic approach to understanding how microbes in soil can affect soil.

3 The Rhizosphere

3.1 Map of the Rhizosphere

The rhizosphere, comprising the endorhizosphere, the rhizoplane and the ectorhizosphere (Fig. 1), defines the narrow region between roots and soil directly influenced by both root exudates and exfoliates, and associated microorganisms (Jones 1998). At the heart of the rhizosphere is the root of a plant that is undergoing symbiosis. The root surface (epidermis and outer cortex) and its adhering soil are collectively termed the rhizoplane: the interface where both microbial population and biochemical plant–microbe interactions are at their maximum. The root systems of plants serve critical roles in the provision of anchorage, water and mineral absorption and conduction, lateral movement, reproduction, metabolite synthesis and food storage centre (Kenrick 2013; Kramer and Boyer 1995; Selosse and Strullu-Derrien 2015). Roots are linear units composed of multiple regions along the root growth axis. The units—the root cap, root tip, elongation zone, root-hair zone and mature zone—are uniquely differentiable and perform distinct functions (Minz et al. 2013). Each of these units uses different libraries of metabolites for communication to other units while releasing a multitude of metabolites that modulate plant–microbe interactions in the rhizosphere (Huang et al. 2014). Root exudates, used as both substrates and signalling molecules by soil microbes, comprise of both low (e.g. amino acids, organic acids, carbohydrates, phenolics and

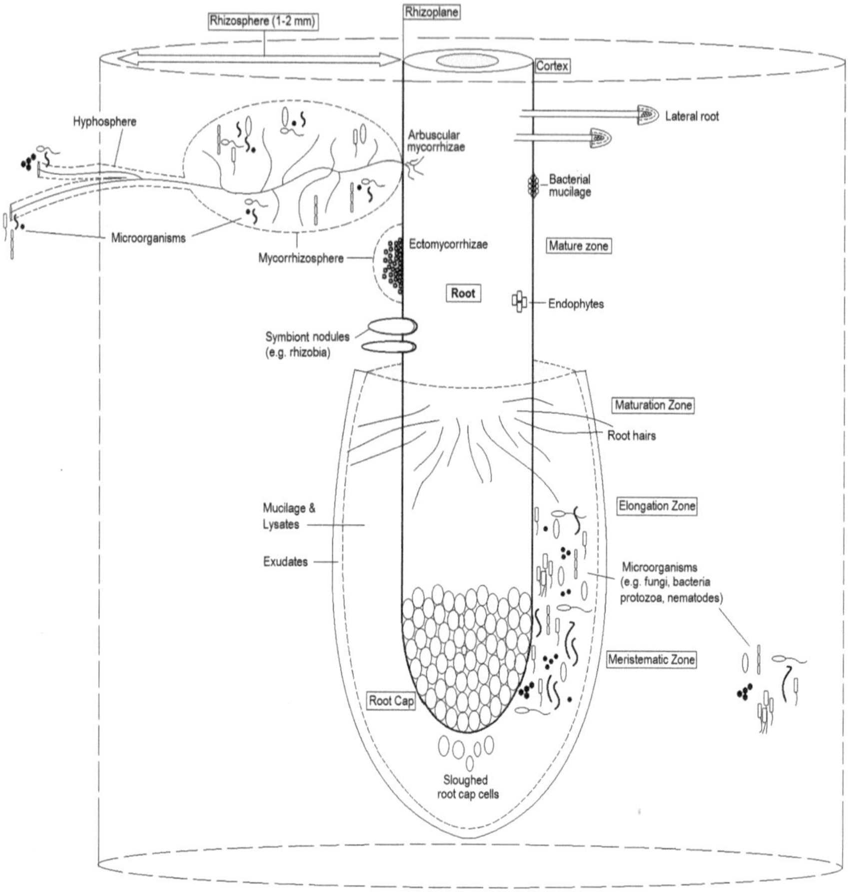

Fig. 1 Schematic representation of the rhizosphere showing commonly associated microbial associations with plant root systems

other secondary metabolites) and high molecular weight compounds (e.g. mucilage and proteins) (Bais et al. 2008; Walker et al. 2003; Ziegler et al. 2013). The nature of root exudates is observed to vary significantly between plant species. Consequently, the rhizosphere microbiome differs with both plant species and soil type (Tate 2000; Wieland et al. 2001). The influence of the rhizosphere can even extend beyond the immediate area of the plant, as discovered with metabolomic analyses of leaf litter around different tree species (Wallenstein et al. 2013; Wallenstein et al. 2010).

Apart from the root, the rhizosphere's main constituents are eukaryotic and prokaryotic microbial species, with a diversity of microbes that can range from thousands (Berendsen et al. 2012) to millions (Nihorimbere et al. 2011). The dynamic microbiome which surrounds the roots is of critical importance to ecosystem function (Bertin et al. 2003; Sardans et al. 2011), below-ground carbon (Mendes et al. 2011) and nutrient cycling (Swenson et al. 2015b; Van Der Heijden et al. 2008), affecting overall plant fitness and soil quality (Barea et al. 2005; Minz et al. 2013). Plants spend up to half of their energy producing exudates into the rhizosphere (van Dam and Bouwmeester 2016), though it has been argued that the source of metabolites (e.g. plant vs. microbe) in the rhizosphere is still to be elucidated convincingly (Dennis et al. 2010). A major role of exudates from roots is the communication to soil-borne microbes, the most dominant class of soil biota (Van Der Heijden et al. 2008). These microbes are phylogenetically diverse (e.g. bacteria, archaea and fungi), and comprise symbionts, pathogens and saprotrophs (Balestrini et al. 2015; Reynolds et al. 2003).

Saprotrophic microbes are crucial to nutrient cycling in terrestrial ecosystems, generating the majority of nutrients required by terrestrial vegetation (Crowther et al. 2012; Schlesinger 1991). The soil microbial community is able to do this primarily via photosynthetically fixed carbon introduced to the soil ecosystem in the form of plant biomass and root exudates (Badri et al. 2013a; Dennis et al. 2010; Tate 2000). While there is competition between plants and microbes for available nutrients (Kaye and Hart 1997; Kuzyakov and Xu 2013), two key plant–microbe symbiotic associations, arbuscular mycorrhizal fungi (AMF) (Van Der Heijden et al. 1998) and root nodule symbiosis (RNS) (Li et al. 2013), impart significant benefits to both microbes and plants. AMF improve the supply of water and nutrients (such as phosphate and nitrogen) to the plant via extraradical hyphae (Kawaguchi and Minamisawa 2010; Parniske 2008). For RNS, nitrogen-fixing bacteria enable enzymatic conversion of atmospheric nitrogen into bioavailable ammonia for plant growth (Brewin 2010). Ecological benefits of symbioses are also contributed as is the case of AMF, which increases host resilience toward drought (Baum et al. 2015), decreased susceptibility to diseases (Saia et al. 2015b), improved heavy metal (Schützendübel and Polle 2002), excess salinity tolerance (Luo et al. 2009) and other abiotic factors (Habib et al. 2013).

3.2 *Rhizosphere Metabolomics*

The rhizosphere has undergone scrutiny using various omic techniques including genomics, metagenomics, proteomics, transcriptomics and more recently metabolomics (van Dam and Bouwmeester 2016). The principle metabolomic tools are NMR spectroscopy [Table 2, (Sardans et al. 2011)], LC–MS [Table 4, (Allwood and Goodacre 2010)], GC–MS [Table 3, (Kusano et al. 2011)], although other techniques such as capillary electrophoresis–time of flight–MS (CE–TOF–MS, Table 1) (Abhilash et al. 2012) have been utilised. For the symbiotic process to occur between microbes and plants, there must be a series of metabolites that are transferred between the two moieties (Rasmussen et al. 2012). Major bioactive metabolites found in the rhizosphere include flavonoids (Cesco et al. 2012; Cheynier et al. 2013; Narasimhan et al. 2003), phenolic compounds (Badri et al. 2013a;

Kakumanu et al. 2013; Rawat et al. 2011), exopolysaccharides (Workentine et al. 2010), antibiotics (Frisvad et al. 2004) and those participating in quorum sensing (QS) signals (De-la-Pena and Loyola-Vergas 2014; Jia et al. 2016). QS describes the process by which bacterial population density, biofilm formation and gene expression (modulating niche persistence and root colonisation) are controlled within the population through production of low-mass signalling molecules [(Braeken et al. 2008; Lima et al. 2014), Table 3]. Bacterial pathogens and symbionts are largely dependent on QS to colonise and infect their respective hosts, as in the case of *Pseudomonas aeruginosa*, which has been shown to have up to 20 % of its genes and proteins regulated via QS (Bauer and Mathesius 2004; Das and Mukherjee 2007). Other abiotic soil factors including nutrient availability (Saia et al. 2015b), soil pH (Dakora and Phillips 2002), salinity (Oikawa et al. 2011) and other environmental stressors are also known to correlate with changes in plant metabolite composition. The physiological impact of the soil microbiome on plant metabolism is receiving increasingly more attention due to changes to the climate and the need to feed an increasing population (Park et al. 2014; Sanchez et al. 2008). While assessment and identification of the full complement of rhizosphere microbes presents significant challenges, rhizosphere microbes have demonstrated a capacity to alter plant morphology, enhance plant growth and increase mineral content (Berendsen et al. 2012; Lakshmanan et al. 2014; Mendes et al. 2011). The following sections demonstrate several notable examples highlighting the use of rhizosphere metabolomics in understanding the plant–microbe interactions that underpin the practical aspects of enhancing plant performance.

4 Exudates

Plant communication to soil microbes is almost solely conducted using metabolites called exudates (Bais et al. 2004; Bronick and Lal 2005). Exudates directly influence the structure and function of the soil microbiome and are strong mediating factors in preferential microbiome selection (Coats and Rumpho 2014). The exudates are commonly amino acids and sugars (Broeckling et al. 2008). Secondary metabolites that may be active in exudates include flavones (Redmond et al. 1986) and terpenes (Hartmann et al. 2008). The secondary metabolites mediate processes to improve nutrient uptake and microbial resistance. To date, most studies have focussed on how a single microbial species reacts to plant exudates, with only recent research taking into account the multitude of microbial species in soil (Swenson et al. 2015a). The manner in which exudates are utilised by and affect microbes varies considerably and has been extensively reviewed (Bais et al. 2006; Bertin et al. 2003; Boyce 2005; Brewin 2010; Huang et al. 2014).

A variety of positive and negative interactions with microbes have been observed occurring directed by exudates from plants (Bais et al. 2006). Positive microbial interactions include the following:

- facilitating plant growth-promoting rhizobacteria (PGPRs) and symbiosis with, among others, acids, sugars and vitamin metabolites (Ahemad and Kibret 2014);
- biocontrol of nutrient fluxes including macronutrients such as P and micronutrients including Fe (Carvalhais et al. 2011, 2013; Dakora and Phillips 2002; Valentinuzzi et al. 2015);
- isoflavanoids (Morandi et al. 1984) and phenolic acids [(Azaizeh et al. 1995; Mandal et al. 2010), Table 1] involved with vesicular-AMF symbiosis;
- alkaloids and nitrogen containing metabolites produced by endophyte symbiosis, often to counteract insect predation of plants (Rasmussen et al. 2012).

Interactions that involve defence or attack against invaders of the rhizosphere include the following:

- QS metabolites such as bradyoxetin from the soybean symbiont, *bradyrhizobium japonicum* (Loh et al. 2002), that appears to help the bacterium fight off invading microbes;
- the use of lactones such as N-(3-oxohexanoyl)homoserine lactone in the rhizosphere of ginger (*Zingiber officinale*) to defend against plant viruses using bacteria such as *Acinetobacter* and *Burkholderia* (Braeken et al. 2008; Chan et al. 2011; Cooper 2007);
- phytotoxin use against invasive plants, including metabolites that include phenolics, coumarins and quinines (Khanh et al. 2005), and antimicrobial agents such as rosmarinic acid (Bais et al. 2008; Haichar et al. 2012; Hartmann et al. 2008).

The research listed identifies that there are still many unknown aspects of exudate processes that have only been hinted at with current research efforts (Bonanomi et al. 2009; Rabie 1998). Further understanding of rhizosphere microbe–exudate interactions would increase our capacity to engineer the rhizosphere to suit particular applications, as for example the use of AMF as biofertilisers (Bonfante and Genre 2010). This strategy entails engineering ideal rhizospheric growth conditions which target particular microbes for their ability to metabolise distinct nutrients.

4.1 Rhizoengineering

Rhizoengineering entails controlling the plant's rhizosphere, a considerable challenge considering the number of different entities involved. This could be to improve plant yield [e.g. wheat (Saia et al. 2015a)] or output (with *Arabidopsis* being a plant of focus in this research) (Kabouw et al. 2012), or to use the plants to improve the surrounding environment (as was found with grasses (*Hyparrhenia hirta)* and beans (*Zygophyllum fabago*) taking up heavy metals from mine tailings in Spain) (Padmavathiamma and Li 2007). A notable example of rhizoengineering used a metabolomics-driven approach to enhance plant–microbe interactions for bioremediation of soil contaminated by polychlorinated biphenyls (PCBs) (Narasimhan et al. 2003). In the study, an *Arabidopsis–Pseudomonas* rhizosphere model was established in which 76 % were phenylpropanoids [e.g. ampelopsin (dihydromyricetin)] of 125 identified compounds (identified by metabolic profiling of *Arabidopsis Thaliana*). The root exudates identified by LC–MS were found to create a nutritional bias for efficient rhizocolonising strain of *Pseudomonas putida* PLM2. The strain was chosen both for its ability to utilise a diverse range of phenylpropanoid compounds, and its PCB-degrading capabilities. Using a gnotobiotic system, the study showed a 90 % reduction in PCBs in flavonoid-producing *Arabidopsis thaliana* strains.

Rhizoengineering need not only be limited to *in planta* studies. Innovative efforts have been made to examine microbial community dynamics through the development of artificial root models using agarose-covered slides amended with various carbon-rich compounds (i.e. glucose, malic acid and serine) simulating root exudate composition (Ziegler et al. 2013). Such novel approaches could theoretically be modified to provide convenient models for the simulation of root–microbe metabolism. Similar research into artificial roots has utilised mucilage (Ahmed et al. 2014), a polymeric gel exuded by plants that includes metabolites xylose, glucose and uronic acids. The authors acquired mucilage from chia seeds (*Salvia hispanica* L.) to emulate maize (*Zea mays* L.) root exudate. Ostensibly, this was to identify how soil in the rhizosphere is often wetter than the roots of the plant producing the exudate. The artificial roots in this system were simplified with the assumption that chia mucilage is similar to maize. The use of artificial roots highlights the difficulties in accurately measuring metabolomics fluxes in the rhizosphere.

Another effort to undertake rhizoengineering was to accentuate microbial consumption of polychlorinated biphenyls (PCB) (Narasimhan et al. 2003). Metabolites in soil were identified and delineated between microbial and those of Arabidopsis. A number of metabolites (e.g. flavonoids, lignins, indoles, anthocyanins) identified in the plant and soil led to the realisation that phenylpropanoids could be target metabolites for rhizoengineering of the soil. This was based on the criteria that phenylpropanoid metabolites are complicated enough to be resistant to microbial degradation, thus allowing them to act as a nutrient source for bacteria that will selectively degrade PCBs.

Rhizoengineering could be an exciting new area of research. As promoted by these papers, it is expected that a mixture of different species of plants and microbes will be required to exude metabolites to the required composition to improve soil productivity. Metabolomics could be a useful approach to characterise and optimise the processes to successfully engineer the rhizosphere to suit society's needs (Abhilash et al. 2012; Bonfante and Genre 2010).

5 Metabolite Coverage Over the Lifespan of Plants

Another aspect yet to be fully elucidated by research and conducive to metabolomics analyses is studying temporal variabilities to metabolites as a plant matures and is affected by external stimuli such as climate change, or is situated in different soil types (Bais et al. 2008; Borisjuk et al. 2012). For instance, the nature of metabolite secretions from the roots of *Arabidopsis thaliana* has been found to differ over the plant lifespan which, by extension, differentially affect root microbes (Chaparro et al. 2013). Analysis of the root exudates by GC–MS identified 57 metabolites from 107 possible compounds. As the plant developed over 31 days, the metabolites showed a comparative decrease in the cumulative secretions of sugars (e.g. fructose, glucose) and sugar alcohols (e.g. glycerol) and an increase in secretion levels of amino acids (e.g. glycine, alanine) and other metabolites (i.e. organic acids, carboxylic acids, FAs and other phenolic compounds). This was noted as being suggestive of a genetically programmed developmental pattern of varied phytochemical root exudation. Rhizosphere mRNA pyrosequencing showed strong correlations between microbial functional genes involved in carbohydrate, amino acid and secondary metabolite metabolism, and metabolites secreted by *Arabidopsis thaliana* at specified developmental stages. Another metabolomic study of interactions over time between potential soil-borne pathogens *Phytophthora infestans* on potatoes showed a similar pattern of amino and organic acid metabolites concentration increase and sugar concentration decrease in response to microbial inoculation (Abu-Nada et al. 2007). Similar results have also been seen due to fungal infection of soybean (Scandiani et al. 2015) and strawberries (Valentinuzzi et al. 2015). One interesting study showed how maize uses the soil fungus *Fusarium verticillioides* to attack another fungus, *Ustilago maydis*, over a 7-day period (Jonkers et al. 2012). As time progressed, the battle between the two fungi could be monitored through the changes in metabolite concentrations of compounds such as fusaric acid and a mannosylerythritol lipid (Arutchelvi et al. 2008). Generally, however, the manner in which rhizosphere microbiome function is affected by temporally varied root exudates over the course of plant development remains largely unknown. Advances in analytical technologies that allow for real-time monitoring [e.g. portable MS (Yang et al. 2008)] may allow for a greater interest in using metabolomics to study how metabolites change over time.

6 Microbial Soil Inoculants

The interactions of beneficial rhizosphere soil microbes with root systems have pivotal roles in the growth, development and ecological fitness of their plant hosts. The prevalences of intensive farming practices that are high yield and/or quality centric are traditionally predicated on extensive use of environmentally harmful and costly chemical fertilisers (Riding et al. 2015; Wissuwa et al. 2008). Subsequently, this has led to increased industry interest in the use of sustainable and environmentally ethical farming practices (Nihorimbere et al. 2011; Zahir et al. 2004). The use of soil inoculants or 'biofertilizers' comprising beneficial soil microbes has been observed to strongly fulfil this niche through enhancement of plant growth, biological control of plant pathogens, nutrient supply and promotion of soil productivity [(Cappellari et al. 2013), Table 1]. Examples of soil inoculants are mycorrhizal fungi, the filamentous fungi *Trichoderma* spp. and plant growth-promoting rhizobacteria (PGPR) (including, but not limited to, genera *Acetobacter*, *Azospirillum*, *Bacillus*, *Burkholderia*, *Herbaspirillum*, *Paenibacillus*, *Phyllobacterium* and *Pseudomonas*) [(Chamam et al. 2013; Saharan and Nehra 2011), Table 1].

For example, the effect of soil inoculation with PGPRs in the commercially valuable wild marigold (*Tagetes minuta*) has been assessed (Cappellari et al. 2013). Wild marigold produces an essential oil (EO) known as "Tagetes oil" sought for the preparation of high-grade perfumes. The effect of single and co-inoculation of *Tagetes minuta* with *Pseudomonas fluorescens* and *Azospirillum brasilense* on plant growth parameters and essential oil production was assessed under glasshouse conditions. *Azospirillum* has been used for growth and yield promotion in cereal plants as rice, maize and wheat (Chamam et al. 2013). Both single and co-inoculations showed an increase in shoot fresh weight by approximately 50 %. Total phenolic content of shoots was upregulated by up to twofold with total EO yield increased by 70 % in single and co-inoculated plants. Major components of the EO that significantly increased ($p < 0.05$) included nine types of terpenoids, such as tagetone and ocimenone, identified by GC–MS. While individual inoculation with either *Azospirillum brasilense* or *Pseudomonas fluorescens* increased plant growth and EO production, significant enhancement to both metrics was observed in co-inoculated plants suggesting synergy between the two bacterial genera.

PGPRs demonstrate significant capacity for plant enhancement. In some cases, the magnitude of enhancement may be both bacterial strain and plant cultivar-dependant (Chamam et al. 2013). This study showed a strain-dependent effect in the association between two types of nitrogen-fixing bacterium *Azospirillum* and Asian rice (*Oryza sativa*). The bacteria *Azospirillum lipoferum* 4B (rhizosphere-colonising strain isolated from *Oryza sativa* cultivar Cigalon) and *Azospirillum* sp. B510 (an endophytic strain isolated from *Oryza sativa* cultivar Nipponbare) were observed to significantly increase ($p < 0.05$) growth of the cultivar by up to 1.5 mg/plant over 10 days if they were used to inoculate the rice strain from which they were isolated. Metabolic profiling data from reverse-phase LC–MS demonstrated significant modification in rice secondary metabolites in PGPR-inoculated plants, with 17 flavonoids, 10 hydroxycinnamic acid derivatives and four alkylresorcinols the most affected metabolite classes. Moreover, the metabolites were unique to the cultivar with only a few compounds, such as tryptophan, found common to all cultivars. This research stands as a strong example of how metabolomics techniques can be used to directly assess the nature of host–PGPR symbiotic interactions, here being able to distinguish the physiological responses of both rice cultivars to specific *Azospirillum* strains with opposing root colonisation strategies. In another case, similar changes in plant secondary metabolism were observed in a study investigating maize-*Azospirillum* interactions (Walker et al. 2011). Major secondary metabolic changes were exclusively observed in the roots of the Cigalon-4B pairing, predominantly for benzoxazinones and benzoxazolinones, while the endophytic B510 strain elicited a systemic response inducing metabolic shifts in both shoots and roots of both rice cultivars tested.

Compared to bacterial soil inoculants, fungal soil inoculants including *Trichoderma* spp., soil fungi that are often associated with plant root ecosystems (Vinale et al. 2008), and AMF can impart significant benefit to host plants (Contreras-Cornejo et al. 2011). Considered beneficial to plants, they protect them from disease by attacking phytopathogenic fungi. Metabolomic studies have identified that *Trichoderma* detect fungi due to the exudation of metabolites from cell wall degrading enzymes and sensing the sugars released by degradation. More importantly, *Trichoderma* releases secondary metabolites that are antifungal, including antibiotics, water soluble acids and peptaibols. The non-polar nature of the antibiotics (e.g. 6-pentyl-α-pyrone) suggests that these are used as a long-range defence, while more polar metabolites (i.e. peptaibols) are used to attack fungi at 'close quarters'. The positive relationship between *Trichoderma* and plants, which includes increasing crop productivity by up to 300 %, along with activation of the plant's defences (Woo et al. 2006), has led to the use of this fungi as a natural biocontrol agent.

Other research that has identified microbially related metabolites involved in systemic acquired resistance and pathogen protection include the following:

- salicylic (SA) and jasmonic acid (JA) (Segarra et al. 2007) in cucumbers;
- SA, JA and indole-3-carboxaldehyde (Contreras-Cornejo et al. 2011), sugars, amino acids, ethanolamine, tagatose, oxoproline, GABA and urea (Chaparro et al. 2013) in *Arabidopsis thaliana*;
- the increased production of phytoalexins (e.g. flavonoids, terpenoids, indoles) after inoculation by *Trichoderma* that showed increased root development and biomass of *Arabidopsis thaliana* over 8 days (Contreras-Cornejo et al. 2011);
- 2,4-diacetylphloroglucinol, from AMF (Wehner et al. 2010);

Another reason for inoculation has been to counteract stressors on plants such as drought or salinity. Abiotic stress adaptation has been shown to involve a range of metabolites including the following:

- rhizospheric fungi *Trichoderma harzianum* Rifai strain T-22 using lipid peroxides in tomatoes as stress biomarkers (Mastouri et al. 2010);
- drought protection of plant through the increased production of hormones such as indole-3-acetic acid, JA, SA, ethylene, auxins, cytokinins and gibberellins and amino acids like proline (Azcón et al. 2013);
- reduced concentrations of malondialdehyde and increased concentrations in proline and phenol metabolites in wheat due to salinity (Rawat et al. 2011); and,

- in the case of *Trichoderma* spp., the regulation of root system architecture that was made possible with reactive oxygen species that convert into hydrogen peroxide (Contreras-Cornejo et al. 2013; Mastouri et al. 2010; Samolski et al. 2012).

Metabolomics can also be used as a tool to find soil microbial metabolites that are a signal of diseases occurring within the soil (Bundy et al. 2009). For example, the fungi *Trichoderma* spp. (Lorito et al. 2010; Vinale et al. 2008) have been found to improve plant defences through secondary metabolite communication. A pathogen attacking a plant results in *Trichoderma* expressing amino acids to create small proteins called hydrophobins that act to coat the plant as a barrier. Other metabolites used for defence included heptelidic (koningic) acid (Itoh et al. 1980), along with isocyanide derivatives, proteins with α-aminoisobutyric acid and a range of FAs (e.g. C8, C10, C10:1).

7 Humus

Another major component within the rhizosphere is humus. Humus is a complex, amorphous and colloidal substance of natural organic matter, a condensation of phenolics and nitrogen containing compounds derived from plant and animal tissue decay (Paul 2016; Sutton and Sposito 2005). Humic substances, which until recently (Lehmann and Kleber 2015) have been considered largely the products of microbial metabolism (Manlay et al. 2007), are recognised as promoting both microbial and plant growth (Coûteaux et al. 1995; Ponge 2013). A recent study examined the changes in the metabolic profile of maize seedling root exudates by ^{1}H NMR and GC–MS following co-inoculation with the PGPR diazotrophic β-proteobacterium *Herbaspirillum seropedicae* and humic acid (Lima et al. 2014). The classes of compounds detected in maize exudates included nitrogenous compounds, FAs, organic acids, steroids and terpenoid derivatives. Substantial changes in the exudation patterns 14 days post-inoculation were observed. For instance, root exudates from seedlings exclusively treated with humic acids demonstrated differing quantities of FAs, phenols and organic acids from that of the controls. Those seedlings treated singularly with *Herbaspirillum seropedicae* or in combination with humic acids exclusively exuded a diversity of heterocyclic, nitrogenous benzilamines and polyamines. The study showed that enhanced root colonisation of *Herbaspirillum seropedicae* in the presence of humic acids could be explained by the interplay between increased endophytic colonisation of *Herbaspirillum seropedicae* and sorption of humic acids to plant cell wall surfaces (Canellas et al. 2013). Interestingly, compounds identified as possible QS-inducing agents (e.g. substituted γ-butyrolactones and 3-hydroxypalmitic acid methyl ester) were also detected.

There is some controversy with the metabolomic characterisation of humus, with some researchers claiming that extraction methods are creating the larger "biomolecules" from smaller soil-sourced metabolites (Schmidt et al. 2011). For example, they argue that the well-known but difficult to characterise humics are actually synthesised from smaller soil metabolites such as carboxylates that condense due to the extraction process, as evidenced by the use of non-destructive analytical techniques, and in situ observations (e.g. near-edge X-ray fine structure spectroscopy combined with scanning transmission X-ray microscopy).

8 Terroir

In the viticulture industry, the word 'terroir' describes regional influences associated with climate, soil factors and plant genotype which strongly affect varietal flavours and aromas (Styger et al. 2011). Biochemically, the same term relates to modifications to the metabolome of a plant that impart a set of discrete desired qualities, such as increased plant growth or reduced feeding by larvae (e.g. cabbage "worm" caterpillar *Trichoplusia ni*) on plants (Badri et al. 2013b). The 'terroir effect' is shown to manifest itself in cultivated grape varieties that have been resident in the same location for extended time periods (van Leeuwen and Seguin 2006), and this concept has begun to be used to describe 'microbial terroir' (del Mónaco et al. 2016). It has been suggested that this effect could likely be attributed to soil microbes or the overall soil microbiome actively adjusting metabolite fluxes in response to signals from plants (Badri et al. 2013b). In one study, GC–MS was used to assess the effect of diverse soil microbiomes applied to the roots of *Arabidopsis thaliana* on the leaf metabolome and, by extension, whether the changes in leaf metabolome influenced the feeding behaviour of *Trichoplusia ni* larvae. The study demonstrated that variations in soil microbiome composition (primarily comprised >60 % actinobacteria and proteobacteria) produced a differential response to canopy and root biomass accumulation in *A. thaliana* plants. In comparison to control soil slurries (absence of soil microbiome), all *A. thaliana* plants showed an upregulation of amino acids, phenolics, sugars and sugar alcohols in leaf material. In addition, presented evidence suggested a strong capacity of soil microbial communities to either modulate above-ground feeding behaviour of *Trichoplusia ni* or to enhance the herbivory resistance of 4-week-old *A. thaliana* plants. It is surprising that in examining what makes a quality grape for wines that soil and the microbial life has until recently been largely ignored (Burns et al. 2015). The reason behind this is that soil microbial effects are subtle versus climate [(Pereira et al. 2006), Table 2], rainfall [(Lee et al. 2009), Table 2] and soil texture (Pereira et al. 2006) which have been found to be larger drivers of terroir.

9 Nutrients in the Rhizosphere

Nutrient availability is at the heart of producing a healthy crop or pasture, and changes to the macro- (e.g. N, P and K) and micronutrients (e.g. S, Mg, Ca) affect plants and soil, with soil microbes adjusting accordingly. As the soil environment changes due to season, moisture content, pH, soil texture, etc., nutrient fluxes in soil result in changes in metabolite concentrations as microbes adapt or suffer from these changes (Lorenz and Wackernagel 1994).

Phosphorus, due to its low solubility (≤ 10 μM), remains one of the most challenging soil nutrients for plants to acquire (Smith et al. 2011). For instance, organic P is thought to be made up of predominantly inositols, DNA, RNA and phospholipids (Nash et al. 2015). Often complexed to the minerals that make up the soil (e.g. iron and aluminium), they can be difficult for soil microbes to access (Jones 1998). One study examining organophosphorus metabolites used glucose-6-phosphate with labelled ^{33}P or ^{14}C atoms to identify how microbes are affected by deficiencies in nutrients (Heuck et al. 2015). The authors determined that soil microbes prefer utilisation of this metabolite as a C rather than P source with even complete uptake of the sugar resulting in excretion of P. The addition of the additional organophosphorus compound revealed an increase in microbial activity, which seemed to level off after 66 h, although this was not further examined to see if this was due to other nutrient deficiencies (i.e. N) or specific to fungi or bacteria. A metabolomic analyses might have also revealed the type of C containing compounds that were being taken up. Increase in P availability to the host plant is one reported benefit of mycorrhizal root colonisation, which plants do through the expenditure of energy and organic metabolites to organisms such as AMF (Smith and Smith 2015). The trade off to acquire P has been observed in tomato roots from AMF *Funneliformis mosseae* and *Rhizophagus irregularis* colonisation experiments [(Rivero et al. 2015), Table 4]. The authors observed that in acquiring P, the use of AMF led to significantly greater N in mycorrhizal roots ($p < 0.05$). However, shoot and root biomass, root/shoot ratio and total C were not significantly altered. The cost to the plant appeared to be the reduction in phenyl alcohols and vitamins, along with some amino acids (i.e. tryptophan, tyrosine, phenylalanine, alanine and leucine). However, increases in root concentration of intermediaries to amino acid (i.e. phenylalanine and tyrosine), sugar, carboxylic acids and fatty acid metabolites revealed the benefit of the interaction between plant and fungi, from the increased uptake of nutrients to the improved stress response of the colonised plants.

Nitrogen has been the major driver of increases in pasture and crop production since the 1950s. Due to the success of inorganic N used to increase plant biomass, our knowledge of organic N is surprisingly limited (Warren 2013, 2014). Characterising organic N metabolites that are produced by microbes is also a nascent field, and soil metabolomic data would certainly advance the knowledge base of this important nutrient. To date, characterisation of amino acids in soil has been the most studied group of metabolites that are associated with microbes, with fumigation of samples often used to determine microbial metabolites (Heuck et al. 2015; Swenson et al. 2015b; Warren 2015). However, there has been progress in identifying key metabolites that microbes use in the soil.

AMF are also important soil microbes that have been studied for understanding the movement of organic N. However, the study of N fluxes between plants and AMF has been limited compared to the current focus of understanding the transfer of C from plants (Hodge and Fitter 2010). Until recently, it was thought that AMF received most N from its host plant, with transport generally occurring through the metabolite arginine. However, it has been shown that AMF also seek out and promote decomposition of organic compounds for the acquisition of N, with a substantial concentration retained within arbuscular mycorrhizae structures for their own growth. As N limitation is reported to reduce the benefits of AMF symbioses (particularly with excess P) (Huang et al. 2014), a comparable study examined the effect of AMF field inoculation (either single or co-inoculated with PGPRs) on the root metabolome of durum wheat (*Triticum durum* Desf.) under N-limited, P-rich conditions [(Saia et al. 2015b), Table 4]. Metabolomics was used to determine how AMF or plant growth-promoting rhizobacteria (PGPR) affect wheat growth. Despite low N and high P soil conditions, plant growth doubled when adding AMF at the expense of amino acids and saturated fatty acid concentrations in roots. However, with the addition of PGPR, the acids were retained. Overall, 118 metabolites were identified and using the Kegg database (http://www.genome.jp/kegg/) metabolic pathways were delineated from 83 identified compounds. Multivariate analyses showed separation by treatment effects, with amines and unsaturated FA conserved in the AMF treatments. Organic acids correlated with AMF + PGPR treatments versus the control samples, which were also correlated with P-containing compounds, saturated FA, carbohydrates and amino acids. An interesting observation was the increase in xylitol, indicating a strong interaction between AMF and wheat.

A recent paper has shown how a multi-omics approach can lead to an improved understanding of fungi and plant interactions (Larsen et al. 2016). The experiments were designed to identify the signalling metabolites between aspen trees (*Populus tremuloides*) and the mycorrhizal fungi *Laccaria bicolour*. Combining transcriptomics, metabolomics and genomics, the authors were able to identify a number of biochemical processes related to communication between the plants and fungi. Metabolites identified from the fungi and correlated to plant gene regulation included amino acids and phosphor sugars related to biochemical pathways for the biosynthesis of aromatic compounds, plant hormones, plant metabolites including quinines and their precursor, and metabolites related to plant terpene biosynthesis. These metabolites were identified as being involved with fungal communication with the aspen to modulate cell adhesion, defence response and cell wall modification, presumably to facilitate the symbiosis between the two. A similar multi-omics approach has also been reported for novel compounds such as nanoparticles that can potentially exhibit unknown and potentially hazardous effects (MacCormack and Goss 2008).

Other examples of enhancement of nutrient availability and the metabolites involved include the following:

- the suppression of soil pathogens and production of auxins, peptides, ketones and terpenes (López-Bucio et al. 2015) by *Trichoderma* spp. protecting chick-peas (Rudresh et al. 2005) so as to increase P uptake; and,
- interactions between fungi and plants that express lipids and trehalose that result in microbe (e.g. *Glomus versiforme*) mediated exchange of nitrates, phosphates and amino acids [particularly arginine (Smith and Smith 2015)] from soil to plant (Bonfante and Genre 2010). This two-way interaction includes the release of metabolites such as strigolactones from plants into the rhizosphere.

Tree litter is another important source of nutrients on which soil microbes can feed. Research has shown that microbial communities adapt to the tree they are under (Ayres et al. 2009). A study of leaf litter revealed that different tree species litter, monotypic stands of trembling aspen (*Populus tremuloides*), lodgepole pine (*Pinus contorta*), and Engelmann spruce (*Picea engelmannii*), resulted in different populations of microbes (Wallenstein et al. 2010). A number of metabolites were identified that varied significantly between the tree species. Although the metabolites were only identified by retention time and mass, it was suspected with comparison to other studies that the metabolites were products of soil microbes or fungi. A followup study using pyrolysis molecular beam MS (py-MBMS) was able to determine that the metabolites were a mixture of lower mass (<137 m/z) carbohydrates, phenols and lignin monomers combined with higher mass (m/z = 252–706), lipids, alkanes, alkenes and FA (Wallenstein et al. 2013).

This research was expanded to analyses of typical oak, beech and grassland soils, and likewise identified that soil microbes were adapting to tree species differences, expressed through changes to a range of soil metabolites [(Liebeke et al. 2009), Table 2]. Grassland and beech forest soils were found to be less diverse than oak forest. This research contrasted that with improvements to soil quality when using soil that the microbes were collected from compared with artificial substrates, due to the variety of metabolites were available for each of the three groups of soil. Glutamic acid was the predominant amino acid (approximately 70 μM in concentration) in oak and in the top three for the other two soil types (although at lower concentration). Leucine and valine were also common to all three, and of the other amino acids, only the oak soil sample had sulphur containing methionine. Amino acids were attributed to degradation of proteins in the soil from microbial decomposition. The most concentrated sugar was trehalose, again being an order of magnitude greater in concentration in oak compared to the other soil samples. Other sugars and organic acids were identified with the total mass of these compounds revealing the richness of the oak soil metabolites (136 μM vs. 7 μM for grassland and beech), a similar ratio as found for the physicochemical analyses (e.g. soil organic carbon). Gram-positive and Gram-negative bacteria grew exponentially in the oak soils, and the types of soil metabolites seemed to indicate what biochemical processes were being used by the bacteria that were found.

10 Extracellular Enzymes

As the previous sections attest, there is no doubt that it can be difficult to identify microbial metabolites from the myriad of chemical compounds in soil (Ponomarova and Patil 2015). This becomes more problematic when sampling away from the microbial powerhouse of the rhizosphere. One example is extracellular enzymes (EE), enzymes released by microbes or plants into the soil to facilitate biochemical processes (Wallenstein and Weintraub 2008). These processes include degrading recalcitrant fractions of SOM for uptake by the microbes, secretion of metabolites to "sense" what predators or prey are in the immediate environment, and release of antibiotics to attack other microbes (Burns et al. 2013; González-Fernández et al. 2015; Wallenstein and Weintraub 2008). For example, EE from fungi produce metabolites such as the lactone-based botcinolides and the terpene-based botrydial compounds (González-Fernández et al. 2015). These metabolites are thought to be excreted to attack plants through decomposition of plant cell walls followed by nutrient acquisition from the plants. EEs from soil bacteria and fungi are known to consume carbon-rich biomolecules such as chitin (Roberts and Selitrennikoff 1988), lignin (Burns et al. 2013), tannins (Joanisse et al. 2007) and pectins (González-Fernández et al. 2015; Tepper and Anderson 1990). Sugars and amino acids from glycoproteins are also found on microbial adhesives (Wang et al. 2014) from EE-producing microorganisms after consumption of the plant and microbial debris. Plant litter is also a rich source of nutrients for EE-producing microorganisms. Monitoring of lignin decomposition has revealed the production of metabolites such as quinines and radical lipids that may potentially form humic compounds in soils with access to phenols, peptides and carbohydrates (Schmidt et al. 2011).

Quantitatively and qualitatively, there is still ambiguity in the sources and consumption of metabolites in soil. Whether the source is via a combination of EE excretion, plant decomposition or experimental error, metabolomics techniques may be amenable to characterising the macromolecules produced by EEs (e.g. polysomes). A potential solution described in the literature involves attaching coloured marker molecules to EEs in a dilute soil slurry, as the enzymes tend to stay fixed to the soil and unavailable for analyses through typical extraction techniques (Burns et al. 2013). Limitations of the current methodology include no knowledge of enzyme turnover rates, limited number of fluorescent markers available to attach to a limited number of functional groups and selection of only those enzymes capable of being stabilised in the slurry, versus in situ soil samples.

Metabolomics has been suggested as a way of removing these limitations by detecting the entire metabolome of a soil sample. Both pyrolysis-GC-IRMS and LC–MS analyses have provided examples of how this might work. For instance, a general survey of EE metabolites involved soil from the USA and Germany (Liebeke et al. 2009). Analyses of soil involved both untargeted metabolomics of GC–MS and ^{1}H-NMR of SOM, while targeting metabolites of microbe *Bacillus licheniformis* for comparison. A range of FAs, sugars, amines, amino acids and organic acids were identified. Most metabolites were species dependent though metabolites acetic, fumaric, aspartic and glutamic acid, glycine, proline, serine, glucose and sucrose were found ubiquitously across the varied agricultural types.

11 Anthropogenic Effects on Soil Microbes

It is argued that we are in the Anthropocene age (Bundy et al. 2009; Desai et al. 2010; Rockström et al. 2009), so it is not surprising that soil microbes have had to adapt to human endeavours, some of which can be deleterious to the environment. Estimates made in 1998 include up to 100 000 chemicals which were available for purchase (Rockström et al. 2009), many of which will end up in the environment. Since then, material synthesis and engineering has advanced greatly, and so it should be no surprise that many of these compounds end up affecting the soil microbial food chain [(Simpson and McKelvie 2009), Table 2]. Indictor animals such as earthworms are often used to determine soil health (Rochfort et al. 2009; Whitfield et al. 2013), but there is an increasing amount of research into the effects of synthetic chemicals on soil microbes (Hernandez-Soriano and Jimenez-Lopez 2014). The use of microbes to monitor or remediate contaminated sites is one of particular interest (Desai et al. 2010; Jones et al. 2014). A soil metabolome would allow for the ability to look across a range of biogeochemical factors that may affect soil health. The discussion here will focus solely on metabolomics research as it relates to microbes.

Agricultural changes of soil from pristine forest to farm land has been a major change to the environment that has been occurring for thousands of years (Foley et al. 2005). Changes occurring in soil management have been described through the lens of metabolomics (Singh 2006). For instance, land use provided an opportunity to use NMR metabolomics combined with mid-infrared spectroscopy (MIR) to identify the effect of land management across four different regions of Victoria, Australia (Rochfort et al. 2015). Comparing soil samples from relatively untouched native land and adjacent farm land of oats or wheat, ^{1}H-NMR was able to identify a series of signals from lipid, terpene and sugar. The study identified that these metabolites could be differentiated by NMR due to different concentrations depending on land use, whereas soil location was differentiated using MIR. This is similar to a ^{1}H-NMR study of various mine sites across England (Jones et al. 2014) which identified a similar series of metabolites, with differences between compounds thought to have occurred due to a different solvent extraction system (methanol for this study versus deuterium oxide). In both cases, the assumed microbial source of these metabolites was explored, using a target microbe, *Bacillus subtilis*, for the Australian study and the identification that most soil metabolites were microbially based for the English analysis. Labelling techniques may give enhanced information, as was shown in the effects of microbes feeding on mine waste. Metabolite confirmation of mine waste and its effects on microbes was conducted using stable isotope labelling (Mosier et al. 2013). Using this method with ^{15}N labelling allowed for the identification of 80 metabolites from 3500 metabolite features that included artefacts, non-biological metabolites, adducts, etc., the latter of which were considered predominantly microbial.

How soil microbes are able to adapt to new environments and unusual metabolites has found use in regions away from their original habitat, as for example in recovering petroleum oil (Arora et al. 2014). For example, the use of mutant bacteria known to be resistant to metalloids, *Pseudomonas pseudoalcaligene*, was used to remove polychlorinated biphenyls (Tremaroli et al. 2009). To identify the mode of action, metabolomic studies identified that thiols were oxidised when the microbe reacted to the metal. ^{1}H-NMR combined with multivariate analyses showed that wild-type and mutant bacteria resulted in changes to concentrations of amino acids (i.e. glutamate, aspartate, glycine, histidine, tryptophan and tyrosine), betaine and NAD+. *Pseudomonas pseudoalcaligene* was found to be resistant to other toxic compounds, including caffeine, sulphates, streptomycin and chlorinated compounds.

11.1 Engineered Nanomaterials (ENM)

Novel compounds, often of the size or smaller than the microbes themselves (e.g. nanometre), are increasingly being developed, used and disposed of, on soils. The design and use of engineered nanomaterials (ENM) is in part because they have unusual and useful properties (e.g. strength and conductivity) that are different from the bulk compound found in nature (Dinesh et al. 2012). ENM, so called to distinguish them from natural soil nanoparticles (e.g. colloids), are receiving increasing attention in agricultural and environmental literature. As production of ENM increases to meet the demand of high-tech materials, these products are more likely to end up in soil. Products with ENM include sunscreens, cleaning products and therapeutic goods. Some ENM, such as zero-valent Fe, are widely used in parts of the world for cleaning up toxic chemicals (Lee et al. 2008). Until recently, the effect of ENM on the environment had not been systematically studied, in part due to the cost of manufacture, the similarity in size and composition of natural colloids (Klaine et al. 2008), as well as the physicochemical aspects of the soil (MacCormack and Goss 2008). Various studies have shown that ENM are potentially hazardous in a laboratory setting, but there are few field studies (Dinesh et al. 2012; Johansen et al. 2008). To date, ENM have been predicted to be in the environment (i.e. surface waters) in concentrations of 0.8 ng/L for carbon-based ENM up to 10 μg/L for Ag-, Ti- and Zn-based compounds (Maurer-Jones et al. 2013). No such study has been conducted for concentrations of ENM in soil.

Adding ENM to soil to see how they affect the biosphere has been the main method of determining their effects [(Jin et al. 2014; Johansen et al. 2008; Shah and Belozerova 2009), Table 1]. This has enabled studies of how the dosage of ENM affects the soil environment. As ENM are in a solid matrix, it can be difficult to ascertain the dose actually received by microbes, and so studies may overestimate their soil concentration. Although it is well recognised that they have the potential to cause pollution in soil through accumulation, ENM have particular properties that can make it difficult to determine their interactions with denizens of the soil matrix, including microbes. This includes aggregation into larger particles and adsorption to minerals within the soil. It is inevitable that nanoparticles are found in soil due to accidental release as they have already been detected in marine and airborne environments.

This lack of knowledge means that these compounds are an unknown threat to the soil and its microbial community. A recent review of 10,000 papers on ENM found that despite the explosion of research on human health, the consensus on their toxicity is at best, weak and often misleading (Krug 2014). As to be expected, research on ENM and their effects on soil and its inhabitants are even less clear. Beyond health and environmental aspects, ENM are also being explored as intermediaries and markers in microbe communication to help researchers identify disease rates in soil (MacCormack and Goss 2008).

Carbon fullerenes and nanotubes are arguably the most well-known ENM. They are also increasingly finding their way into the environment, and into soil (Berry et al. 2016). However, studies to date find limited interactions with soil microbes (Pettibone and Louie 2015). Two papers by researchers in South Korea highlighted how microbes are affected by carbon nanotubes in their environment (Jin et al. 2014; Jin et al. 2013). The metabolic profile of the microbes, revealed through phospholipid fatty acid analyses (PLFA), showed that microbial FA typically changed abundance depending on whether the soil were treated with powdered SWCNT (single-walled carbon nanotubes) or solvent suspended SWCNT. Generally, as the concentration of SWCNT increased, the biomarker fatty acid (i.e. odd chained and/or hydroxyl grouped and/or cyclic unsaturated) metabolite concentration for Gram-positive bacteria, Gram-negative bacteria and fungi significantly decreased. This effect on microbes occurred for at least 25 days. Increases in other types of FA (e.g. iso-branched) led the authors to believe that microbes change their lipid composition to defend against SWCNT. Another study of bacteria and protozoa and how they are affected by C60 fullerene (Johansen et al. 2008) noted that systematic biases may be an issue leading to difficulty in comparing results with other research. In particular, using pure solvents versus soil samples meant that other chemical characteristics of fullerenes are not taken into account when the effects on microbes are analysed. However, as was found in studies with cleaner conditions, fast-growing microbes suffered significantly upon addition of C60 to soil ($p = 0.004$–0.033). The bacteria eventually recovered from fullerene exposure. Suggested reasons for this included absorption of C60 to soil or other particles that coated the ENM and minimised contact with microbes.

Metal ENM have a longer history than carbon ENM, and have been used, sometimes unknowingly, since ancient times [i.e. gold nanoparticles for decoration of ancient Roman sculpture (http://phys.org/news/2013-08-goblet-ancient-romans-nanotechnology.html)]. Despite their more extensive history, research into their interactions with microbes in the soil is limited. Like other ENM, metal ENM analyses are complicated by the number of naturally occurring nanoparticles or similar minerals already present in the soil. It has been suggested that labelling metal ENM with unusual isotopes could improve the analyses of these particles (Klaine et al. 2008). Most research of metal ENM has shown that they act as bactericides (Lee et al. 2008). As these compounds have been used for water treatment and fabric formation, some of the pollution may end up in soil (Stefaniuk et al. 2016). Measuring microbial responses to metal ENM have usually focussed on their toxicity through the use of fatty acid methyl esters (FAMEs) analysis. For instance, a study on microbial response to Pd, Cu, Si and Au ENM used FAMEs typical to microbes to determine how microbes survived in their presence [(Sasser 2006; Shah and Belozerova 2009), Table 1]. Despite using two comparative concentrations of metal ENM in soil over 15 days (0.013 and 0.066 % w/w), no significant change in these microbial biomarkers metabolites was seen. However, it was speculated that ENM may exhibit an indirect effect through interaction of compounds required by microbes. This was shown in a concurrent experiment where the growth of lettuce seeds was reduced in the presence of Pd and Au ENMs. One positive review discussed how microbes could be used to create metal ENM for commercial use, based on microbes being able to create metal organic complexes such as iron oxide magnets, metal phosphate medicines and transition metal catalysts (Lloyd et al. 2008).

11.2 Heavy Metal Contamination

Microbes such as AMF can be affected by heavy metal contamination (Karimi et al. 2011), and it has been shown that the metals reduce microbial biomass. Metal wastes reaching soils are another environmental issue (Maurer-Jones et al. 2013; Simpson and McKelvie 2009). Similarity in atomic size to nutrients, some heavy metals can access microbes through channels designed to diffuse and transport cations (Singh et al. 2016). This similarity has been put to use with Cu used as a fungicide that has been shown to improve crop yield (Dhawi et al. 2015). As metals cannot be degraded, microbial action tends to involve immobilising the contamination so that its toxic effects are mitigated (Azcón et al. 2013). The success of microbes and plants in resisting radiation effects (Stone 2009) and reducing concentrations of toxic metals has been identified in places of recent disasters such as the Chernobyl and Fukishima nuclear accidents (Aung et al. 2015; Geras'kin et al. 2008). It has been noted that continuous exposure to heavy metals can result in tolerance forming in microbes as they employ metabolic strategies to reduce metal toxicity though reduction of the metal's oxidation state or coordination of metabolites to soluble metals ions (Jones 1998; Karimi et al. 2011). For instance, the microbe *Klebsiella mobilis* CIAM 880 was able to release metabolites that bind with Cd to promote plant growth in soil with high cadmium concentrations, precipitating the metal (Nies and Silver 1995; Pishchik et al. 2002). This was seen in the plant having a larger root system and increased exudate concentrations released into the soil.

The compound that probably typifies pollution since the industrial era is lead (Nriagu and Pacyna 1988). One metabolomics study identified metabolites that changed in concentration due to mycorrhizal microbe influence attenuating the plants response to Pb contamination (Souza et al. 2014). The analyses identified a series of amino acids that may complex heavy metal and remove the metal's ability to affect the plants. The amino acids include asparagine, histidine, proline and glutamine. Using the AMF *Glomus etunicatum*, it was possible to identify potential microbial interactions with plants (e.g. N and carbohydrate metabolism) that offer protection from Pb. Pb was also identified as a factor in metabolomics analyses of a series of mines in the UK [see the "Community Metabolomics" chapter by Jones et al. (2014) this book]. Using ^{1}H-NMR, metabolomics was used across 11 sites that mined Pb and Zn (Jones et al. 2014). A simple methodology with minimal extraction meant that, along with the microbial community, soil and invertebrates were also sampled. The authors examined specific groups of metabolites (nucleotides, sugars, lactate and amino acids) using multivariate statistics. They were able to show that there were sites that had similar metabolic profiles and lower Fe concentrations even though the mines were otherwise different in geochemical makeup. The authors proposed that monitoring metabolomics could act as an early warning to hazardous pollutant levels before any visible effects were seen. PLFA was used in the Czech Republic to determine how mining waste materials fly ash and mine digestate that contained high Pb, Si and Zn concentrations affected soil microbes (Garcia-Sánchez et al. 2015). Organic compounds' concentrations, including phenolics, were identified as increasing under both contaminants, with the researchers identifying this via increased microbial activity, particularly fungi, under the digestate treatment (9.3 ± 1.4 μg/kg after 60 days, $p \leq 0.05$). Other compounds found due to contamination included carbohydrates, carboxylic and amino acids, amines and polymers, although the researchers cautioned that the metabolomics and genomics analyses could not differentiate between active and passive microbes in the soil. A timed study showed that after two months, amine compounds had become preferentially consumed. Digestate contaminated soil was shown to result in a preference for microbes to consume carboxylic acids taken from the SOM. The PLFA analyses found that metabolites associated with fungi were most attenuated by the digestate and fly ash. Metabolites associated with Gram-negative bacteria were also affected. The digestate increased the concentration of Gram-positive bacteria after two months. Fly ash was found to be beneficial to soil community structure, with the metabolites associated with most types of microbes increasing. Biofilms in mines have also proven to be a rich source of metabolomic data on how microbes are affected by various metal wastes that included sulphate, iron, zinc, copper and arsenic (Mosier et al. 2013).

Cadmium contamination is also a concern in soil as it affects the life cycles of most species [(Sarry et al. 2006), Table 2]. Cd has been found to affect people through kidney damage after consumption of contaminated food, with the suspicion that the metal's affinity to thiols is deleterious to organisms. This includes yeasts and fungi that are found in a variety of forms in agriculture (Slávikova and Vadkertiová 2003). Yeast under the effects of cadmium was studied using metabolomic and proteomic techniques (Lafaye et al. 2005). The research identified that the typical sulphur pathway in yeast for producing glutathione increased in rate and the sulphur amino acid concentration reduced by 30 % as the microbe attempted to remove cadmium. Proteome and metabolomic results were correlated when Cd was used, but other treatments (e.g. sulphur starvation) that also result in increased glutathione production were not, indicating an independent pathway is initiated when Cd contamination is present.

A controlled study of increasing Fe concentrations was used to determine if the microbe *Pseudomonas stutzeri* RCH2 was affected by metal-poor or metal-rich soils (Swenson et al. 2015a). The idea was that Fe will affect the microbe's metabolite output as the competition for sites on either the metal or carbon of SOM in soil is changed. As expected, increasing the Fe concentration increased the sorption of all metabolites including those with phosphate, N-containing and carboxylate functional groups. The research identified that concentration change in metabolites was correlated to the charge of the anion (e.g. $Phosphate^{-} - Fe^{+}$) when it came to sorption to Fe. The authors felt that it was important to conduct metabolomic analyses to ascertain the rates of sorption due to a typical mix of microbial metabolites, which may differ due to competing interactions compared to when separate metabolites being tested. While the analyses did not bring many surprising results (i.e. phosphates and dicarboxylates adsorb strongest to Fe), this paper is a rare example of the application of metabolomics to understand the holistic system of soil–microbe interactions.

11.3 Organic Contaminants

Research on organic compound contamination has been presented that offers not just problems but solutions using metabolomics of soil microbes. For instance, metabolomics was used to determine how microbes might be able to recover petroleum from reservoirs that are not cost effective by traditional extraction techniques (Arora et al. 2014). This study used microbial enhanced oil recovery of an oil well on soil samples from India to determine how efficient the microbes were and what decomposition products they produced during the extraction. Using indigenous hyperthermophilic *Clostridium* sp., they tested how these microbes may extract oil at high temperature sites (>91 °C). The use of water from the oil well site ensured no contamination of microbes from elsewhere in the soil. Metabolites collected included biosurfactants, organic acids, solvents, exopolysaccharides and volatile FAs. Following a targeted analysis of the metabolite mix from different groups of bacteria, the researchers were able to optimise conditions to increase the concentration of metabolites that would be suitable for extraction, including sugars (particularly sucrose), nitrates and ammonium metabolites (particularly urea). Metabolomics has also been used to better understand the bioremediation of soils or soil models (Singh 2006). NMR studies have generally used soil samples rather than solutions, although there are metabolomic studies using liquid-state NMR to monitor microbial degradation of xenobiotics. For instance, in one study on how *Mycobacterium* biodegrades morpholine, piperidine and thiomorpholine, scientists were able to use enriched carbon and nitrogen compounds to identify the metabolites formed as the bacterium consumed the antibiotics [(Delort and Combourieu 2001), Table 2].

Persistent organic pollutants (POP) are another source of contamination that can affect all species, including soil microbes (Wania and Mackay 1996). These are small compounds with aromatic rings and potentially halogenated functional groups. While it has been thought that condensation of these compounds into soil is a better alternative to POP being airborne, the effect on organisms is still a problem. Polycyclic aromatic hydrocarbons (PAH) are one target of active research, particularly on contaminated sites such as oil fields and fires (Seo et al. 2009). One study identified a series of metabolites isolated from the microbe *Sinorhizobium* sp. C4 that was extracted from soils contaminated with another PAH, phenanthrene (Keum et al. 2008). After determining that the mode of degradation was ring opening, the authors used untargeted analyses to monitor polar metabolites such as FAs and polyhydroxyalkanoates as the bacteria were fed phenanthrene. This was detrimental to the microbe, with more than 70 % of these metabolites decreasing in concentration after being fed phenanthrene. POP eradication via soil microbes is another area of active research. Polychlorinated biphenyls (PCBs) have had a 90 % removal rate over one month when placed in the rhizosphere of Arabidopsis (Narasimhan et al. 2003). The expression of phenylpropanoids increased by over 100-fold compared to control samples which the authors linked to the breakdown of PCBs. It was proposed that by adding to the soil, 10–100 fold the current concentration of the microbe *Pseudomonas* spp., the current estimates of PCBs in soil [328 000 tonnes (est. 1988)] could be significantly removed. A consensus is that soil microbes in the rhizosphere have adapted to using PCBs as a source of carbon and energy, utilising exudates to degrade them, through the use of metabolites such as biotin, thiamine, amino acids and isoflavanoids (Jha et al. 2015). The two major metabolic processes identified from this research were anaerobic dechlorination and aerobic biodegradation.

Sometimes, as in the case of pesticides and antibiotics, chemicals are deliberately applied to soil. How soil microbes are affected is often overlooked. Pesticides have a long history in agriculture and so their effects on non-target organisms have been overshadowed by the benefit to plant yield and productivity (Imfeld and Vuilleumier 2012). Antibiotics in agriculture, either by accident or design, are becoming a real concern due the resistance of many bacteria that infect humans (Di Marco et al. 2014), even when their primary use is intended to promote animal growth (Horrigan et al. 2002).

Pesticide reduction through improved efficiency of soil microbes such as AMF has been proposed (Baum et al. 2015) with *Trichoderma* spp. providing an excellent example of how this can be achieved (Vinale et al. 2008). It should be noted that studies have shown that the replacement of conventional pesticides with "biopesticides" must be a gradual process as mycorrhizae concentrations have decreased while synthetic fertilisers were applied (Imfeld and Vuilleumier 2012;

Ruzicka et al. 2012). Metabolomics can monitor how biopesticides function by observing concentrations of carbohydrates, lipids and n-acetylglucosamine, among others (Baum et al. 2015). Other secondary metabolites from plants that also reduce the need for pesticides include phenolics, alkaloids and coumarins. For instance, phenolics have been shown to reduce weed growth (Khanh et al. 2005). Another example is the effect of 2,4-dichlorophenoxyacetic acid (2,4-D) on *Escherichia coli* (Bhat et al. 2015). Previous conflicting research identified that there were significant disturbances to soil microbe communities. Using GC–MS for the analysis [combined with analyses using scanning electron microscopy (SEM) and atomic force microscopy (AFM)] the authors were able to determine a specific pathway attributed to 2,4-D exposure. This involved a combination of attenuation of oxidative phosphorylation, ABC transporters, peptidoglycan biosynthesis, glutathione metabolism and purine/pyrimidine metabolism, and an increase in amino acid, protein, sugar and starch metabolism. There was also a significant reduction in the concentration of metabolites associated with membranes and cell walls.

In regards to antibiotics, the decreasing effectiveness of current medicines can be balanced with the discovery of new antibiotics in soil created by microbial action, as in the example of teixobactin (Ling et al. 2015). For instance, one compound that is rapidly losing effectiveness as a medicine, tetracycline, has been examined and found to affect the rhizosphere leading to a loss of exuded metabolites (e.g. phenols, flavonoids) by up to 48 % (Di Marco et al. 2014). Similar research looking at maize and its interactions with AMF found up to sixfold increases in carbohydrates, amino acids and phenolics in the soil when the antibiotics Cefotaxime and Trimethoprim were added, showing how deleterious antibiotics can be (Azaizeh et al. 1995). Research detailing how microbes and plants can themselves act as antibiotics is demonstrated by a recent study showing that *Bacillus subtilis* was inhibited when placed in soil samples from wheat farms and untouched forest (Rochfort et al. 2015). Correlating data between the lipids, terpenes and sugars in the soil was able to be matched with the antibiotic effects against the microbe. Metabolites such as fulvic acids, isochromantoxins, organic acids and xanthocillins have been identified in a wide ranging study of microbes including *Aspergilloides*, *Furcatum* and *Penicillium* (Frisvad et al. 2004). It should be noted that due to technological limitations, the metabolites of these well-known microbes may be misidentified.

11.4 Climatic Change Effects

Potentially, the greatest anthropogenic effect of the current century is how the planet responds to an increasingly variable climatic pattern of weather (Edenhofer et al. 2014). Metabolomics has shown promise as one way to quickly quantitate metabolites that microbes exude in response to climate stressors (Simpson et al. 2012). Expected changes in weather and climate have led to the call and use of metabolomics techniques to understand how the denizens of the land will adapt (Ahuja et al. 2010). Climatic effects on soil microbes are expected as changes in temperature along with water and nutrient concentrations become more variable (Rennenberg et al. 2009).

Salinity is one climate change issue that is already common. Salinity is increasing and the replenishment of water tables is difficult. Exacerbating this problem is research showing that replenishment of water tables is more difficult as temperature rises (Lozupone and Knight 2007). Metabolomics approaches have predominantly examined plants, but microbial interactions with plants due to salinity have also been examined. Studies of a variety of plants (i.e. *Arabidopsis thaliana*, *Lotus japonicus* and *Oryza sativa)* seem to suggest metabolites used to communicate between microbes and plants in the rhizosphere will be significantly affected (Sanchez et al. 2008). These metabolites include organic and amino acids, along with sugars. This may be mitigated depending on the plant species, as was found in the case of the highly salt-tolerant Brassicaceae, *Thellungiella salsuginea*, where there appeared to be little change regardless of salinity levels [(Lugan et al. 2010), Table 3].

One issue with salinity is that the high salt concentration in these soils can interfere with the sample, leading to loss of data through ion suppression (Oikawa et al. 2011). To address this issue, the authors used capillary electrophoresis–mass spectrometry (CE–MS) in combination with solid-phase extraction (SPE) to selectively remove up to 17 cations from soil solution. They were also able to differentiate between plant and microbial cations. Optimising the method with a model soil solution of 78 organic and 12 inorganic compounds, the method was then used on soil solutions from rice farms. Significant differences between soil with and without rice were obtained, with microbial metabolites such as histamine and tyrosine present only in the absence of plants, and leucine, isoleucine, phenylalanine and serine significantly more concentrated without plants.

Climate change will probably result in increased periods of drought, and irrigation is one method to ensure crop survivability. This results in a dry–wet event that can lead to various stresses to both plant and microbial life (Kakumanu et al. 2013). In particular, a change in osmotic potential between the soil and intracellular contents of microorganism can result in reduced life expectancy of the microbial biomass. Besides K^+, which acts as a regulator of ionic strength in cells, amino acids, carbohydrates, quaternary amines and tetrahydropyrimidine are regulated and maintained within microbial cells when the amount of water available is reduced. Fungi have been found to preferentially accumulate polyols while bacteria have a preference for amino acids and sugars. Other factors that may improve drought resistance are a robust AMF that will increase N uptake to plants (Baum et al. 2015),

At the end of drought when rains have come, the sudden influx of fresh water generally results in a flush of microbial activity (Kakumanu et al. 2013). It has still not been determined whether this is due to the lysis of microbes, a change in equilibrium of microbes as the concentration gradient changes, or other reasons. The authors reasoned that they could quantify metabolites that are released when the concentration gradient changes. It was found that the accumulation of metabolites appeared more related to keeping energy-rich compounds at hand than to regulate osmotic pressure. Metabolite composition of sugars and polyols in drought and non-drought prone areas revealed that microbial lysis was unlikely to have occurred when soils were dry. The authors speculated that metabolite accumulation during droughts had more to do with energy and nutrient conservation, and production of survival metabolites, such as exopolysaccharides.

12 Tools for Microbial Soil Metabolomics

When looking for microbial metabolites, it is often necessary to utilise as many methods as possible to identify the source of metabolites. Attempts at a broad understanding of microbes in soil have also been elucidated through model communities (Ahmed et al. 2014; Ziegler et al. 2013) or plants (Ahmed et al. 2014). However, extrapolating laboratory results to the field can be risky and increasing the types of data collected can be useful. Correlations with other parameters such as genomic data or physicochemical parameters can be helpful in understanding why metabolites are present (Larsen et al. 2016; Rochfort et al. 2015). Recent combined analyses have shown that a single analytical technique can give an incomplete picture of microbial fluxes (Larsen et al. 2016).

One common enabling technology compatible with metabolomics methods is stable isotope labelling, which has been shown to identify a variety of soil functions of soil microbes, from rhizosphere signalling molecules to nutrient-associated metabolites [(Gunina et al. 2014; Haichar et al. 2012; Watzinger 2015), Tables 3 and 4]. Probably, the most common method to identify metabolites of microbial origin is to feed labelled substrates which produce digested products that are easy to identify (Heuck et al. 2015). Sugars are an important energy source for microbes, and this has led to metabolomic studies to determine how sugars are taken up by soil microbes [(Apostel et al. 2015), Table 3]. Using ^{13}C-labelled glucose and ribose with PLFA to identify microbe type, a loamy soil in Germany was dosed to determine sugar uptake by the soil. Initial decomposition of the sugars occurred within 3 days. Glucose concentration then decreased by 50 % between 3 and 10 days, while ribose remained relatively constant. The position of the carbon on the sugars was important, with the majority of carbon being incorporated from glucose C-2 (approximately 90 %) and ribose C-5 (approximately 70 %). PLFA identified the majority of detected microbes taking up sugars were Gram-negative bacteria while Gram-positive bacteria and actinomycetes incorporated the greatest concentration of labelled sugars into their PLFA (0.2–0.4 %). Other microbes found to take up labelled sugars included protozoa, VA-mycorrhiza, anaerobes and fungi. Gram-negative bacteria were also found to take up the greatest concentrations of labelled carbon, while Gram-positive bacteria appeared to prefer sugars from older SOM. A similar study sought to measure the use of plant versus soil organic carbon by microbes in maize, wheat and rye farms, also in Germany (Kramer and Gleixner 2006). Following the rate and mechanism of these metabolites allowed the researchers to determine that glycolysis and the pentose phosphate pathway were parallel processes and characteristic to the soils studied. Initially, both glucose and ribose were consumed at similar rates. After 10 days, concentration decreases in the C-5 labelled ribose showing the preference by microbes to incorporate this metabolite over glucose, presumably due to the formation of RNA and DNA. Data for other labelled sugars showed a complex series of reactions occurring in the soil that include glycolysis, pentose phosphate and gluconeogenesis metabolic pathways. PLFA data revealed a preference for the ^{13}C substrates to be taken up by Gram-negative bacteria, known to be the dominant species around the rhizosphere, while methanotrophs were found to consume non-plant material, despite its proximity.

The flux of a series of amino acids was determined using labelled compounds (Gunina et al. 2014). Compared to fungi, sugars were more efficiently taken up by bacteria, especially glucose and sucrose, if the concentrations of these metabolites were low in the soil (Gunina et al. 2014). Ribose uptake was similar to glucose for Gram-negative bacteria. The pentose phosphate pathway for these bacteria was also characterised by the uptake of xylose. In contrast, fungi appeared to have a preference for larger, more complex sugars along with decomposition of glucose to form triacylglycerols. Acetate, a common metabolite in soils due to it being sourced in plant litter and cattle slurry, was found to be incorporated preferentially into Gram-negative bacteria.

Using these labelled metabolites helped the authors to identify that Gram-negative bacteria were the most efficient at using low molecular weight metabolites, due to the preference of this type of bacteria to use the anabolic pentose phosphate pathway. Fungi and filamentous microorganisms were found to be better utilisers of acidic and complex organic compounds like palmitate and double-bonded FA. Carbon-13 labelling allowed for comparison between microbial and soil metabolites in a series of experiments investigating SOM accessibility of nutrients (Swenson et al. 2015b). Comparing samples of soil that had been fumigated versus untouched soil samples, it was possible to identify the labelled metabolites of the microbe *Pseudomonas stutzeri* RCH2. Extracellular metabolites were found to be more likely to be detected than intracellular ones, indicating that additional steps such as adding salts to reduce osmosis may not be required. As minimal processing was a goal of this research, water was used as an extractant, a solution the authors point out is not ideal for the study of metabolites such as FA or sterols. Despite this, the labelled metabolites identified included amino acids and analogues, nucleobases such as uracil, and a series of organic acids.

13 Conclusion

Soil is a rich, complex ground for metabolomics research. The majority of microbes within soil are still a mystery, but metabolomics is beginning to reveal their secrets. Even with the advances in understanding how soil microbes interact with plants

(Mendes et al. 2011), significant challenges remain in characterising the spatial separation and metabolic compartmentalization of differing metabolic pathways and metabolites to their respective source—plant, microbe or systemic metabolic response. Metabolomics techniques will help to holistically understanding how humans have changed the soil environment; hopefully we will learn how to sustain not just the soil but the environment that microbes interact with, whether it be in the ground (Holland 2004), waterways (Lozupone and Knight 2007) or sky (Conrad 1996). As advances in research occur, many of the mysterious interactions between biology, geology and chemistry will become apparent.

The knowledge acquired from the interactions between soil microbes will then allow for specific manipulation of soil from improving crop yield to soil remediation (Mosa et al. 2016). Soil microbes are potential vectors that could be manipulated and eventually synthesised to help with soil management. As humanity comes to master concepts like rhizoengineering, metabolomic techniques may be needed to monitor the efficiency of designed microbial consortiums, as they are used to inoculate soil for improved soil health and productivity (Jia et al. 2016; Zhang et al. 2015). While by no means the only technique to understand microbial life (Desai et al. 2010), metabolomics has the advantage of being able to cast a wide net over the biological, chemical and geological interactions occurring in soil.

References

Abhilash PC, Powell JR, Singh HB, Singh BK (2012) Plant-microbe interactions: novel applications for exploitation in multipurpose remediation technologies. Trends Biotechnol 30:416–420. doi:10.1016/j.tibtech.2012.04.004

Abu-Nada Y, Kushalappa AC, Marshall WD, Al-Mughrabi K, Murphy A (2007) Temporal dynamics of pathogenesis-related metabolites and their plausible pathways of induction in potato leaves following inoculation with *Phytophthora infestans*. Eur J Plant Pathol 118:375–391. doi:10.1007/s10658-007-9150-8

Aguilar R, Quesada M, Ashworth L, Herrerias-Diego Y, Lobo J (2008) Genetic consequences of habitat fragmentation in plant populations: susceptible signals in plant traits and methodological approaches. Mol Ecol 17:5177–5188. doi:10.1111/j.1365-294X.2008.03971.x

Ahemad M, Kibret M (2014) Mechanisms and applications of plant growth promoting rhizobacteria: current perspective. J King Saud Univ Sci 26:1–20. doi:10.1016/j.jksus.2013.05.001

Ahmed MA, Kroener E, Holz M, Zarebanadkouki M, Carminati A (2014) Mucilage exudation facilitates root water uptake in dry soils. Funct Plant Biol 41:1129–1137. doi:10.1071/FP13330

Ahuja I, de Vos RCH, Bones AM, Hall RD (2010) Plant molecular stress responses face climate change. Trends Plant Sci 15:664–674. doi:10.1016/j.tplants.2010.08.002

Alivisatos AP, Blaser MJ, Brodie EL, Chun M, Dangl JL, Donohue TJ, Dorrestein PC, Gilbert JA, Green JL, Jansson JK, Knight R, Maxon ME, McFall-Ngai MJ, Miller JF, Pollard KS, Ruby EG, Taha SA (2015) A unified initiative to harness Earth's microbiomes. Science 350:507–508. doi:10.1126/science.aac8480

Allwood JW, Goodacre R (2010) An introduction to liquid chromatography-mass spectrometry instrumentation applied in plant metabolomic analyses. Phytochem Anal 21:33–47. doi:10.1002/pca.1187

Apostel C, Dippold M, Kuzyakov Y (2015) Biochemistry of hexose and pentose transformations in soil analyzed by position-specific labeling and 13C-PLFA. Soil Biol Biochem 80:199–208. doi:10.1016/j.soilbio.2014.09.005

Arora P, Ranade DR, Dhakephalkar PK (2014) Development of a microbial process for the recovery of petroleum oil from depleted reservoirs at 91–96°C. Bioresour Technol 165:274–278. doi:10.1016/j.biortech.2014.03.109

Arutchelvi JI, Bhaduri S, Uppara PV, Doble M (2008) Mannosylerythritol lipids: a review. J Ind Microbiol Biotechnol 35:1559–1570. doi:10.1007/s10295-008-0460-4

Aung HP, Djedidi S, Oo AZ, Aye YS, Yokoyama T, Suzuki S, Sekimoto H, Bellingrath-Kimura SD (2015) Growth and 137Cs uptake of four *Brassica* species influenced by inoculation with a plant growth-promoting rhizobacterium *Bacillus pumilus* in three contaminated farmlands in Fukushima prefecture, Japan. Sci Total Environ 521–522:261–269. doi:10.1016/j.scitotenv.2015.03.109

Ayres E, Steltzer H, Simmons BL, Simpson RT, Steinweg JM, Wallenstein MD, Mellor N, Parton WJ, Moore JC, Wall DH (2009) Home-field advantage accelerates leaf litter decomposition in forests. Soil Biol Biochem 41:606–610. doi:10.1016/j.soilbio.2008.12.022

Azaizeh HA, Marschner H, Römheld V, Wittenmayer L (1995) Effects of a vesicular-arbuscular mycorrhizal fungus and other soil microorganisms on growth, mineral nutrient acquisition and root exudation of soil-grown maize plants. Mycorrhiza 5:321–327. doi:10.1007/bf00207404

Azcón R, Medina A, Aroca R, Ruiz-Lozano JM (2013) Abiotic stress remediation by the arbuscular mycorrhizal symbiosis and rhizosphere bacteria/yeast interactions. In: de Bruijn FJ (ed) Molecular microbial ecology of the rhizosphere, vol 1 & 2. Wiley, Hoboken

Badri DV, Chaparro JM, Zhang R, Shen Q, Vivanco JM (2013a) Application of natural blends of phytochemicals derived from the root exudates of *arabidopsis* to the soil reveal that phenolic-related compounds predominantly modulate the soil microbiome. J Biol Chem 288:4502–4512. doi:10.1074/jbc.M112.433300

Badri DV, Zolla G, Bakker MG, Manter DK, Vivanco JM (2013b) Potential impact of soil microbiomes on the leaf metabolome and on herbivore feeding behavior. New Phytol 198 :264–273. doi:10.1111/nph.12124

Bais HP, Broeckling CD, Vivanco JM (2008) Root exudates modulate plant–microbe interactions in the rhizosphere. In: Karlovsky P (ed) secondary metabolites in soil ecology soil biology 14. Springer, Berlin

Bais HP, Park SW, Weir TL, Callaway RM, Vivanco JM (2004) How plants communicate using the underground information superhighway. Trends Plant Sci 9:26–32. doi:10.1016/j.tplants.2003.11.008

Bais HP, Tiffany L, Weir T, Perry LG, Gilroy S, Vivanco JM (2006) The role of root exudates in the rhizosphere interactions with plants and other organisms. Annu Rev Plant Biol 57:233–266

Balestrini R, Lumini E, Borriello R, Bianciotto V (2015) Plant-soil biota interactions. In: Paul E (ed) Soil microbiology, ecology and biochemistry, 4th edn. Academic Press, London

Barea J-M, Pozo MJ, Azcon R, Azcon-Aguilar C (2005) Microbial co-operation in the rhizosphere. J Exp Bot 56:1761–1778

Bauer WD, Mathesius U (2004) Plant responses to bacterial quorum sensing signals. Curr Opin Plant Biol 7:429–433

Baum C, El-Tohamy W, Gruda N (2015) Increasing the productivity and product quality of vegetable crops using arbuscular mycorrhizal fungi: a review. Sci Hortic 187:131–141. doi:10.1016/j.scienta.2015.03.002

Berendsen RL, Pieterse CMJ, Bakker PAHM (2012) The rhizosphere microbiome and plant health. Trends Plant Sci 17:478–486. doi:10.1016/j.tplants.2012.04.001

Berry TD, Clavijo AP, Zhao Y, Jafvert CT, Turco RF, Filley TR (2016) Soil microbial response to photo-degraded C60 fullerenes. Environ Pollut 211:338–345. doi:10.1016/j.envpol.2015.12.025

Bertin C, Yang X, Weston LA (2003) The role of root exudates and allelochemicals in the rhizosphere. Plant Soil 256:67–83. doi:10.1023/a:1026290508166

Bertram HC, Weisbjerg MR, Jensen CS, Pedersen MG, Didion T, Petersen BO, Duus JØ, Larsen MK, Nielsen JH (2010) Seasonal changes in the metabolic fingerprint of 21 grass and legume cultivars studied by nuclear magnetic resonance-based metabolomics. J Agric Food Chem 58:4336–4341. doi:10.1021/jf904321p

Bhat SV, Booth SC, Vantomme EAN, Afroj S, Yost CK, Dahms TES (2015) Oxidative stress and metabolic perturbations in *Escherichia coli* exposed to sublethal levels of 2,4-dichlorophenoxyacetic acid. Chemosphere 135:453–461. doi:10.1016/j.chemosphere.2014.12.035

Bommarco R, Kleijn D, Potts SG (2013) Ecological intensification: harnessing ecosystem services for food security. Trends Ecol Evol 28:230–238. doi:10.1016/j.tree.2012.10.012

Bonanomi G, Vinale F, Scala F (2009) The role of natural products in plant-microbe interactions. In: Osbourn EA, Lanzotti V (eds) Plant-derived natural products: synthesis, function, and application. Springer, New York

Bonfante P, Genre A (2010) Mechanisms underlying beneficial plant–fungus interactions in mycorrhizal symbiosis. Nat Commun 1:48

Borisjuk L, Rolletschek H, Neuberger T (2012) Surveying the plant's world by magnetic resonance imaging. Plant J 70:129–146. doi:10.1111/j.1365-313X.2012.04927.x

Boyce KC (2005) The evolutionary history of roots and leaves. In: Holbrook NM, Zwieniecki MA (eds) Vascular transport in plants. Elsevier, San Diego

Braeken K, Daniels R, Ndayizeye M, Vanderleyden J, Michiels J (2008) Quorum sensing in bacteria-plant interactions. In: Nautiyal CS, Dion P (eds) Molecular mechanisms of plant and microbe coexistence. Springer, Berlin

Brewin NJ (2010) Root nodules (legume–rhizobium symbiosis). eLS, Nov 2010

Broeckling CD, Broz AK, Bergelson J, Manter DK, Vivanco JM (2008) Root exudates regulate soil fungal community composition and diversity. Appl Environ Microbiol 74:738–744. doi:10.1128/aem.02188-07

Bronick CJ, Lal R (2005) Soil structure and management: a review. Geoderma 124:3–22. doi:10.1016/j.geoderma.2004.03.005

Bundy JG, Davey MP, Viant MR (2009) Environmental metabolomics: a critical review and future perspectives. Metabolomics 5:3–21. doi:10.1007/s11306-008-0152-0

Burns KN, Kluepfel DA, Strauss SL, Bokulich NA, Cantu D, Steenwerth KL (2015) Vineyard soil bacterial diversity and composition revealed by 16S rRNA genes: differentiation by geographic features. Soil Biol Biochem 91:232–247. doi:10.1016/j.soilbio.2015.09.002

Burns RG, DeForest JL, Marxsen J, Sinsabaugh RL, Stromberger ME, Wallenstein MD, Weintraub MN, Zoppini A (2013) Soil enzymes in a changing environment: current knowledge and future directions. Soil Biol Biochem 58:216–234. doi:10.1016/j.soilbio.2012.11.009

Canellas LP, Balmori DM, Médici LO, Aguiar NO, Campostrini E, Rosa RCC, Façanha AR, Olivares FL (2013) A combination of humic substances and *Herbaspirillum seropedicae* inoculation enhances the growth of maize (Zea mays L.). Plant Soil 366:119–132

Cappellari LDR, Santoro MV, Nievas F, Giordano W, Banchio E (2013) Increase of secondary metabolite content in marigold by inoculation with plant growth-promoting rhizobacteria. Appl Soil Ecol 70:16–22

Carvalhais LC, Dennis PG, Fan B, Fedoseyenko D, Kierul K, Becker A, von Wiren N, Borriss R (2013) Linking plant nutritional status to plant-microbe interactions. PLoS ONE 8:e68555

Carvalhais LC, Dennis PG, Fedoseyenko D, Hajirezaei MR, Borriss R, von Wirén N (2011) Root exudation of sugars, amino acids, and organic acids by maize as affected by nitrogen, phosphorus, potassium, and iron deficiency. J Plant Nutr Soil Sci 174:3–11. doi:10.1002/jpln.201000085

Cassman KG (1999) Ecological intensification of cereal production systems: yield potential, soil quality, and precision agriculture. Proc Natl Acad Sci 96:5952–5959. doi:10.1073/pnas.96.11.5952

Cesco S, Mimmo T, Tonon G, Tomasi N, Pinton R, Terzano R, Neumann G, Weisskopf L, Renella G, Landi L, Nannipieri P (2012) Plant-borne flavonoids released into the rhizosphere: impact on soil bio-activities related to plant nutrition. A review. Biol Fertil Soils 48:123–149. doi:10.1007/s00374-011-0653-2

Chamam A, Sanguin H, Bellvert F, Meiffren G, Comte G, Wisniewski-Dye F, Bertrand C, Prigent-Combaret C (2013) Plant secondary metabolite profiling evidences strain-dependent effect in the *Azospirillum-Oryza sativa* association. Phytochemistry 87:65–77

Chan KG, Atkinson S, Mathee K, Sam CK, Chhabra SR, Cámara M, Koh CL, Williams P (2011) Characterization of N-acyl-homoserine lactone-degrading bacteria associated with the *Zingiber officinale* (ginger) rhizosphere: co-existence of quorum quenching and quorum sensing in *Acinetobacter* and *Burkholderia*. BMC Microbiol 11:1–14. doi:10.1186/1471-2180-11-51

Chaparro JM, Badri DV, Bakker MG, Sugiyama A, Manter DK, Vivanco JM (2013) Root exudation of phytochemicals in *Arabidopsis* follows specific patterns that are developmentally programmed and correlate with soil microbial functions. PLoS ONE 8:e55731

Cheynier V, Comte G, Davies KM, Lattanzio V, Martens S (2013) Plant phenolics: recent advances on their biosynthesis, genetics, and ecophysiology. Plant Physiol Biochem 72:1–20. doi:10.1016/j.plaphy.2013.05.009

Clair SB, Lynch JP (2010) The opening of pandora's box: climate change impacts on soil fertility and crop nutrition in developing countries. Plant Soil 335:101–115

Coats VC, Rumpho ME (2014) The rhizosphere microbiota of plant invaders: an overview of recent advances in the microbiomics of invasive plants. Front Microbiol 5:368

Conrad R (1996) Soil microorganisms as controllers of atmospheric trace gases (H2, CO, CH4, OCS, N2O, and NO). Microbiol Rev 60:609–640

Contreras-Cornejo HA, Macías-Rodríguez L, Beltrán-Peña E, Herrera-Estrella A, López-Bucio J (2011) *Trichoderma*-induced plant immunity likely involves both hormonal and camalexin dependent mechanisms in *Arabidopsis thaliana* and confers resistance against necrotrophic fungi Botrytis cinerea. Plant Signaling Behav 6:1554–1563

Contreras-Cornejo HA, Ortiz-Castro R, Lopez-Bucio J (2013) Promotion of plant growth and the induction of systemic defence by *Trichoderma*: physiology, genetics and gene expression. In: Mukherjee PK, Horwitz BA, Singh US, Mukherjee M, Schmoll M (eds) *Trichoderma*: biology and applications. CABI, Oxfordshire

Cooper JE (2007) Early interactions between legumes and rhizobia: disclosing complexity in a molecular dialogue. J Appl Microbiol 103:1355–1365. doi:10.1111/j.1365-2672.2007.03366.x

Coûteaux MM, Bottner P, Berg B (1995) Litter decomposition, climate and liter quality. Trends Ecol Evol 10:63–66. doi:10.1016/S0169-5347(00)88978-8

Crowther TW, Boddy L, Jones TH (2012) Functional and ecological consequences of saprotrophic fungus–grazer interactions. ISME J 6:1992–2001

Dakora FD, Phillips DA (2002) Root exudates as mediators of mineral acquisition in low-nutrient environments. Plant Soil 245:35–47. doi:10.1023/a:1020809400075

Das K, Mukherjee AK (2007) Crude petroleum-oil biodegradation efficiency of *Bacillus subtilis* and *Pseudomonas aeruginosa* strains isolated from a petroleum-oil contaminated soil from north-east India. Bioresour Technol 98:1339–1345. doi:10.1016/j.biortech.2006.05.032

De-la-Pena C, Loyola-Vergas VM (2014) Biotic interactions in the rhizosphere: a diverse cooperative enterprise for plant productivity. Plant Physiol 166:701–719

del Mónaco SM, Rodríguez ME, Lopes CA (2016) *Pichia kudriavzevii* as a representative yeast of north patagonian winemaking terroir. Int J Food Microbiol 230:31–39. doi:10.1016/j.ijfoodmicro.2016.04.017

Delort AM, Combourieu B (2001) In situ 1H NMR study of the biodegradation of xenobiotics: application to heterocyclic compounds. J Ind Microbiol Biotech 26:2–8. doi:10.1038/sj.jim.7000022

Dennis PG, Miller AJ, Hirsch PR (2010) Are root exudates more important than other sources of rhizodeposits in structuring rhizosphere bacterial communities? FEMS Microbiol Ecol 72:313–327. doi:10.1111/j.1574-6941.2010.00860.x

Desai C, Pathak H, Madamwar D (2010) Advances in molecular and "-omics" technologies to gauge microbial communities and bioremediation at xenobiotic/anthropogen contaminated sites. Bioresour Technol 101:1558–1569. doi:10.1016/j.biortech.2009.10.080

Dhawi F, Datta R, Ramakrishna W (2015) Mycorrhiza and PGPB modulate maize biomass, nutrient uptake and metabolic pathways in maize grown in mining-impacted soil. Plant Physiol Biochem 97:390–399. doi:10.1016/j.plaphy.2015.10.028

Di Marco G, Gismondi A, Canuti L, Scimeca M, Volpe A, Canini A (2014) Tetracycline accumulates in *Iberis sempervirens L.* through apoplastic transport inducing oxidative stress and growth inhibition. Plant Biol 16:792–800. doi:10.1111/plb.12102

Dinesh R, Anandaraj M, Srinivasan V, Hamza S (2012) Engineered nanoparticles in the soil and their potential implications to microbial activity. Geoderma 173–174:19–27. doi:10.1016/j.geoderma.2011.12.018

Edenhofer OR, Pichs-Madruga Y, Sokona E, Farahani S, Kadner K, Seyboth A, Adler I, Baum S, Brunner P, Eickemeier B, Kriemann J, Savolainen S, Schlömer, C von Stechow, Zwickel T (eds) IPCC, 2014: climate change 2014: mitigation of climate change. Contribution of working group III to the fifth assessment report of the Intergovernmental Panel on Climate Change, Cambridge, UK and New York, NY, USA

Ewert F, Rounsevell MDA, Reginster I, Metzger MJ, Leemans R (2005) Future scenarios of European agricultural land use: I. estimating changes in crop productivity. Agric Ecosyst Environ 107:101–116. doi:10.1016/j.agee.2004.12.003

Foley JA, DeFries R, Asner GP, Barford C, Bonan G, Carpenter SR, Chapin FS, Coe MT, Daily GC, Gibbs HK, Helkowski JH, Holloway T, Howard EA, Kucharik CJ, Monfreda C, Patz JA, Prentice IC, Ramankutty N, Snyder PK (2005) Global consequences of land use. Science 309:570–574. doi:10.1126/science.1111772

Food and Agriculture Organization (FAO) (2002) World agriculture towards 2015/2030: Summary report. In: Food and Agriculture Organization of the United Nations (ed). Food and Agriculture Organization (FAO), Rome

Food and Agriculture Organization of the United Nations (2013) World agriculture towards 2030/2050. In: Alexandratos N, Bruinsma J (ed) ESA Working paper No 12–03. Rome, Italy

Frisvad JC, Smedsgaard J, Larsen TO, Samson RA (2004) Mycotoxins, drugs and other extrolites produced by species in *Penicillium* subgenus *Penicillium*. Stud Mycol 2004:201–241

Gamir J, Pastor V, Kaever A, Cerezo M, Flors V (2014) Targeting novel chemical and constitutive primed metabolites against *Plectosphaerella cucumerina*. Plant J 78:227–240. doi:10.1111/tpj.12465

Garcia-Sánchez M, Garcia-Romera I, Cajthaml T, Tlustoš P, Száková J (2015) Changes in soil microbial community functionality and structure in a metal-polluted site: the effect of digestate and fly ash applications. J Environ Manage 162:63–73. doi:10.1016/j.jenvman.2015.07.042

Geras'kin SA, Fesenko SV, Alexakhin RM (2008) Effects of non-human species irradiation after the chernobyl NPP accident. Environ Int 34:880–897. doi:10.1016/j.envint.2007.12.012

González-Fernández R, Valero-Galván J, Gómez-Gálvez FJ, Jorrín-Novo JV (2015) Unraveling the in vitro secretome of the phytopathogen *Botrytis cinerea* to understand the interaction with its hosts. Front Plant Sci 6. doi:10.3389/fpls.2015.00839

Gunina A, Dippold MA, Glaser B, Kuzyakov Y (2014) Fate of low molecular weight organic substances in an arable soil: from microbial uptake to utilisation and stabilisation. Soil Biol Biochem 77:304–313. doi:10.1016/j.soilbio.2014.06.029

Habib MT, Heller T, Polle A (2013) Molecular physiology of tree ectomycorrhizal interactions. In: Eshel A, Beckman T (eds) Plant roots: the hidden half, 4th edn. CRC Press, Boca Raton

FEZ Haichar, Roncato MA, Achouak W (2012) Stable isotope probing of bacterial community structure and gene expression in the rhizosphere of *Arabidopsis thaliana*. FEMS Microbiol Ecol 81:291–302. doi:10.1111/j.1574-6941.2012.01345.x

Hartmann A, Schmid M, Dv Tuinen, Berg G (2008) Plant-driven selection of microbes. Plant Soil 321:235–257. doi:10.1007/s11104-008-9814-y

Hernandez-Soriano MC, Jimenez-Lopez JC (2014) Metabolomics for soil contamination assessment

Heuck C, Weig A, Spohn M (2015) Soil microbial biomass C:N: P stoichiometry and microbial use of organic phosphorus. Soil Biol Biochem 85:119–129. doi:10.1016/j.soilbio.2015.02.029

Hodge A, Fitter AH (2010) Substantial nitrogen acquisition by arbuscular mycorrhizal fungi from organic material has implications for N cycling. Proc Nat Acad Sci USA 107:13754–13759

Holland JM (2004) The environmental consequences of adopting conservation tillage in Europe: reviewing the evidence. Agric Ecosyst Environ 103:1–25. doi:10.1016/j.agee.2003.12.018

Horrigan L, Lawrence RS, Walker P (2002) How sustainable agriculture can address the environmental and human health harms of industrial agriculture. Environ Health Perspect 110:445–456

Huang X-F, Chaparro JM, Reardon KF, Zhang R, Shen Q, Vivanco JM (2014) Rhizosphere interactions: root exudates, microbes, and microbial communities. Botany 92:267–275

Imfeld G, Vuilleumier S (2012) Measuring the effects of pesticides on bacterial communities in soil: a critical review. Eur J Soil Biology 49:22–30. doi:10.1016/j.ejsobi.2011.11.010

Itoh Y, Kodama K, Furuya K, Takahashi S, Haneishi T, Takiguchi Y, Arai M (1980) A new sesquiterpene antibiotic, heptelidic acid producing organisms, fermentation, isolation and characterization. J Antibiot 33:468–473

Jha P, Panwar J, Jha PN (2015) Secondary plant metabolites and root exudates: guiding tools for polychlorinated biphenyl biodegradation. Int J Environ Sci Technol 12:789–802. doi:10.1007/s13762-014-0515-1

Jia X, Liu C, Song H, Ding M, Du J, Ma Q, Yuan Y (2016) Design, analysis and application of synthetic microbial consortia. Synth Syst Biotechnol 1:109–117. doi:10.1016/j.synbio.2016.02.001

Jin L, Son Y, DeForest JL, Kang YJ, Kim W, Chung H (2014) Single-walled carbon nanotubes alter soil microbial community composition. Sci Total Environ 466–467:533–538. doi:10.1016/j.scitotenv.2013.07.035

Jin L, Son Y, Yoon TK, Kang YJ, Kim W, Chung H (2013) High concentrations of single-walled carbon nanotubes lower soil enzyme activity and microbial biomass. Ecotoxicol Environ Saf 88:9–15. doi:10.1016/j.ecoenv.2012.10.031

Joanisse GD, Bradley RL, Preston CM, Munson AD (2007) Soil enzyme inhibition by condensed litter tannins may drive ecosystem structure and processes: the case of *Kalmia angustifolia*. New Phytol 175:535–546. doi:10.1111/j.1469-8137.2007.02113.x

Johansen A, Pedersen AL, Jensen KA, Karlson U, Hansen BM, Scott-Fordsmand JJ, Winding A (2008) Effects of C60 fullerene nanoparticles on soil bacteria and protozoans. Environ Toxicol Chem 27:1895–1903. doi:10.1897/07-375.1

Jones DL (1998) Organic acids in the rhizosphere—a critical review. Plant Soil 205:25–44. doi:10.1023/a:1004356007312

Jones OAH, Sdepanian S, Lofts S, Svendsen C, Spurgeon DJ, Maguire ML, Griffin JL (2014) Metabolomic analysis of soil communities can be used for pollution assessment. Environ Toxicol Chem 33:61–64. doi:10.1002/etc.2418

Jonkers W, Rodriguez Estrada AE, Lee K, Breakspear A, May G, Kistler HC (2012) Metabolome and transcriptome of the Interaction between *Ustilago maydis* and *Fusarium verticillioides* In Vitro. Appl Environ Microbiol 78:3656–3667. doi:10.1128/aem.07841-11

Kabouw P, van Dam NM, van der Putten WH, Biere A (2012) How genetic modification of roots affects rhizosphere processes and plant performance. J Exp Bot 63:3475–3483. doi:10.1093/jxb/err399

Kakumanu ML, Cantrell CL, Williams MA (2013) Microbial community response to varying magnitudes of desiccation in soil: a test of the osmolyte accumulation hypothesis. Soil Biol Biochem 57:644–653. doi:10.1016/j.soilbio.2012.08.014

Karimi A, Khodaverdiloo H, Sepehri M, Sadaghiani MR (2011) Arbuscular mycorrhizal fungi and heavy metal contaminated soils. Afr J Micro Res 5:1571–1576

Kawaguchi M, Minamisawa K (2010) Plant-microbe communications for symbiosis. Plant Cell Physiol 51:1377–1380
Kaye JP, Hart SC (1997) Competition for nitrogen between plants and soil microorganisms. Trends Ecol Evol 12:139–143
Kenrick P (2013) The origin of roots. In: Eshel A, Beckman T (eds) Plant roots: the hidden half, 4th edn. CRC Press, Boca Raton
Keum YS, Seo JS, Li QX, Kim JH (2008) Comparative metabolomic analysis of *Sinorhizobium* sp. C4 during the degradation of phenanthrene. Appl Microbiol Biotechnol 80:863–872. doi:10.1007/s00253-008-1581-4
Khanh TD, Chung MI, Xuan TD, Tawata S (2005) The exploitation of crop allelopathy in sustainable agricultural production. J Agron Crop Sci 191:172–184. doi:10.1111/j.1439-037X.2005.00172.x
Kirk JL, Beaudette LA, Hart M, Moutoglis P, Klironomos JN, Lee H, Trevors JT (2004) Methods of studying soil microbial diversity. J Microbiol Methods 58:169–188. doi:10.1016/j.mimet.2004.04.006
Klaine SJ, Alvarez PJJ, Batley GE, Fernandes TF, Handy RD, Lyon DY, Mahendra S, McLaughlin MJ, Lead JR (2008) Nanomaterials in the environment: Behavior, fate, bioavailability, and effects. Environ Toxicol Chem 27:1825–1851. doi:10.1897/08-090.1
Kramer C, Gleixner G (2006) Variable use of plant-and soil-derived carbon by microorganisms in agricultural soils. Soil Biol Biochem 38:3267–3278. doi:10.1016/j.soilbio.2006.04.006
Kramer PJ, Boyer JS (1995) Water relations of plants and soil. Academic Press, New York
Krug HF (2014) Nanosafety research—are we on the right track? Angew Chem Int Ed 53:12304–12319. doi:10.1002/anie.201403367
Kusano M, Fukushima A, Redestig H, Saito K (2011) Metabolomic approaches toward understanding nitrogen metabolism in plants. J Exp Bot 62:1439–1453. doi:10.1093/jxb/erq417
Kuzyakov Y, Xu X (2013) Competition between roots and microorganisms for nitrogen: mechanisms and ecological relevance. New Phytol 198:656–669
Lafaye A, Junot C, Pereira Y, Lagniel G, Tabet JC, Ezan E, Labarre J (2005) Combined proteome and metabolite-profiling analyses reveal surprising insights into yeast sulfur metabolism. J Biol Chem 280:24723–24730. doi:10.1074/jbc.M502285200
Lakshmanan V, Selvaraj G, Bais HP (2014) Functional soil microbiome: belowground solutions to an aboveground problem. Plant Physiol 166:689–700
Larsen PE, Sreedasyam A, Trivedi G, Shinde SV, Dai Y, Cseke LJ, Collart FR (2016) Multi-omics approach identifies molecular mechanisms of plant-fungus mycorrhizal interaction. Front Plant Sci 6. doi:10.3389/fpls.2015.01061
Lee C, Kim JY, Lee WI, Nelson KL, Yoon J, Sedlak DL (2008) Bactericidal effect of zero-valent iron nanoparticles on *Escherichia coli*. Environ Sci Technol 42:4927–4933. doi:10.1021/es800408u
Lee J-E, Hwang G-S, Van Den Berg F, Lee C-H, Hong Y-S (2009) Evidence of vintage effects on grape wines using 1H NMR-based metabolomic study. Anal Chim Acta 648:71–76. doi:10.1016/j.aca.2009.06.039
Lehmann J, Kleber M (2015) The contentious nature of soil organic matter. Nature 528:60–68. doi:10.1038/nature16069
Li Q-G, Zhang L, Li C, Dunwell JM, Zhang Y-M (2013) Comparative genomics suggests that an ancestral polyploidy event leads to enhanced root nodule symbiosis in the papilionoideae. Mol Biol Evol 30:2602–2611. doi:10.1093/molbev/mst152
Liebeke M, Brözel V, Hecker M, Lalk M (2009) Chemical characterization of soil extract as growth media for the ecophysiological study of bacteria. Appl Microbiol Biotechnol 83:161–173. doi:10.1007/s00253-009-1965-0
Lima LDS, Olivares FL, de Oliveira RR, Vega MRG, Aguiar NO, Canellas LP (2014) Root exudate profiling of maize seedlings inoculated with *Herbaspirillum seropedicae* and humic acids. Chem Biol Technol Agric 1:1–18

Lin CY, Viant MR, Tjeerdema RS (2006) Metabolomics: methodologies and applications in the environmental sciences. J Pestic Sci 31:245–251. doi:10.1584/jpestics.31.245

Ling LL, Schneider T, Peoples AJ, Spoering AL, Engels I, Conlon BP, Mueller A, Schaberle TF, Hughes DE, Epstein S, Jones M, Lazarides L, Steadman VA, Cohen DR, Felix CR, Fetterman KA, Millett WP, Nitti AG, Zullo AM, Chen C, Lewis K (2015) A new antibiotic kills pathogens without detectable resistance. Nature 517:455–459. doi:10.1038/nature14098

Lloyd JR, Pearce CI, Coker VS, Pattrick RAD, Van Der Laan G, Cutting R, Vaughan DJ, Paterson-Beedle M, Mikheenko IP, Yong P, Macaskie LE (2008) Biomineralization: linking the fossil record to the production of high value functional materials. Geobiology 6:285–297. doi:10.1111/j.1472-4669.2008.00162.x

Loh J, Carlson RW, York WS, Stacey G (2002) Bradyoxetin, a unique chemical signal involved in symbiotic gene regulation. Proc Nat Aca Sci 99:14446–14451. doi:10.1073/pnas.222336799

López-Bucio J, Pelagio-Flores R, Herrera-Estrella A (2015) Trichoderma as biostimulant: exploiting the multilevel properties of a plant beneficial fungus. Sci Hortic 196:109–123. doi:10.1016/j.scienta.2015.08.043

Lorenz MG, Wackernagel W (1994) Bacterial gene transfer by natural genetic transformation in the environment. Microbiol Rev 58:563–602

Lorito M, Woo SL, Harman Gary E, Monte E (2010) Translational research on trichoderma: from 'omics to the field. Annu Rev Phytopathol 48:395–417. doi:10.1146/annurev-phyto-073009-114314

Lozupone CA, Knight R (2007) Global patterns in bacterial diversity. Proc Nat Aca Sci 104:11436–11440. doi:10.1073/pnas.0611525104

Lugan R, Niogret M-F, Leport L, Guégan J-P, Larher FR, Savouré A, Kopka J, Bouchereau A (2010) Metabolome and water homeostasis analysis of Thellungiella salsuginea suggests that dehydration tolerance is a key response to osmotic stress in this halophyte. Plant J 64:215–229. doi:10.1111/j.1365-313X.2010.04323.x

Lumbsch HT, Leavitt SD (2011) Goodbye morphology? a paradigm shift in the delimitation of species in lichenized fungi. Fungal Divers 50:59–72. doi:10.1007/s13225-011-0123-z

Luo ZB, Janz D, Jiang XN, Gobel C, Wildhagen H, Tan Y, Rennenberg H, Feussner I, Polle A (2009) Upgrading root physiology for stress tolerance by ectomycorrhizas: insights from metabolite and transcriptional profiling into reprogramming for stress anticipation. Plant Physiol 151:1902–1917

Lynch JP (2007) Roots of the second green revolution. Aust J Bot 55:493–512. doi:10.1071/BT06118

MacCormack TJ, Goss GG (2008) Identifying and predicting biological risks associated with manufactured nanoparticles in aquatic ecosystems. J Ind Ecol 12:286–296. doi:10.1111/j.1530-9290.2008.00041.x

Mandal SM, Chakraborty D, Dey S (2010) Phenolic acids act as signaling molecules in plant-microbe symbioses. Plant Signal Behav 5:359–368. doi:10.4161/psb.5.4.10871

Manlay RJ, Feller C, Swift MJ (2007) Historical evolution of soil organic matter concepts and their relationships with the fertility and sustainability of cropping systems. Agric Ecosyst Environ 119:217–233. doi:10.1016/j.agee.2006.07.011

Maron P-A, Mougel C, Ranjard L (2011) Soil microbial diversity: methodological strategy, spatial overview and functional interest. C R Biol 334:403–411. doi:10.1016/j.crvi.2010.12.003

Marshall TJ, Holmes JW (1980) Soil physics. Cambridge University Press, Cambridge

Mastouri F, Björkman T, Harman GE (2010) Seed treatment with *Trichoderma harzianum* alleviates biotic, abiotic, and physiological stresses in germinating seeds and seedlings. Phytopathology 100:1213–1221

Matson PA, Parton WJ, Power AG, Swift MJ (1997) Agricultural Intensification and Ecosystem Properties. Science 277:504–509. doi:10.1126/science.277.5325.504

Maurer-Jones MA, Gunsolus IL, Murphy CJ, Haynes CL (2013) Toxicity of engineered nanoparticles in the environment. Anal Chem 85:3036–3049. doi:10.1021/ac303636s

McBride MB (1994) Environmental chemistry of soils. Oxford university press, Oxford

Mendes R, Garbeva P, Raaijmakers JM (2013) The rhizosphere microbiome: significance of plant beneficial, plant pathogenic, and human pathogenic microorganisms. FEMS Microbiol Rev 37:634–663. doi:10.1111/1574-6976.12028

Mendes R, Kruijt M, de Bruijn I, Dekkers E, van der Voort M, Schneider JHM, Piceno YM, DeSantis TZ, Andersen GL, Bakker PAHM, Raaijmakers JM (2011) Deciphering the rhizosphere microbiome for disease-suppressive bacteria. Science 332:1097–1100. doi:10.1126/science.1203980

Minz D, Ofek M, Yitzhak H (2013) Plant rhizosphere microbial communities. In: Rosenberg E, DeLong EF, Lory S, Stackebrandt E, Thompson F (eds) The prokaryotes: prokaryotic communities and ecophysiology. Springer, Berlin

Morandi D, Bailey JA, Gianinazzi-Pearson V (1984) Isoflavonoid accumulation in soybean roots infected with vesicular-arbuscular mycorrhizal fungi. Physiolog Plant Pathol 24:357–364. doi:10.1016/0048-4059(84)90009-2

Mosa KA, Saadoun I, Kumar K, Helmy M, Dhankher OP (2016) Potential biotechnological strategies for the cleanup of heavy metals and metalloids. Front Plant Sci 7. doi:10.3389/fpls.2016.00303

Mosier AC, Justice NB, Bowen BP, Baran R, Thomas BC, Northen TR, Banfield JF (2013) Metabolites associated with adaptation of microorganisms to an acidophilic, metal-rich environment identified by stable-isotope-enabled metabolomics. MBio 4. doi:10.1128/mBio.00484-12

Narasimhan K, Basheer C, Bajic VB, Swarup S (2003) Enhancement of plant-microbe interactions using a rhizosphere metabolomics-driven approach and its application in the removal of polychlorinated biphenyls. Plant Physiol 132:146–153. doi:10.1104/pp.102.016295

Nash DM, Watkins M, Heaven MW, Hannah M, Robertson F, McDowell R (2015) Effects of cultivation on soil and soil water under different fertiliser regimes. Soil Tillage Res 145:37–46. doi:10.1016/j.still.2014.08.006

Nies DH, Silver S (1995) Ion efflux systems involved in bacterial metal resistances. J Ind Microbiol 14:186–199. doi:10.1007/bf01569902

Nihorimbere V, Ongena M, Smargiassi M, Thonart P (2011) Beneficial effect of the rhizosphere microbial community for plant growth and health. Biotechnol Agron Soc Environ 15:327–337

Nriagu JO, Pacyna JM (1988) Quantitative assessment of worldwide contamination of air, water and soils by trace metals. Nature 333:134–139

Oikawa A, Fujita N, Horie R, Saito K, Tawaraya K (2011) Solid-phase extraction for metabolomic analysis of high-salinity samples by capillary electrophoresis–mass spectrometry. J Sep Sci 34:1063–1068. doi:10.1002/jssc.201000890

Padmavathiamma PK, Li LY (2007) Phytoremediation technology: hyper-accumulation metals in plants. Water Air Soil Pollut 184:105–126. doi:10.1007/s11270-007-9401-5

Pang Q, Zhang A, Zang W, Wei L, Yan X (2016) Integrated proteomics and metabolomics for dissecting the mechanism of global responses to salt and alkali stress in Suaeda corniculata. Plant Soil:1–16. doi:10.1007/s11104-015-2774-0

Park S, Seo YS, Hegeman AD (2014) Plant metabolomics for plant chemical responses to belowground community change by climate change. J Plant Biol 57:137–149. doi:10.1007/s12374-014-0110-5

Parniske M (2008) Arbuscular mycorrhiza: the mother of plant root endosymbioses. Nat Rev Microbiol 6:763–775

Paul EA (2016) The nature and dynamics of soil organic matter: plant inputs, microbial transformations, and organic matter stabilization. Soil Biol Biochem 98:109–126. doi:10.1016/j.soilbio.2016.04.001

Pereira GE, Gaudillere J-P, Cv Leeuwen, Hilbert G, Maucourt M, Deborde C, Moing A, Rolin D (2006) 1H NMR metabolite fingerprints of grape berry: comparison of vintage and soil effects in Bordeaux grapevine growing areas. Anal Chim Acta 563:346–352. doi:10.1016/j.aca.2005.11.007

Pettibone JM, Louie SM (2015) Research highlights: detecting, characterizing and quantifying the presence and impact of carbon nanomaterials in environmental systems. Environ Sci Nano 2:308–311. doi:10.1039/c5en90014j

Pishchik VN, Vorobyev NI, Chernyaeva II, Timofeeva SV, Kozhemyakov AP, Alexeev YV, Lukin SM (2002) Experimental and mathematical simulation of plant growth promoting rhizobacteria and plant interaction under cadmium stress. Plant Soil 243:173–186. doi:10.1023/a:1019941525758

Ponge J-F (2013) Plant–soil feedbacks mediated by humus forms: a review. Soil Biol Biochem 57:1048–1060. doi:10.1016/j.soilbio.2012.07.019

Ponomarova O, Patil KR (2015) Metabolic interactions in microbial communities: untangling the Gordian knot. Curr Opin Microbiol 27:37–44. doi:10.1016/j.mib.2015.06.014

Rabie GH (1998) Induction of fungal disease resistance in Vicia faba by dual inoculation with *Rhizobium leguminosarum* and vesicular-arbuscular mycorrhizal fungi. Mycopathologia 141:159–166. doi:10.1023/a:1006937821777

Rasmussen S, Parsons AJ, Jones CS (2012) Metabolomics of forage plants: a review. Ann Bot 110:1281–1290. doi:10.1093/aob/mcs023

Rastogi G, Sani R (2011) Molecular techniques to assess microbial community structure, function, and dynamics in the environment. In: Ahmad I, Ahmad F, Pichtel J (eds) Microbes and microbial technology. Springer, New York

Rawat L, Singh Y, Shukla N, Kumar J (2011) Alleviation of the adverse effects of salinity stress in wheat (*Triticum aestivum L.*) by seed biopriming with salinity tolerant isolates of Trichoderma harzianum. Plant Soil 347:387–400. doi:10.1007/s11104-011-0858-z

Redmond JW, Batley M, Djordjevic MA, Innes RW, Kuempel PL, Rolfe BG (1986) Flavones induce expression of nodulation genes in Rhizobium. Nature 323:632–635

Rennenberg H, Dannenmann M, Gessler A, Kreuzwieser J, Simon J, Papen H (2009) Nitrogen balance in forest soils: nutritional limitation of plants under climate change stresses. Plant Biol 11:4–23. doi:10.1111/j.1438-8677.2009.00241.x

Reynolds HL, Packer A, Bever JD, Clay K (2003) Grassroots ecology: plant-microbe-soil interactions as drivers of plant community structure and dynamics. Ecology 84:2281–2291

Řezanka T, Sigler K (2009) Odd-numbered very-long-chain fatty acids from the microbial, animal and plant kingdoms. Prog Lipid Res 48:206–238. doi:10.1016/j.plipres.2009.03.003

Riding MJ, Herbert BMJ, Ricketts L, Dodd I, Ostle N, Semple KT (2015) Harmonising conflicts between science, regulation, perception and environmental impact: the case of soil conditioners from bioenergy. Environ Int 75:52–67. doi:10.1016/j.envint.2014.10.025

Rivero J, Gamir J, Aroca R, Pozo MJ, Flors V (2015) Metabolic transition in mycorrhizal tomato roots. Front Microbiol 6:598

Roberts WK, Selitrennikoff CP (1988) Plant and bacterial chitinases differ in antifungal activity. Microbiology 134:169–176. doi:10.1099/00221287-134-1-169

Rochfort S, Ezernieks V, Mele P, Kitching M (2015) NMR metabolomics for soil analysis provide complementary, orthogonal data to MIR and traditional soil chemistry approaches—a land use study. Magn Reson Chem 53:719–725. doi:10.1002/mrc.4187

Rochfort S, Ezernieks V, Yen A (2009) NMR-based metabolomics using earthworms as potential indicators for soil health. Metabolomics 5:95–107. doi:10.1007/s11306-008-0140-4

Rockström J, Steffen W, Noone K, Persson A, Chapin Iii FS, Lambin E, Lenton TM, Scheffer M, Folke C, Schellnhuber HJ, Nykvist B, de Wit CA, Hughes T, van der Leeuw S, Rodhe H, Sörlin S, Snyder PK, Costanza R, Svedin U, Falkenmark M, Karlberg L, Corell RW, Fabry VJ, Hansen J, Walker B, Liverman D, Richardson K, Crutzen P, Foley J (2009) Planetary boundaries: exploring the safe operating space for humanity. Ecol Soc 14

Rudresh DL, Shivaprakash MK, Prasad RD (2005) Tricalcium phosphate solubilizing abilities of *Trichoderma spp.* in relation to P uptake and growth and yield parameters of chickpea (*Cicer arietinum L.*). Can J Microbiol 51:217–222

Ruzicka D, Hausmann N, Barrios-Masias F, Jackson L, Schachtman D (2012) Transcriptomic and metabolic responses of mycorrhizal roots to nitrogen patches under field conditions. Plant Soil 350:145–162. doi:10.1007/s11104-011-0890-z

Saharan BS, Nehra V (2011) Plant growth promoting rhizobacteria: a critical review. Life Sci Med Res 21:1–30

Saia S, Rappa V, Ruisi P, Abenavoli MR, Sunseri F, Giambalvo D, Frenda AS, Martinelli F (2015a) Soil inoculation with symbiotic microorganisms promotes plant growth and nutrient transporter genes expression in durum wheat. Front Plant Sci 6:815. doi:10.3389/fpls.2015.00815

Saia S, Ruisi P, Fileccia V, Di Miceli G, Amato G, Martinelli F (2015b) Metabolomics suggests that soil inoculation with arbuscular mycorrhizal fungi decreased free amino acid content in roots of durum wheat grown under n-limited. P-Rich Field Cond. PLoS ONE 10:e0129591. doi:10.1371/journal.pone.0129591

Samolski I, Rincón AM, Pinzón LM, Viterbo A, Monte E (2012) The qid74 gene from *Trichoderma harzianum* has a role in root architecture and plant biofertilization. Microbiology 158:129–138

Sanchez DH, Siahpoosh MR, Roessner U, Udvardi M, Kopka J (2008) Plant metabolomics reveals conserved and divergent metabolic responses to salinity. Physiol Plant 132:209–219. doi:10.1111/j.1399-3054.2007.00993.x

Sardans J, Peñuelas J, Rivas-Ubach A (2011) Ecological metabolomics: overview of current developments and future challenges. Chemoecology 21:191–225. doi:10.1007/s00049-011-0083-5

Sarry J-E, Kuhn L, Ducruix C, Lafaye A, Junot C, Hugouvieux V, Jourdain A, Bastien O, Fievet JB, Vailhen D, Amekraz B, Moulin C, Ezan E, Garin J, Bourguignon J (2006) The early responses of *Arabidopsis thaliana* cells to cadmium exposure explored by protein and metabolite profiling analyses. Proteomics 6:2180–2198. doi:10.1002/pmic.200500543

Sasser M (2006) Bacterial identification by gas chromatographic analysis of fatty acids methyl esters (GC-FAME). Microbial ID Inc

Scandiani MM, Luque AG, Razori MV, Ciancio Casalini L, Aoki T, O'Donnell K, Cervigni GDL, Spampinato CP (2015) Metabolic profiles of soybean roots during early stages of Fusarium tucumaniae infection. J Exp Bot 66:391–402. doi:10.1093/jxb/eru432

Schlesinger WH (1991) Biogeochemistry: an analysis of global change. Academic Press, San Diego

Schmidt MWI, Torn MS, Abiven S, Dittmar T, Guggenberger G, Janssens IA, Kleber M, Kogel-Knabner I, Lehmann J, Manning DAC, Nannipieri P, Rasse DP, Weiner S, Trumbore SE (2011) Persistence of soil organic matter as an ecosystem property. Nature 478:49–56

Schützendübel A, Polle A (2002) Plant responses to abiotic stresses: heavy metal-induced oxidative stress and protection by mycorrhization. J Exp Bot 53:1351–1365

Segarra G, Casanova E, Bellido D, Odena MA, Oliveira E, Trillas I (2007) Proteome, salicylic acid and jasmonic acid changes in cucumber plants inoculated with Trichoderma asperellum strain T34. Proteomics 7:3943–3952

Selosse M-A, Strullu-Derrien C (2015) Origins of terrestrial flora: a symbiosis with fungi? BIO Web Conf 4:1–12

Seo J-S, Keum Y-S, Li Q (2009) Bacterial degradation of aromatic compounds. Int J Environ Res Pub Health 6:278

Shah V, Belozerova I (2009) Influence of metal nanoparticles on the soil microbial community and germination of lettuce seeds. Water Air Soil Pollut 197:143–148. doi:10.1007/s11270-008-9797-6

Simpson AJ, Simpson MJ, Soong R (2012) Nuclear magnetic resonance spectroscopy and its key role in environmental research. Environ Sci Technol 46:11488–11496. doi:10.1021/es302154w

Simpson M, McKelvie J (2009) Environmental metabolomics: new insights into earthworm ecotoxicity and contaminant bioavailability in soil. Anal Bioanal Chem 394:137–149. doi:10.1007/s00216-009-2612-4

Singh OV (2006) Proteomics and metabolomics: the molecular make-up of toxic aromatic pollutant bioremediation. Proteomics 6:5481–5492. doi:10.1002/pmic.200600200

Singh S, Parihar P, Singh R, Singh VP, Prasad SM (2016) Heavy metal tolerance in plants: role of transcriptomics, proteomics, metabolomics and ionomics. Front Plant Sci 6. doi:10.3389/fpls.2015.01143

Sinsabaugh RL, Turner BL, Talbot JM, Waring BG, Powers JS, Kuske CR, Moorhead DL, Follstad Shah JJ (2016) Stoichiometry of microbial carbon use efficiency in soils. Ecol Monogr 86:172–189. doi:10.1890/15-2110.1

Sláviková E, Vadkertiová R (2003) The diversity of yeasts in the agricultural soil. J Basic Microbiol 43:430–436. doi:10.1002/jobm.200310277

Smith FA, Smith SE (2015) How harmonious are arbuscular mycorrhizal symbioses? inconsistent concepts reflect different mindsets as well as results. New Phytol 205:1381–1384

Smith SE, Jakobsen I, Gronlund M, Smith FA (2011) Roles of arbuscular mycorrhizas in plant phosphorus nutrition: interactions between pathways of phosphorus uptake in arbuscular mycorrhizal roots have important implications for understanding and manipulating plant phosphorus acquisition. Plant Physiol 156:1050–1057

Souza LA, Camargos LS, Schiavinato MA, Andrade SAL (2014) Mycorrhization alters foliar soluble amino acid composition and influences tolerance to Pb in Calopogonium mucunoides. Theoret Exp Plant Physiol 26:211–216. doi:10.1007/s40626-014-0019-x

Stefaniuk M, Oleszczuk P, Ok YS (2016) Review on nano zerovalent iron (nZVI): from synthesis to environmental applications. Chem Eng J 287:618–632. doi:10.1016/j.cej.2015.11.046

Stone M (2009) Metal balance helps explain survival of microbial "superhero". Microbe 4: 164–165

Styger G, Prior B, Bauer FF (2011) Wine flavour and aroma. J Ind Microbiol Biotech 38: 1145–1159

Sutton R, Sposito G (2005) Molecular structure in soil humic substances: the new view. Environ Sci Technol 39:9009–9015. doi:10.1021/es050778q

Swenson TL, Bowen BP, Nico PS, Northen TR (2015a) Competitive sorption of microbial metabolites on an iron oxide mineral. Soil Biol Biochem 90:34–41. doi:10.1016/j.soilbio.2015.07.022

Swenson TL, Jenkins S, Bowen BP, Northen TR (2015b) Untargeted soil metabolomics methods for analysis of extractable organic matter. Soil Biol Biochem 80:189–198. doi:10.1016/j.soilbio.2014.10.007

Tate RL (2000) Soil Microbiology. Wiley, New York

Tepper CS, Anderson AJ (1990) Interactions between pectic fragments and extracellular components from the fungal pathogen Colletotrichum lindemuthianum. Phys Mol Plant Pathol 36:147–158. doi:10.1016/0885-5765(90)90103-5

Tilman D, Cassman KG, Matson PA, Naylor R, Polasky S (2002) Agricultural sustainability and intensive production practices. Nature 418:671–677

Tittonell P (2014) Ecological intensification of agriculture—sustainable by nature. Curr Opin Environ Sustain 8:53–61

Torsvik V, Øvreås L (2002) Microbial diversity and function in soil: from genes to ecosystems. Curr Opin Microbiol 5:240–245. doi:10.1016/S1369-5274(02)00324-7

Tremaroli V, Workentine ML, Weljie AM, Vogel HJ, Ceri H, Viti C, Tatti E, Zhang P, Hynes AP, Turner RJ, Zannoni D (2009) Metabolomic investigation of the bacterial response to a metal challenge. Appl Environ Microbiol 75:719–728. doi:10.1128/aem.01771-08

Valentinuzzi F, Pii Y, Vigani G, Lehmann M, Cesco S, Mimmo T (2015) Phosphorus and iron deficiencies induce a metabolic reprogramming and affect the exudation traits of the woody plant Fragaria × ananassa. J Exp Bot 66:6483–6495. doi:10.1093/jxb/erv364

van Dam NM, Bouwmeester HJ (2016) Metabolomics in the rhizosphere: tapping into belowground chemical communication. Trends Plant Sci 21:256–265. doi:10.1016/j.tplants.2016.01.008

van den Berg EM, van Dongen U, Abbas B, van Loosdrecht MCM (2015) Enrichment of DNRA bacteria in a continuous culture. ISME J 9:2153–2161. doi:10.1038/ismej.2015.26

Van Der Heijden MGA, Bardgett RD, Van Straalen NM (2008) The unseen majority: soil microbes as drivers of plant diversity and productivity in terrestrial ecosystems. Ecol Lett 11:296–310. doi:10.1111/j.1461-0248.2007.01139.x

Van Der Heijden MGA, Klironomos JN, Ursic M, Moutoglis P, Streitwolf-Engel R, Boller T, Wiemken A, Sanders IR (1998) Mycorrhizal fungal diversity determines plant biodiversity, ecosystem variability and productivity. Nature 396:69–72. doi:10.1038/23932

van Leeuwen C, Seguin G (2006) The concept of terroirs in viticulture. Wine Res 17:1–10

Vinale F, Sivasithamparam K, Ghisalberti EL, Marra R, Woo SL, Lorito M (2008) Trichoderma–plant–pathogen interactions. Soil Biol Biochem 40:1–10. doi:10.1016/j.soilbio.2007.07.002

Walker TS, Bais HP, Grotewold E, Vivanco JM (2003) Root exudation and rhizosphere biology. Plant Physiol 132:44–51. doi:10.1104/pp.102.019661

Walker V, Bertrand C, Bellvert F, Moënne-Loccoz Y, Bally R, Comte G (2011) Host plant secondary metabolite profiling shows a complex, strain-dependent response of maize to plant growth-promoting rhizobacteria of the genus *Azospirillum*. New Phytol 189:494–506

Wallenstein MD, Haddix ML, Ayres E, Steltzer H, Magrini-Bair KA, Paul EA (2013) Litter chemistry changes more rapidly when decomposed at home but converges during decomposition–transformation. Soil Biol Biochem 57:311–319. doi:10.1016/j.soilbio.2012.09.027

Wallenstein MD, Hess AM, Lewis MR, Steltzer H, Ayres E (2010) Decomposition of aspen leaf litter results in unique metabolomes when decomposed under different tree species. Soil Biol Biochem 42:484–490. doi:10.1016/j.soilbio.2009.12.001

Wallenstein MD, Weintraub MN (2008) Emerging tools for measuring and modeling the in situ activity of soil extracellular enzymes. Soil Biol Biochem 40:2098–2106. doi:10.1016/j.soilbio.2008.01.024

Wang W, Li Y, Wang H, Zu Y (2014) Differences in the activities of eight enzymes from ten soil fungi and their possible influences on the surface structure, functional groups, and element composition of soil colloids. PLoS ONE 9:e111740. doi:10.1371/journal.pone.0111740

Wania F, Mackay D (1996) Tracking the distribution of persistent organic pollutants. Environ Sci Technol 30:390A–397A

Warren CR (2013) High diversity of small organic N observed in soil water. Soil Biol Biochem 57:444–450. doi:10.1016/j.soilbio.2012.09.025

Warren CR (2014) Organic N molecules in the soil solution: what is known, what is unknown and the path forwards. Plant Soil 375:1–19. doi:10.1007/s11104-013-1939-y

Warren CR (2015) Comparison of methods for extraction of organic N monomers from soil microbial biomass. Soil Biol Biochem 81:67–76. doi:10.1016/j.soilbio.2014.11.005

Watzinger A (2015) Microbial phospholipid biomarkers and stable isotope methods help reveal soil functions. Soil Biol Biochem 86:98–107. doi:10.1016/j.soilbio.2015.03.019

Wehner J, Antunes PM, Powell JR, Mazukatow J, Rillig MC (2010) Plant pathogen protection by arbuscular mycorrhizas: a role for fungal diversity? Pedobiologia 53:197–201

Whitfield MÅ, Stephenson GL, Simpson AJ, Simpson MJ (2013) Comparison of earthworm responses to petroleum hydrocarbon exposure in aged field contaminated soil using traditional ecotoxicity endpoints and 1H NMR-based metabolomics. Environ Pollut 182:263–268. doi:10.1016/j.envpol.2013.07.026

Wieland G, Neumann R, Backhaus H (2001) Variation of microbial communities in soil, rhizosphere, and rhizoplane in response to crop species, soil type, and crop development. Appl Environ Microbiol 67:5849–5854

Wissuwa M, Mazzola M, Picard C (2008) Novel approaches in plant breeding for rhizosphere-related traits. Plant Soil 321:409–430. doi:10.1007/s11104-008-9693-2

Woo SL, Scala F, Ruocco M, Lorito M (2006) The molecular biology of the interactions between trichoderma spp., phytopathogenic fungi, and plants. Phytopathology 96:181–185. doi:10.1094/phyto-96-0181

Workentine ML, Harrison JJ, Weljie AM, Tran VA, Stenroos PU, Tremaroli V, Vogel HJ, Ceri H, Turner RJ (2010) Phenotypic and metabolic profiling of colony morphology variants evolved from *Pseudomonas fluorescens* biofilms. Environ Microbiol 12:1565–1577. doi:10.1111/j.1462-2920.2010.02185.x

Yang M, Kim T-Y, Hwang H-C, Yi S-K, Kim D-H (2008) Development of a palm portable mass spectrometer. J Am Soc Mass Spectrom 19:1442–1448. doi:10.1016/j.jasms.2008.05.011
Zahir AZ, Arshad M, Frankenberger ET Jr (2004) Plant growth promoting rhizobacteria: application and respectives in agriculture. Adv Agron 81:97–168
Zhang Y, Ruyter-Spira C, Bouwmeester HJ (2015) Engineering the plant rhizosphere. Curr Opin Biotechnol 32:136–142. doi:10.1016/j.copbio.2014.12.006
Ziegler M, Engel M, Welzl G, Schloter M (2013) Development of a simple root model to study the effects of single exudates on the development of bacterial community structure. J Microbiol Methods 94:30–36

Chapter 7
Community Metabolomics in Environmental Microbiology

Oliver A.H. Jones, Gavin Lear, Aalim M. Welji, Gavin Collins and Christopher Quince

1 Introduction

Metabolomics may be defined as the analysis of thousands of naturally occurring small biomolecules (metabolites) that are the substrates and products of both primary and secondary cellular metabolism (Jones and Cheung 2007). Such molecules include sugars, organic acids, amino acids, flavonoids, lipids and nucleotides amongst many others. An organism's "metabolome" is its full complement of metabolites, in the same way that its genome comprises its complete genetic content

O.A.H. Jones (✉)
Australian Centre for Research on Separation Science (ACROSS),
School of Science, RMIT University, GPO Box 2476, Melbourne, VIC 3001, Australia
e-mail: oliver.jones@rmit.edu.au

G. Lear
School of Biological Sciences, University of Auckland, Private Bag 92019,
Auckland 1010, New Zealand
e-mail: g.lear@auckland.ac.nz

A.M. Welji
10-113 Smilow Centre for Translational Research, University of Pennsylvania,
3400 Civic Centre Blvd, Bldg. 421, Philadelphia, PA 19104, USA
e-mail: aalim@upenn.edu

G. Collins · C. Quince
Department of Civil Engineering (Rankine Building), Glasgow University,
76 Oakfield Avenue, Glasgow G12 8LT, UK
e-mail: gavin.collins@nuigalway.ie

C. Quince
e-mail: quince@civil.gla.ac.uk

G. Collins
Microbial Ecology Laboratory, Department of Microbiology, School of Natural
Sciences, National University of Ireland, University Road, Galway, Ireland

D.J. Beale et al. (eds.), *Microbial Metabolomics*,
DOI 10.1007/978-3-319-46326-1_7

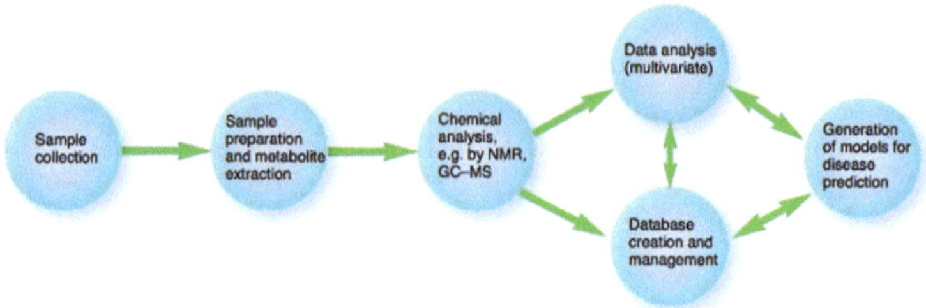

Fig. 1 Simplified overview of a metabolomics experiment

and a proteome entails the complete set of proteins expressed and modified following their expression by the genome (Wilkins 2009). A very basic outline of a metabolomics experimental setup is shown in Fig. 1.

The metabolome can be used to study the underlying biochemical responses of an organism/population/community to an environmental stimulus or combination of stimuli. However, because both populations and communities of microorganisms are often difficult to culture, metabolomic analysis of such assemblages must take a different approach from that taken by studies using tissue or biofluid(s) from larger organisms such as humans. A new and effective way to do this is to borrow techniques and approaches from the field of metagenomics.

Metagenomics is a rapidly growing area of the genome sciences founded by Venter et al. (2004) who hit upon the idea of undertaking the analysis of the total complement of microbial DNA extracted directly from entire microbial communities in their native habitats without worrying about which species it came from. This approach allowed them to sequence such communities in their entirety, regardless of the ability of individual component members to be cultured in the laboratory (Handelsman 2004). Early studies had great success. For instance, when the technique was applied to the analysis of samples from the Sargasso sea over 1.2 million previously unknown genes were identified (Venter et al. 2004).

Metagenomics allows one to see, in ever-increasing detail, the vast microbial and metabolic diversity that exists in the biosphere. It has been utilised in several high-profile publications in studies ranging from the aforementioned study of the genetic diversity of microbes in seawater (Venter et al. 2004) to phosphorus and carbon management in sewage sludge (Garcia et al. 2006; Sales and Lee 2015), to the effects of warming on carbon sequestration and lignin oxidation in soil (Feng et al. 2008).

It is of note, however, that the other omic sciences have yet to replicate this success. The first detailed study on the NMR-based metabolomics of microbial communities in environmental samples was published by Jones et al. (2014) who looked at the effects of pollutants on soil metabolic profiles in old mine sites in the UK. There have since been a few similar studies including Swenson et al. (2015) and Rochfort et al. (2015) who both looked at metabolites in the extractable organic matter in soil, and Llewellyn et al. (2015) who used community metabolomics as a new approach to discriminate marine microbial particulate organic matter in sediments from various locations in the English Channel. Plant-associated fungal communities assessed by meta-omics techniques were also the subject of a recent review by Peršoh (2015).

Much remains to be learnt from this area of research through what might be termed "Meta-Metabolomics" or, as proposed by Jones et al. (2014), "*Community Metabolomics*". Under this definition, in the same way as metagenomics indicates the analyses of all the DNA from a given sample, community metabolomic analysis uses all of the thousands of naturally occurring metabolites from the meta-population of a sample of a given environment such as soil or water, and perhaps even air (Castillo et al. 2012), as opposed to those from simple, laboratory based monoclonal cultures (Eisen 2007) or clinical isolates (Bundy et al. 2005).

Community metabolomics has the potential to influence a range of "bio" related disciplines, including medicine. For instance, Bundy et al. (2005) showed that one can use metabolomics to classify pathogens on the basis of their metabolic profile, even when it is not possible to infer a direct link to known virulence factors. This could also have a wide range of possible applications in both general microbiology and microbial ecology for distinguishing and identifying different functional/physiological ecotypes of bacterial strains or species. In a similar vein, pioneering work by Nicholson et al. (2005) at Imperial College London proposed that the appropriate consideration of individual human gut microbial activities is a necessary part of future personalised healthcare. This is because the gut microbiota of most species interacts extensively with the host through metabolic exchanges and cometabolism of substrates. Such interactions are presently poorly understood but are highly likely to be involved in the aetiology of many human diseases as well as the fate and toxicity of drugs taken to alleviate said diseases (Nicholson et al. 2005). This topic is expanded upon elsewhere in this book, so will not be explored further here.

2 Soil Science

Environmental science is another area where microbial metabolomics has great potential and it is this area on which this chapter will focus. Whilst most work carried out on this topic has been, thus far, relatively simple, it illustrates the potential of the approach. For instance, Fourier Transform-Infrared Spectroscopy (FT-IR) has been demonstrated to allow the chemically based discrimination of microbial genotypes, and to produce biochemical fingerprints which are both reproducible and distinct for different bacteria and fungi (Timmins et al. 1998). Scullion et al. (2003) also tested the potential of FT-IR spectroscopy for investigating microbial communities and their activities in soil. A range of samples including laboratory cultures of similar soil bacteria, plant materials and earthworm casts from worms of various ages and feeding regimes, were analysed using cluster analyses and this proved capable of differentiating between different bacterial, litter and cast samples. However, this work has not been followed up in detail and the amount of diagnostic information that can be obtained from FT-IR is limited.

A more complicated study was that mentioned above by Jones et al. (2014), which utilised ^{1}H-nuclear magnetic resonance spectroscopy (NMR)-based community metabolic profiling to assess the changes in biochemical profiles of communities living in contaminated soils from various sites in the UK, each with very different physicochemical characteristics (levels of metal contamination, underlying geology and soil type). Each site could be clearly distinguished on the basis of the metabolic profile of its microbial community. While some of these site differences may also have been caused by additional abiotic factors (such as soil type or pH), pattern recognition analysis of the data showed that both site- and contaminant-specific effects on the metabolic profiles could be discerned. The study therefore acts as a proof of principle for the use of community metabolomics of microbial populations from whole soil samples (rather than single isolates) as a diagnostic tool for pollution assessment. Assigning peaks in NMR spectra of soil extracts is difficult but software tools such as Chenomx NMR Suite that overlay library spectra of known compounds over the sample spectra can help (an example is shown in Fig. 2). Recently developments in NMR software such as Bayesil, (http://bayesil.ca/) a web based system that automatically identifies and quantifies metabolites developed at the University of Alberta (Canada) may also be of help in future.

Rochfort et al. (2015) also used NMR to assess metabolites in soils. They found similar results to Jones et al. (2014), although in the Rochfort study there were more

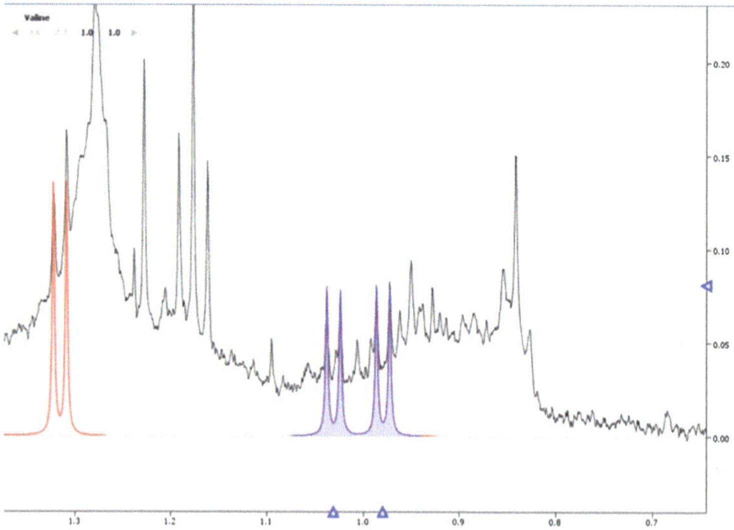

Fig. 2 Section of a ^{1}H NMR of soil extract (*black*) overlain with NMR spectra of lactate (*red*) and valine (*blue*)

resonances as a result of lipids and terpenes in the NMR spectra. The authors put this down to differences between the extraction methodologies employed in each study, although it seems equally possible that soil type, condition or land use could have been responsible. Both the Jones and Rochfort studies demonstrated that soil extracts could be measured by NMR, and there were no obvious interference or line broadening as a result of paramagnetic materials that might have been present in the samples. It is interesting to note that Rochfort et al. also used mid-infrared spectroscopy (MIR) to analyse their soil samples and found some differences. Whilst the NMR metabolomic data were more associated with land use rather than location, the MIR data were correlated more to location and inorganic chemical analysis. The results may reflect the differences between NMR and MIR data (the latter technique is far less sensitive) but may also indicate that it may be beneficial to combine two or more analytical methods in such studies to ensure a comprehensive soil analysis.

It should also be noted that some microbial metabolites rapidly adsorb to soil and thus one should also keep in mind that soil physical properties such as mineral composition, surface area, shape and porosity present a notably different set of challenges in soil analysis compared to the cell and tissue extractions more common in metabolomics. Swenson et al. (2015) showed that even after a short incubation, with water, many metabolites, for example amines and carboxylic acids, could be adsorbed to soil and/or chelated by metals or other ions in the soil or extraction medium. Cationic and anionic metabolites seem to be most affected, whilst the majority of neutral metabolites were recovered. This is an important finding since metabolites not recovered in the extraction procedure could be overlooked in the subsequent analysis. Other factors that might affect such binding include soil type, metabolite-soil contact time, temperature, moisture levels, and which extraction solvents are used. Swenson et al. (2015) point out that there is, however, a silver lining to this problem as understanding how specific metabolites bind to soil particles under differing environmental conditions may also increase our understanding of how such compounds will behave in response to future climatic changes (e.g. heat or precipitation) and thereby affect substrate availability and the structure of soil microbial communities in the future.

Analysing small molecule metabolites, in addition to more abundant macromolecules, could also help to increase knowledge of biogeochemical cycles. For example, ^{13}C-labelled carbon sources could be fed to plants or pumped into/through the soil and the bacterial metabolome studied to assess where the carbon is transported or to detect subtle shifts in microbial populations interacting with plant litter. Indeed, the use of NMR techniques has already provided valuable data supporting the occurrence, diversity and extent of carbon cycling in the carbohydrate metabolism of microorganisms; for a review see Portais and Delort (2002).

Targeted studies have also been undertaken in order to measure flux rates of specific metabolic pathways, for example through the use of isotope labelling studies. Scullion et al. (2003) used FT-IR-based metabolic analysis to assess several potential foods (oat grain, and fresh and aged tobacco) for the earthworm species *Lubricus terrestris* and *L. rubellus*. As casts aged, there was a predictable microbial succession (from bacteria to fungi) in both earthworm species. Each species was also found to differ in terms of their cast microbiology in response to food type. Analysis of ageing casts by FT-IR spectroscopy indicated greater chemical changes in casts of *L. terrestris* than *L. rubellus* irrespective of food type.

Community metabolomics of soils is potentially very useful since the identification of biomarkers indicative of a defined response to a pollutant or pollutants, before major outward changes become apparent, would be very useful in preventing damage to a variety of sensitive systems (e.g. eutrophication). This could potentially have applications in testing of contaminated land prior to redevelopment (e.g. house-building on old industrial sites). It could also be useful for the water industry since the discharge of toxic substances to the sewage network can have negative effects on wastewater treatment plants, which rely heavily on bacteria and other microorganisms to function effectively. Similar research could also identify microorganisms with favourable metabolisms for bioremediation work.

Community metabolomics is not limited to the terrestrial environment. Llewellyn et al. (2015) used 'non-targeted' community metabolomics to determine patterns in metabolite profiles associated with particulate organic matter (POM) at four locations from two long-term monitoring stations in the western English Channel. They identified a range of compounds including amino-acid derivatives, diacylglyceryltrimethylhomoserine (DGTS) lipids, oxidised fatty acids (oxylipins), various glycosylated compounds, oligohexoses, phospholipids, triacylglycerides (TAGs) and oxidised TAGs. Metabolic profiles varied significantly across the four locations with the largest differences for both the polar and lipid fractions again being due to geographic location (and/or time). Smaller differences were also associated with depth.

The work demonstrated that community metabolomics is not only applicable to the aquatic, as well as the terrestrial, environment but that it has the potential to contribute significantly to the comprehensive and unbiased characterisation of marine microbial populations; particularly if linked to metagenomic data. The authors also speculated that the oxylipins could have a possible link to the formation of oxygenated volatile organic compounds produced by microbial species that are important in atmospheric chemistry.

3 Case Studies in Community Metabolomics

3.1 Bacteria and Metal Toxicity

As microorganisms form an extremely important, sessile component of ecosystems, it is little surprise that they have evolved numerous strategies to deal with environmental stressors. Laboratory investigations of individual bacterial strains provide well-controlled conditions from which to study mechanistic behaviour in response to stress. The ability of microbial strains to survive in the presence of relatively high levels of metal ions is an important environmental adaptation to toxicological stress. Recently, two specific approaches to microbial adaptation have been examined using metabolomics methods: (i) an innate metabolic 'priming' which confers adaptive advantages to bacteria via specific pathways, and (ii) morphological adaptation into biofilm communities which completely alter the bacterial metabolism and consequent response to metal toxicity.

The negative environmental consequences of heavy metal toxicity have long been known, and recent political and regulatory awareness has sparked a renewed vigour in attacking this problem scientifically. In living systems, such as microbial communities, heavy metal toxicity manifests itself through numerous mechanisms (reviewed in Harrison et al. 2007 amongst others). Briefly, toxic effects include: substitution for other essential inorganic elements, interaction with protein and metabolite thiol groups, production of reactive oxygen species through Fenton-type reactions, competitive inhibition of the membrane transport process and siphoning electrons directly from respiratory pathways.

3.1.1 Tellurite Resistance

Incorporation of Tellurite (tellurium dioxide—TeO_2) into electronics and industrial applications has led to an increased interest in the study of the effects of environmental exposure (Chasteen and Bentley 2002). A picture of pure Tellurium can be seen in Fig. 3 but this element is not often found in its native form. A common, soluble oxyanion form of this metal, however, is tellurium oxide (TeO_3^-), which is highly toxic to most organisms even at low concentrations. *Pseudomonas pseudoalcaligenes* KF707 is a naturally Te-resistant (TeR) bacterial strain. The exact mechanisms by which KF707 is resistant to Te are poorly understood, although there are known interactions with thiol groups that mediate the bacterial interaction with Te (Zannoni et al. 2008).

Fig. 3 Pure tellurium

In a metabolomics study aimed at understanding the metabolic shifts associated with Te exposure, a hyper-resistant KF707 strain (T5) was isolated and studied using an NMR-based approach (Tremaroli et al. 2009). In an initial experiment using an established biochemical assay (Ellman's reagent, 5,5′-dithiobis-2-nitrobenzoic acid) to measure cellular thiol content (RSH), it was found that exposure to 25 μg/ml K_2TeO_3 for 30 min resulted in an equivalent depletion of RSH content in both the wild-type KF707 and hyper-resistant T5 strains. Intriguingly, the viability of T5 cells did not decrease further with extended exposure, and infact, the cells were able to replenish RSH. Similarly, T5 cells demonstrated resistance to perturbations of the membrane potential induced by metal exposure.

In order to better understand the metabolic adaptations of the T5 strain to counteract the effects of Te, an untargeted metabolite analysis was performed on both the wild-type and T5 strains $\pm TeO_3^-$ (Tremaroli et al. 2009). This approach consisted of quantitatively profiling 28 metabolites using the targeted approach which accounted for the majority of peaks in NMR spectra of the samples. Orthogonal partial least square-discriminant analysis (OPLS-DA) comparing two sample types at a time revealed that there were unique metabolite profiles for baseline KF707 versus. (a) T5 cells not exposed to TeO_3^-, (b) KF707 cells $\pm TeO_3^-$ and (c) T5 cells $\pm TeO_3^-$. Notably, there was no significant difference between KF707 and T5 cells exposed to TeO_3^-, indicating that the post-exposure profile of the two strains were similar, and that it is the pre-exposure metabolism effects which confers the selective advantage to the T5 strain. One of the compounds elevated in the baseline T5 profile compared to wild-type was glutathione, which is important in responding to cellular oxidative stress. Levels of betaine, which plays a role in response to hyperosmolar stress (Lisa et al. 1994), were also found to be elevated .

Collectively, the metabolomics experiments described above lead to the conclusion that the hyper-resistance of the T5 cells could be attributed to a series of basal adaptations that 'prime' the cells for exposure. This conclusion was supported by further experimental evidence using phenotype microarrays which provide a series of chemical stressors to evaluate the response of exposure to a host of toxic chemicals (Bochner et al. 2001). In addition to tellurite, T5 cells exhibited an increased tolerance to a number of other ions, including selenite, chromium, aluminium and caesium, while there was no altered response to other metals. Also noteworthy was the fact that T5 demonstrated increased resistance to compounds which interact with glutathione, glutathione-related proteins or RSH groups (namely diamide, 1-chloro-2,4-dinitrobenzene and phenylarsine oxide). This result provides a demonstrable link between the higher basal levels of glutathione in T5 observed in the metabolomics experiment and the phenotypic response of the bacteria to toxicants. This is a useful demonstration of the power of community metabolomics and the work could easily be extended to a range of other metals and metalloids.

3.1.2 Investigations of Biofilms and Morphological Variants

Biofilms are increasingly recognised as potentially important morphological communities by which bacteria are able to survive under conditions that would be otherwise detrimental. Recent evidence suggests that this transformation is accompanied by specialisation, analogous to differentiation in higher organisms (Nadell et al. 2009). Such adaptation into synergistically functional communities confers a wide range of physical and chemical characteristics unique to the biofilms compared to planktonic cultures of the same strain.

One of the interesting features of biofilms in both laboratory and 'real-world' settings is heterogeneity in colony morphologies when the bacteria are grown on solid media (Boles et al. 2004). Within *Pseudomonas*, sp., two of these so-called phenotypic variants have been well characterised, namely small-colony variants (SCVs) and rugose small-colony variants [RSCVs, or wrinkly spreaders (WS)]. These morphological variants exhibit increased resistance to anti-microbial compounds and, under laboratory settings, they are selected by exposure to environmental stress agents such as oxidative agents, anti-microbials and metals.

One of the key genetic pathways associated with phenotypic morphological variation is the global activator of cyanide biosynthesis/regulator of the secondary metabolism (*gac*/*rsm*) signal transduction system (Petrova and Sauer 2009). In experiments examining the morphological response to metal ion exposure of *P. fluorescens*, four distinct populations were observed: (1) the parental wild-type (CHA0), which did not give rise to any morphological variants; (2) a *gacS* (CHA19) strain which gave rise to (3) SCV and (4) WS morphological variants upon exposure to toxicants. Using an NMR metabolomics approach, the metabolic difference between these four population types was characterised (Workentine et al. 2010).

Metabolites were extracted from the four groups from cultures grown to mid-log phase and subjected to 1D ^{1}H-NMR analysis and the results analysed in detail by both principle component analysis (PCA) and partial least squares discriminant analysis (PLS-DA). In this case 32 metabolites were identified and quantified, revealed significant differences between all four populations, both in a combined PLS-DA approach, and via comparisons of the various strains/morphological types. Clear metabolic differences were observed between the SCV and WS morphological variants compared to both the wild-type and CHA19 strains (SCV and WS were different to both wild-type and CHA19 strains). For example, valine, phenylalanine and glycine were uniquely found to be important in the WS models, whereas acetate, pyruvate, aspartate, proline and glutamate were important to explaining the SCV metabolic differences. Other metabolites, such as tryptophan, were also found to be important in both variants.

It is not entirely clear that what types of advantages or disadvantages are conferred to morphological variants compared to the parental strains. Metal susceptibility ($AgNO_3$, $CuSO_4$ and $NiSO_4$) was used to evaluate the quantitative survival of the bacteria in response to exposure. The *gacS* clone was susceptible to all three metals, while the SCV and WS variants were both tolerant to $NiSO_4$; however, SCV was uniquely tolerant to $CuSO_4$ and WS was uniquely tolerant to $AgNO_3$. These results suggest that the morphological variants are imbued with unique properties. To further explore the link between the metabolite profiles and metal tolerance, another set of PLS-DA models were built which related the specific metabolites responsible to metal tolerance across all three strains (in other words, the basal metabolite states across all strains was considered together). In this analysis, tryptophan, glutathione, methionine, adenosine and glucose were elevated in sensitivity for both copper and silver, whereas proline was strongly correlated to sensitivity for copper only and lactate and NAD^+ increased with sensitivity to silver.

Overall, the results from this study indicate that the biofilms under study had distinct metabolic states that conferred some advantages to stress, and that morphological variants further allow the bacteria to develop unique profiles. These laboratory results are particularly important to understanding the environmental survival and viability of bacterial strains.

3.1.3 Biofilm Versus Planktonic Response to Copper

One of the proposed mechanisms by which biofilms are known to differ from planktonic cultures is in sugar metabolism. In particular, exo-polysaccharide matrices are thought to provide structure for formation of the biofilms (Harrison et al. 2007). The NMR approach commonly employed for metabolomics analysis is advantageous in that it is robust, quantitative and highly reproducible. One disadvantage, however, is the relatively small number of metabolites ($\sim$30) that can be characterised from extracts, with specific metabolites related to sugar metabolism being poorly characterised. For this reason, in a study of *P. fluorescens* response to copper (Booth et al. 2011) a more sensitive gas chromatography-mass spectrometry (GC-MS) method was employed in addition to NMR. The two analytical platforms provided identification of 79 unique metabolites, and hence a much larger coverage of metabolic space. Pairwise multivariate comparisons indicated that while copper exposure induces a significant metabolic response in both planktonic and biofilm cultures, the nature of this response is highly different. Only three metabolites (NAD+, phosphoric acid and glutathione) were common to both responses.

Overall, the planktonic cultures were found to be far more ‘reactive’ to exposure, characterised by an oxidative stress response with changes to the tricarboxylic acid (TCA) cycle, glycolysis, pyruvate and nicotinate and niacinamide metabolism. On the other hand, the biofilm response was dominated by shifts in exo-polysaccharide metabolism, suggesting a ‘protective’ response. While alterations in levels of glutathione indicate that there was still oxidative stress in the biofilms, the lack of involvement of energy pathways suggests that the biofilms have alternate methods for enhancing protection, which is consistent with previous observations (González et al. 2010).

3.1.4 Methodological Considerations and Conclusions

Given the obvious advantages of being able to tightly control growth and exposure conditions in a laboratory setting, it is no surprise that this is one of the most popular methods used to study the effect of exposure to environmental toxicants. Studies of bacterial metabolism are exquisitely sensitive to a multitude of factors, including sampling conditions, sample processing, and in the context of metabolomics, data acquisition and analysis. Numerous studies have been conducted on appropriate methods to quench bacterial metabolism, which has recently been assessed in the context of bacterial metabolomics (van Gulik 2010). The studies described above were accomplished using cold methanol quenching, which is the most common technique, but may suffer from leakage of metabolites during the process. As the metabolomic scientists interest is often not in absolute quantification of metabolites in the cells, but relative quantification between different populations, an inherent assumption in these experiments is that the quenching and metabolite extraction processes impact the different populations in the same way.

In spite of these limitations, our understanding of the highly unique and specific alterations in microbial metabolism in response to metals has been greatly enhanced by metabolomics methods. The studies described here provide some clues as to how bacteria can adapt, or exist in a uniquely 'primed' state for exposure to toxicants. There remains much to do in the laboratory, including further testing of mixed microbial communities, testing of complex toxicity profiles consisting of multiple toxicants more closely reflective of real-world situations under chronic exposure conditions. From a technical perspective, further expansion of the metabolome coverage would also be advantageous, for example using liquid chromatography-mass spectrometry (LC-MS) methods. In summary, metabolomics experiments in the laboratory have contributed significantly to our understanding of the mechanisms by which microbial populations are metabolically differentiated and, in fact, these types of studies could have important industrial applications.

4 Potential Industrial Uses of Community Metabolomics

4.1 Using Community Metabolomics to Elucidate Metabolic Pathways in Microbial Fuel Cells and Anaerobic Waste Conversion Applications

Effective waste treatment is critical to maintaining ecosystems and human health. Furthermore, in the context of increasing energy costs and the mitigation of greenhouse gas emissions, the development of efficient waste conversion systems is critical. Two related technologies, which promise to help address the challenge of efficient treatment of wastes and wastewaters, are anaerobic digestion (AD) and microbial fuel cells (MFCs).

Digestion (both anaerobic and aerobic) is an established waste treatment technology for residues from various sources, including industrial processes and agriculture (Lester and Birkett 1999). Anaerobic digestion can be defined as the biological mineralisation of organic material to biogas, in the absence of oxygen, by the sequential activity of several microbial groups (Lester and Birkett 1999). It has several advantages over aerobic biological treatment, including lower operating costs and, notably, biogas production (Lettinga 1995). Biogas is similar to natural gas and consists of a mixture of 50–85 % methane and 15–50 % carbon dioxide with trace amounts of other gases. It is a renewable energy source and is used for the generation of heat and electricity, or as vehicle fuel. A very interesting study by Beale et al. (2016), combined metagenomics and community metabolomics to characterise the microbiota in AD digesters, and investigate the resilience of said microbial populations when exposed to operational shocks in the form of temperature and the addition of additional substrates in the form of fats oils and grease. The results provided a good deal of useful information of the structure and function of microbe communities in AD units. It was also found that AD performance was not greatly affected by temperature shocks, but that the addition of fats oils and grease led to significantly promote biogas production.

The anaerobic reactors used for full-scale wastewater treatment and biogas production are high-rate systems based on self-immobilised, or granular, biofilms, e.g. the upflow anaerobic sludge blanket (UASB) reactor (Lettinga et al. 1980; Lettinga 1995). Biofilms are a potentially multi-species community of microbes embedded in a self-generated matrix of extracellular polymeric substance (EPS). They are often found on surfaces but sometimes exist in granules comprising solely EPS and microbes. In the case of the AD applications, each granule theoretically contains all of the trophic groups required for the waste conversion. Increased application of AD could provide benefits, both economically—by the reduction of energy and operational costs—and environmentally—by preserving fossil fuels and reducing emissions for CO_2, particulate matter and other pollutants.

In the recent years AD research has moved towards: (i) energy-efficient, low-temperature (psychrophilic) waste treatment applications (McHugh et al. 2003), (ii) biorefinery applications for the conversion of a range of non-food feedstocks—such as grass and solid wastes—to valuable products, including organic acids with industrial value, bioplastics and alcohols and (iii) anaerobic fermentation processes for biohydrogen production (Oh and Logan 2005) and electricity generation in MFCs (Chaudhuri and Lovley 2003).

MFCs are bioreactors, divided into two compartments separated by means of a membrane that is permeable to cations but is impermeable to oxygen; electrodes are inserted into the compartments and connected through an electrical circuit (Min et al. 2005). Typically, anaerobic microbial consortia oxidise an organic substrate (e.g. glucose or secondary wastewater) in the anodic compartment; electrons generated in the reaction are transferred to the anode and delivered to the cathodic compartment in which they reduce an oxidised substrate (e.g. oxygen or oxidised metal). Thus, the MFC converts chemical energy directly into electrical energy.

The energy potentially available in MFCs is significant, but their efficiency is limited by several factors. These include the presence of electron acceptors in the medium, the incomplete oxidation of the substrate and the kinetic limitations in electron transfer from microorganisms to the anode. This latter process was previously thought to occur strictly through redox mediators that shuttle the electrons. However, only a decade ago, microbes capable of direct electron transfer have been discovered (Min et al. 2005). Although not currently capable of large-scale electricity generation, MFCs present an attractive technology for self-sufficient wastewater treatment with modest electricity generation.

Complex microbial communities underpin AD and MFC applications but the metabolic roles of the individual microbes are still largely undetermined. Moreover, the species are strongly connected through syntrophies where the waste products of one provide resources for another. Consequently, to understand, exploit and extend the application of these systems the 'ecophysiological' roles of the individual populations and species must be determined. Questions, such as *how do the resulting syntrophic interactions structure the system-level behaviour?* can then be addressed to support full-scale engineering and operation. We propose that a unification of metabolomics and other 'omics approaches—particularly metagenomics—could provide a high-throughput solution to link taxonomy with function. This approach could also enable the construction and validation of 'ecosystems biology' models operating at the level of the whole community.

4.2 State of the Science

The vast majority of organisms in anaerobic reactor biofilms and MFCs have not yet been cultured. To study these organisms requires the application of techniques from the 'molecular toolbox' to characterize the complex, mixed microbial consortia present. Much prior research has focused on extracting DNA from these communities and polymerase chain reaction (PCR)-amplifying 16S rRNA genes. This gene is present in all prokaryotes and can be used as a marker for the taxa present (Clarridge 2004). To date, most 16S rRNA studies have used either gel-based, fingerprinting techniques, such as DGGE, which provides a 'barcode' of community diversity but without direct sequence information, or clone libraries, which do provide direct sequence reads but through cost and time constraints are limited in size (Phung et al. 2004). Both methods will fail to resolve rare taxa. Quantitative PCR assays can be used to obtain absolute abundances of even rare organisms but only for a predetermined target.

Until recently there was no economical method capable of resolving the complete community structure. Next-generation sequencing technologies have transformed this situation by providing orders-of-magnitude more sequence data at the same cost (Margulies et al. 2005). This is driving the 'omics revolution', allowing the sequencing of complete genomes from isolated organisms and environmental DNA.

Application to MFCs and AD systems is still in its infancy. The genomes of potentially important species have, however, been sequenced; for example,

Geobacter sulfurreducens (Methe et al. 2003) and *Shewanella oneidensis* (Heidelberg et al. 2002), both of which appear capable of direct mediator-less electron transfer in MFCs. These two genomes have been extensively studied and their genes annotated to functions which have been confirmed by experiments and regulatory networks inferred from transcription profiles. This information has also been integrated into metabolic pathway models. However, even for these two highly studied organisms, a complete systems level understanding is still some way off (Fredrickson et al. 2008). Indeed, to extend this highly reductionist approach to all organisms that could be present in a system that is open to the environment—although potentially desirable—is impossible.

Metagenomics can be considered as the extension of genome sequencing to a community. In shotgun metagenomics, short reads are taken indiscriminately from all organisms and then assembled into longer individual genome fragments or contigs (Handelsman 2004). When only a few genomes are present, it is possible to obtain long contigs, and even complete genomes, from which the same inferences regarding gene function and putative metabolic pathways as for a single genome can be made. This technique can prove very effective for simple communities such as those associated with the breakdown of a particular compound in anaerobic bioreactors (Lykidis et al. 2011). Gene annotation is, however, only the suggestion of potential function; the methods are far from perfect, and that gene may either not be actively expressed or its associated protein could be rendered inactive by post-translational processes. For more complex communities, metagenomics will fail to recover substantially sized contigs unless the sequencing depth is very high. This motivates the amplification of focal genes, most often marker genes such as the 16S rRNA. These amplified gene fragments can then be sequenced to very high depth without dissipating effort through the whole meta-genome and sequences can be recovered from a substantial portion of the community. Unfortunately, the diversity of information associated with the genome fragments of metagenomics is lost and the connection of taxonomy to function is even more tenuous.

Using a careful combination of molecular and analytical techniques, it is possible to link taxa to function in these systems. This can be illustrated with a case study: the detection of *Crenarchaeota* in sludge granules from AD bioreactors (Amann et al. 1990). These organisms were first observed in bioreactors using 16S rRNA gene cloning. Fluorescence in situ hybridisation (FISH), a microscopy technique based on probing mixed-species specimens with fluorescently tagged oligonucleotides, was used to determine the spatial location of *Crenarchaeota* cells which were observed in close association with acetate-consuming methanogens (Collins et al. 2005). Based on this, it was hypothesised that the *Crenarchaeota* in AD bioreactors are H_2-oxidising autotrophs that generate acetate for syntrophic, methanogenic partners. Incubations of granules with radiolabelled acetate (^{14}C-sodium acetate), followed by biofilm cross-sectioning and beta-microimaging to detect isotope label uptake, revealed that the acetate was confined to zones dominated by the co-located *Crenarchaeota* and acetoclastic methanogens (Collins et al. 2005). Stable isotope probing (DNA-SIP) experiments using ^{13}C-bicarbonate and sludge granules, with separate PCR-amplification of 16S rRNA genes from 'heavy' and 'light' DNA fractions, indicated the autotrophic—or at least, mixotrophic—potential of crenarchaeal clones. This is a fascinating example of how culture-independent techniques, and disparate datasets, can together reveal the metabolic function of target groups, but it is not a high-throughput and generically applicable approach while metagenomics is a high-throughput technique, the connection to function is not direct.

4.3 An Integrated Approach for 'Ecosystems Biology'

An integration of 'omics techniques could provide the high-throughput approach required to link taxonomy with function, and to deconstruct the complex microbial communities found in AD bioreactors and MFCs. Next-generation sequencing of marker genes will provide a high-resolution picture of community structure generating relative frequencies of even low-abundance taxa. It is important to sample the minor community constituents, as it has been hypothesised that this so-called 'rare biosphere' may be important in ecosystem functioning (Sogin et al. 2006). This can be coupled to techniques capable of resolving the absolute abundance of higher level taxonomic groups, for example microfluidics cell-counting coupled to quantitative PCR or fluorescence labelling of marker genes. This approach can determine 'who' is there—and in what numbers; metagenomics then allows determination of the functional genes they possess. Annotating those genes can provide a framework of potential metabolic pathways on which further information can be attached. Linking the same marker genes that have been amplified through co-occurrence on contigs to the functional genes could allow the compartmentalisation of metabolic functions to particular species and, consequently, enable a connection to taxa abundance.

The next step is to resolve from these potential pathways the true metabolic processes occurring in the community; central to this is metabolomics. Using methods such as NMR and GC-MS, it is possible to determine the concentrations of small metabolite intermediates along the active pathways. Even more powerful is to couple this with experiments pulsing stable isotope-labelled substrates through the community—through the application of flux balance analysis—to calculate the rates of reaction along targeted pathways. At the same time, mass balance measurements based on resolving the concentrations of input substrates and ions in MFCs and bioreactors, and reaction endpoints, provide a useful corollary. Less obviously, further 'omics information' can be incorporated to resolve ambiguities, or to provide verification of the predictions. For example, transcriptomics can be used to determine which functional genes are expressed and proteomics can be utilised to search for enzymes. There is also a role for more targeted techniques, such as FISH to identify the spatial location of particular groups of organisms.

In summary, metagenomics can be married to community metabolomics to directly elucidate the functional role of whole (but exceedingly diverse) microbial communities. This is an experimental strategy for obtaining the information necessary to resolve the pathways in individual samples and to link the components of those pathways to species. Its success, however, is dependent on coupling it to a computational pipeline capable of extracting the information and an experimental strategy that maximises the statistical power of the complete data set to give meaningful and valid results. The individual components of such a pipeline exist; it is possible to identify microbial taxa from marker genes using resources such as the Ribosomal Database Project (Cole et al. 2009); metagenomics reads can be annotated to functional genes using MG-RAST (Meyer et al. 2008); and potential metabolic pathways can be constructed from the KEGG database (Kanehisa et al. 2004). Similarly, tools exist for converting GC-MS peaks into predictions of molecule identity (Horai et al. 2010; Atherton et al. 2006). The key will be integrating the information in a single pipeline within a single statistical framework that allows for errors, mislabelling and potentially contradictory information.

Deriving the optimal or most likely set of possible pathways and their divisions amongst species will be a highly complex optimisation problem. A Bayesian probabilistic approach will most likely provide the best method to solve this. This will have the advantage of predicting not a single solution but a complete distribution of possible results. Finally, any statistical technique is limited by the quality of the data set. It is vital to link these tools to experiments that maximise the information available. This should be achieved by running replicate systems to determine the variability of the responses and, conversely, by using carefully planned perturbations to expand the range of system states explored. These perturbations could be through changing substrate concentrations or operating conditions, such as temperature. These requirements underpin the importance of the high-throughput nature of 'omics techniques to obtain, at reasonable cost, multiple datasets, each containing huge amounts of information.

The final step of an ecosystems biology approach is integrating this information into mathematical models capable of predicting both community composition and bioreactor function. These will take the metabolic-taxonomic framework elucidated above and add dynamics. The actual mathematical structures to do this are well established: in a well-mixed bioreactor, a system of ordinary differential equations would be appropriate, possibly with the addition of environmental noise; more complex models will be necessary if spatial localisation of groups of organisms in anaerobic granules or on the anodes of MFCs are deemed important. The greater challenge than the model structure will be parameterisation. Once again a Bayesian approach will be vital to propagate that uncertainty into predictions that formally account for it.

4.4 Outlook for the Future

The holistic, polyphasic approach outlined above can be viewed as systems biology at the community level. It is not desirable to reconstruct and model every reaction and process in every microbial cell, but it is desirable to identify those pathways, syntrophies and ecological interactions that are critical for controlling the community—and by extension, process—behaviour. The integration of 'omics' approaches coupled to process monitoring, and advanced microscopy, can generate comprehensive, integrated datasets at microorganism, biofilm and bioreactor level. This will enable the link between processes occurring at microorganism level (scale c. 1 μm–1 mm) and the processes occurring within bioreactors (scale > 1 m). This will, in turn, lead to improved reactor monitoring capability, reactor design, performance and operational stability.

5 Upcoming Challenges: What Can Community Metabolomics Learn from Metagenomics?

5.1 The Challenges Ahead

The past two decades has seen the rapid expansion of 'omics' oriented research, examining the abundance of the vast array of genes (metagenomics), RNA molecules (meta-transcriptomics), proteins (meta-proteomics) and small molecule metabolites (meta/community metabolomics) present in a wide range of different environments. To date, it is genomics that has been the poster child of the omics era, with the global media avidly following the progress of high-profile research projects that have enabled us to map the human genome (Human Genome Sequencing 2004), the genomes of various economically important plants (Goff et al. 2002), animals (Hillier et al. 2004) and disease causing organisms (Heidelberg et al. 2000). More recently, we have begun to describe the vast content of 'metagenomic' DNA, seemingly aiming to catalogue the entirety of life around us using the DNA barcode.

As our analysis of the global meta-genome continues to benefit from recent advances in robotics, DNA sequencing chemistries, analytical and computational procedures, the costs associated with sequencing such large volumes of DNA continue to tumble. For example, the consumable costs associated with sequencing the human genome have decreased from tens of millions of (US) dollars in 2007, to just a few thousand (US) dollars or under today (Drmanac et al. 2010). With so much data now being generated, the major challenges associated with most large-scale metagenomic studies concern data storage and how to make sense of the vast amounts of DNA sequence data generated.

There are several terms in use for the application of metabolomics in ecological/environmental studies. For instance, the term "Ecotoxicogenomics" was proposed by Snape et al. (2004) to describe the integration of genomic-based science into the field of ecotoxicology. "EcoGenomics" (or ecological genomics) was used by Chapman (2001) to describe the application of genomics based techniques to ecology. In both cases the term genomics was taken to encompass all the 'omic sciences' namely genomics (genome sequencing and the annotation of function to genes), transcriptomics (gene expression at the transcription level), proteomics (protein abundance) and metabolomics (metabolite/small molecule levels). In addition, the phrase "Environmental metabolomics" was defined by Viant (2007) as the "application of metabolomics to characterize the metabolism of free-living organisms obtained from a natural environment, and of organisms reared under laboratory conditions, where those conditions serve to mimic scenarios encountered in the natural environment".

Below, we describe how researchers in the field of metagenomics are meeting the new challenges provided by their new 'mega-data sets' and how researchers of community metabolomics might benefit from the many lessons learnt by its 'bigger brother', metagenomics.

5.2 *The Need for Better Reference Samples*

To date, most attempts to 'sequence' the meta-genome of varied environments such as the soil, marine waters or the human intestinal tract have been largely superficial, only exploring as much of the meta-genome as is necessary to enable some descriptions of the diversity and function of the community to be made, and with the volume of data analysed normally dictated by financial or technical constraints. There is therefore an urgent need for better reference data, comprising the complete metagenomic data of various sample media. Several years ago, an international Soil Meta-genome Consortium, 'Terragenome', (http://www.terragenome.org/) was established (Vogel et al. 2009) with the aim of generating the first complete meta-genome sequence of a complex environmental system (soil from Park Grass, Rothamsted, UK). This study will provide reference data, to which data collected from all other soils around the world can be compared. It is only with the provision of these 'full' datasets that we will be able to answer key questions including: What is the extent of microbial diversity and what is the diversity of functional genes within the sample media? What is the abundance and functional significance of each species within the soil, across different domains of life (e.g., bacteria, fungi, protozoa, etc.)? How does the composition and function of the community differ across spatial and temporal scales? And, what 'core' genes are present in most soils?

While it may take many years of data collection and analysis before the complexity of the soil meta-genome is fully understood, no serious efforts are currently underway to catalogue the entire community metabolome of similarly complex environmental samples. The generation of entire reference sets of community metabolomic data from complex communities is essential if we are able to better understand the diversity of functional activities undertaken within them and to prevent community metabolomics being out-shadowed by its metagenomics for years to come.

5.3 The Importance of Meta-Data

Meta-data (or data about data) provides an essential environmental context to metagenomic data, detailing key factors including site and sample descriptions, chemical and physical properties and methods of analysis. To ensure that valid comparisons can be made between the metagenomic data collected by different researchers at different times and locations, and to reduce any unnecessary experimental duplication, it is prudent that researchers should adhere to a standard set of reporting requirements. Possibly inspired by the broadly accepted MIAME (Minimal Information about a Microarray Experiment) standard (Brazma et al. 2001), the Genomics Standard Consortium (GSC, http://gensc.org) has driven the development of MIGS (Minimal Information about a Genome Sequence), MIMS (Minimal Information about a Metagenome Sequence) and MIENS (Minimal Information about an Environmental Sequence) standards. Similar initiatives have also been undertaken to provide comprehensive reporting standards for environmentally derived metabolomics data (Morrison et al. 2007).

However, despite the efforts of such groups, most current studies of the meta-genome (and community metabolome) do not attempt to adhere to such standards. If the vast increases in data generated by metagenomics and community metabolomics are to be used to their greatest potential, it is a matter of urgency that standards for the provision of metadata will be adopted and required for major scientific journals, as they are already required for microarray experiments (e.g., see instructions for authors in the ISME Journal). Such a universal approach would help to avoid unnecessary duplications of effort, whilst also benefiting various fields of ecosystems biology, by helping to improve our understanding of the 'connectivity' between the parts lists generated by the ever growing number of omics approaches (including metagenomics, meta-transcriptomics, meta-proteomics, and community metabolomics).

5.4 Data Storage and Sharing

Modern molecular methods have provided tools for the global scientific community to produce a continual deluge of DNA sequence data. These data, which are generated at an exponential rate, perhaps doubling in volume every 18 months (Lathe et al. 2008), now requires terabyte-scale computation and ever expanding storage facilities to reduce the widening gap between rates of data collection and interpretation. Whilst it is the aim of researchers to increase the volume of useful DNA data stored in their own databases and in online repositories such as NCBI's GenBank (www.ncbi.nlm.nih.gov/genbank/), much of the data being stored by researchers are of limited value. Increasingly, it is being recognised that much, or even most, of the storage space is used to house data that could be thrown away. For this reason, it is now common practice to delete raw data files (e.g.

electropherogram data) immediately after the data have been processed and condensed into far smaller text files containing the DNA sequence data. Due to recent technical advances, the amount of data produced by metabolomic studies is beginning to pose similar challenges and new standards of data storage and interpretation are now required to minimise the cataloguing of redundant information.

5.5 Knowledge Transfer to Applied Research Outcomes

Even once all of our data have been collected, adequately stored and compared to appropriate reference and metadata there remain many challenges to maximise the value of the large datasets characteristic of meta-omic studies. To date, many large-scale metagenomic and community metabolomic studies have sought only to address primarily pure, or fundamental, research aims, cataloguing the complexity of various biological systems and to provide new insights into the functional capabilities of many different microbial communities. Whilst these are worthy research aims, it remains crucial that our new understanding of microbial diversity and function be exploited to its full potential, addressing much broader, applied research questions of benefit to key industrial partners, healthcare providers and environmental protection agencies.

The development and application of new, more sensitive, biological indicators of environmental health will be a key growth area of metagenomics and metabolomics led research. For example, there is an increasing interest in the monitoring and restoration of freshwater systems, aimed at maximising their value for ecological, recreation and economic purposes. Macroinvertebrate communities have been used around the globe as biological indicators of stream health for decades. Extensive research has permitted classification of a wide range of macroinvertebrate taxa into pollution tolerant, sensitive and facultative categories which are used to provide indices of water quality (e.g., Hilsenhoff 1987) and a well-researched body of information now supports their use (Feld and Hering 2007; Trigal et al. 2007).

This macroinvertebrate community data provide little or no information as to the potential causes of any observed declines in freshwater ecological health. However, recent advances in meta-omics centred research mean that it is now possible to rapidly characterise (i) the abundance of ‘functional groups’ of bacteria, such as those conferring resistance to any heavy metal (e.g., cobalt, zinc and cadmium) or the resistance and ability to degrade organic pollutants (e.g., PCBs), (ii) the abundance and activity of genes encoding for these ‘key’ functions and (iii) the presence and abundance of functionally related metabolites. As such, the era of meta-omics research promises to revolutionise the application of biological indicators such that the rapid analysis of microbial communities may not only be used to provide an indication of general ecological stress, but also to identify specific drivers of environmental degradation within any site. This is of particular importance for the restoration of urban ecosystems in which the primarily drivers of environmental changes are frequently hard to distinguish amongst the large background of other potentially harmful anthropogenic influences using traditional chemical analysis.

In contrast, the holistic approach adopted by metagenomic and metabolomic studies aid the identification of key drivers of microbial community function, akin to analysing the total chemical profile within a sample. This information can be used to design highly targeted restoration strategies on a site-specific basis, with no prior knowledge of the site's history required. The field of biomedical sciences similarly stands to benefit from predicted increases in metagenomic and community metabolomic derived data. For example, a number of studies have already sought to catalogue the presence and abundance of microbial genes within the human gut, also identifying key differences in the gut microflora of healthy individuals and patients with inflammatory bowel disease (Qin et al. 2010). By improving our understanding of the links between the diversity and function of microbial communities, and their metabolites, on and in the human body, it is expected that we may devise better treatments and preventative care for a wide range of microbial diseases. For example, could changes in the abundance of certain bacterial genes or metabolites in the human gut be used as an early warning signal to detect the onset of inflammatory bowel disease? It is already evident that the primary legacy of the new 'omics' era of microbial research will not be the catalogue of parts that list diversity of microbial life on earth, but it is possible by applying our new knowledge of the complex relationships between the functional capability and versatility of microbial communities and the complex environments in which they live.

Acknowledgments OAHJ thanks colleagues from the Australia and New Zealand Metabolomics Network (ANZMN), Proteomics and Metabolomics Victoria (PMV), RMIT University (Australia), the University of Cambridge (UK) and the Centre for Ecology and Hydrology Wallingford (UK) for help and support in metabolomics research over the years.

References

Amann RI, Binder BJ, Olson RJ, Chisholm SW, Devereux R, Stahl DA (1990) Combination of 16S ribosomal-RNA-targeted oligonucleotide probes with flow-cytometry for analyzing mixed microbial-populations. Appl Environ Microbiol 56:1919–1925

Atherton H, Bailey N, Zhang W, Taylor J, Major H, Shockcor J, Clarke K, Griffin J (2006) A combined 1H-NMR spectroscopy- and mass spectrometry-based metabolomic study of the PPAR-alpha null mutant mouse defines profound systemic changes in metabolism linked to the metabolic syndrome. Physiol Genomics 27:178–186

Beale DJ, Karpe AV, McLeod JD, Gondalia SV, Muster TH, Othman MZ, Palombo EA, Joshi D (2016) An 'omics' approach towards the characterisation of laboratory scale anaerobic digesters treating municipal sewage sludge. Water Res 88:346–357

Bochner BR, Gadzinski P, Panomitros E (2001) Phenotype microarrays for high-throughput phenotypic testing and assay of gene function. Genome Res 11:1246–1255

Boles BR, Thoendel M, Singh PK (2004) Self-generated diversity produces "insurance effects" in biofilm communities. Proc Nat Aca Sci USA 101:16630–16635

Booth SC, Workentine ML, Wen J, Shaykhutdinov R, Vogel HJ, Ceri H, Turner RJ, Weljie AM (2011) Differences in metabolism between the biofilm and planktonic response to metal stress. J Proteome Res 10:3190–3199

Brazma A, Hingamp P, Quackenbush J, Sherlock G, Spellman P, Stoeckert C, Aach J, Ansorge W, Ball CA, Causton HC, Gaasterland T, Glenisson P, Holstege FCP, Kim IF, Markowitz V, Matese JC, Parkinson H, Robinson A, Sarkans U, Schulze-Kremer S, Stewart J, Taylor RG, Vilo J, Vingron M (2001) Minimal information about a microarray experiment (MIAME)—towards standards for microarray data. Nat Genet 29:365–371

Bundy JG, Willey TL, Castell RS, Ellar DJ, Brindle KM (2005) Discrimination of pathogenic clinical isolates and laboratory strains of *Bacillus cereus* by NMR-based metabolomic profiling. FEMS Microbiol Lett 242:127–136

Castillo JA, Staton SJ, Taylor TJ, Herckes P, Hayes MA (2012) Exploring the feasibility of bioaerosol analysis as a novel fingerprinting technique. Anal Bioanal Chem 403:15–26

Chapman RW (2001) EcoGenomics—a consilience for comparative immunology? Dev Comp Immunol 25:549–551

Chasteen TG, Bentley R (2002) Biomethylation of Selenium and Tellurium: microorganisms and plants. Chem Rev 103:1–26

Chaudhuri SK, Lovley DR (2003) Electricity generation by direct oxidation of glucose in mediatorless microbial fuel cells. Nat Biotechnol 21:1229–1232

Clarridge JE (2004) Impact of 16S rRNA gene sequence analysis for identification of bacteria on clinical microbiology and infectious diseases. Clin Microbiol Rev 17:840–862

Cole JR, Wang Q, Cardenas E, Fish J, Chai B, Farris RJ, Kulam-Syed-Mohideen AS, McGarrell DM, Marsh T, Garrity GM, Tiedje JM (2009) The Ribosomal Database Project: improved alignments and new tools for rRNA analysis. Nucleic Acids Res 37:D141–D145

Collins G, Mahony T, McHugh S, Gieseke A, de Beer D, O'Flaherty V (2005) Distribution, dynamics and in situ ecophysiology of *Crenarchaeota* in anaerobic wastewater treatment granular biofilms. Water Sci Technol 52:233–239

Drmanac R, Sparks AB, Callow MJ, Halpern AL, Burns NL, Kermani BG, Carnevali P, Nazarenko I, Nilsen GB, Yeung G, Dahl F, Fernandez A, Staker B, Pant KP, Baccash J, Borcherding AP, Brownley A, Cedeno R, Chen LS, Chernikoff D, Cheung A, Chirita R, Curson B, Ebert JC, Hacker CR, Hartlage R, Hauser B, Huang S, Jiang Y, Karpinchyk V, Koenig M, Kong C, Landers T, Le C, Liu J, McBride CE, Morenzoni M, Morey RE, Mutch K, Perazich H, Perry K, Peters BA, Peterson J, Pethiyagoda CL, Pothuraju K, Richter C, Rosenbaum AM, Roy S, Shafto J, Sharanhovich U, Shannon KW, Sheppy CG, Sun M, Thakuria JV, Tran A, Vu D, Zaranek AW, Wu XD, Drmanac S, Oliphant AR, Banyai WC, Martin B, Ballinger DG, Church GM, Reid CA (2010) Human genome sequencing using unchained base reads on self-assembling DNA nanoarrays. Science 327:78–81

Eisen JA (2007) Environmental shotgun sequencing: its potential and challenges for studying the hidden world of microbes. PLoS Biol 5:e82

Feld CK, Hering D (2007) Community structure or function: effects of environmental stress on benthic macroinvertebrates at different spatial scales. Freshw Biol 52:1380–1399

Feng X, Simpson AJ, Wilson KP, Dudley Williams D, Simpson MJ (2008) Increased cuticular carbon sequestration and lignin oxidation in response to soil warming. Nat Geosci 1:836–839

Fredrickson JK, Romine MF, Beliaev AS, Auchtung JM, Driscoll ME, Gardner TS, Nealson KH, Osterman AL, Pinchuk G, Reed JL, Rodionov DA, Rodrigues JLM, Saffarini DA, Serres MH, Spormann AM, Zhulin IB, Tiedje JM (2008) Towards environmental systems biology of *Shewanella*. Nat Rev Microbiol 6:592–603

Garcia MH, Ivanova N, Kunin V, Warnecke F, Barry KW, McHardy AC, Yeates C, He S, Salamov AA, Szeto E, Dalin E, Putnam NH, Shapiro HJ, Pangilinan JL, Rigoutsos I, Kyrpides NC, Blackall LL, McMahon KD, Hugenholtz P (2006) Metagenomic analysis of two enhanced biological phosphorus removal (EBPR) sludge communities. Nat Biotechnol 24:1263–1269

Goff SA, Ricke D, Lan TH, Presting G, Wang RL, Dunn M, Glazebrook J, Sessions A, Oeller P, Varma H, Hadley D, Hutchinson D, Martin C, Katagiri F, Lange BM, Moughamer T, Xia Y,

Budworth P, Zhong JP, Miguel T, Paszkowski U, Zhang SP, Colbert M, Sun WL, Chen LL, Cooper B, Park S, Wood TC, Mao L, Quail P, Wing R, Dean R, Yu YS, Zharkikh A, Shen R, Sahasrabudhe S, Thomas A, Cannings R, Gutin A, Pruss D, Reid J, Tavtigian S, Mitchell J, Eldredge G, Scholl T, Miller RM, Bhatnagar S, Adey N, Rubano T, Tusneem N, Robinson R, Feldhaus J, Macalma T, Oliphant A, Briggs S (2002) A draft sequence of the rice genome (*Oryza sativa* L. ssp japonica). Science 296:92–100

González AG, Shirokova LS, Pokrovsky OS, Emnova EE, Martínez RE, Santana-Casiano JM, González-Dávila M, Pokrovski GS (2010) Adsorption of copper on *Pseudomonas aureofaciens*: protective role of surface exopolysaccharides. J Colloid Interface Sci 350:305–314

Handelsman J (2004) Metagenomics: application of genomics to uncultured microorganisms. Microbiol Mol Biol Rev 68:669–685

Harrison JJ, Ceri H, Turner RJ (2007) Multimetal resistance and tolerance in microbial biofilms. Nat Rev Microbiol 5:928–938

Heidelberg JF, Eisen JA, Nelson WC, Clayton RA, Gwinn ML, Dodson RJ, Haft DH, Hickey EK, Peterson JD, Umayam L, Gill SR, Nelson KE, Read TD, Tettelin H, Richardson D, Ermolaeva MD, Vamathevan J, Bass S, Qin HY, Dragoi I, Sellers P, McDonald L, Utterback T, Fleishmann RD, Nierman WC, White O, Salzberg SL, Smith HO, Colwell RR, Mekalanos JJ, Venter JC, Fraser CM (2000) DNA sequence of both chromosomes of the cholera pathogen *Vibrio cholerae*. Nature 406:477–483

Heidelberg JF, Paulsen IT, Nelson KE, Gaidos EJ, Nelson WC, Read TD, Eisen JA, Seshadri R, Ward N, Methe B, Clayton RA, Meyer T, Tsapin A, Scott J, Beanan M, Brinkac L, Daugherty S, DeBoy RT, Dodson RJ, Durkin AS, Haft DH, Kolonay JF, Madupu R, Peterson JD, Umayam LA, White O, Wolf AM, Vamathevan J, Weidman J, Impraim M, Lee K, Berry K, Lee C, Mueller J, Khouri H, Gill J, Utterback TR, McDonald LA, Feldblyum TV, Smith HO, Venter JC, Nealson KH, Fraser CM (2002) Genome sequence of the dissimilatory metal ion-reducing bacterium Shewanella oneidensis. Nat Biotechnol 20:1118–1123

Hillier LW, Miller W, Birney E, Warren W, Hardison RC, Ponting CP, Bork P, Burt DW, Groenen MAM, Delany ME, Dodgson JB, Chinwalla AT, Cliften PF, Clifton SW, Delehaunty KD, Fronick C, Fulton RS, Graves TA, Kremitzki C, Layman D, Magrini V, McPherson JD, Miner TL, Minx P, Nash WE, Nhan MN, Nelson JO, Oddy LG, Pohl CS, Randall-Maher J, Smith SM, Wallis JW, Yang SP, Romanov MN, Rondelli CM, Paton B, Smith J, Morrice D, Daniels L, Tempest HG, Robertson L, Masabanda JS, Griffin DK, Vignal A, Fillon V, Jacobbson L, Kerje S, Andersson L, Crooijmans RPM, Aerts J, van der Poel JJ, Ellegren H, Caldwell RB, Hubbard SJ, Grafham DV, Kierzek AM, McLaren SR, Overton IM, Arakawa H, Beattie KJ, Bezzubov Y, Boardman PE, Bonfield JK, Croning MDR, Davies RM, Francis MD, Humphray SJ, Scott CE, Taylor RG, Tickle C, Brown WRA, Rogers J, Buerstedde JM, Wilson SA, Stubbs L, Ovcharenko I, Gordon L, Lucas S, Miller MM, Inoko H, Shiina T, Kaufman J, Salomonsen J, Skjoedt K, Wong GKS, Wang J, Liu B, Yu J, Yang HM, Nefedov M, Koriabine M, deJong PJ, Goodstadt L, Webber C, Dickens NJ, Letunic I, Suyama M, Torrents D, von Mering C, Zdobnov EM, Makova K, Nekrutenko A, Elnitski L, Eswara P, King DC, Yang S, Tyekucheva S, Radakrishnan A, Harris RS, Chiaromonte F, Taylor J, He JB, Rijnkels M, Griffiths-Jones S, Ureta-Vidal A, Hoffman MM, Severin J, Searle SMJ, Law AS, Speed D, Waddington D, Cheng Z, Tuzun E, Eichler E, Bao ZR, Flicek P, Shteynberg DD, Brent MR, Bye JM, Huckle EJ, Chatterji S, Dewey C, Pachter L, Kouranov A, Mourelatos Z, Hatzigeorgiou AG, Paterson AH, Ivarie R, Brandstrom M, Axelsson E, Backstrom N, Berlin S, Webster MT, Pourquie O, Reymond A, Ucla C, Antonarakis SE, Long MY, Emerson JJ, Betran E, Dupanloup I, Kaessmann H, Hinrichs AS, Bejerano G, Furey TS, Harte RA, Raney B, Siepel A, Kent WJ, Haussler D, Eyras E, Castelo R, Abril JF, Castellano S, Camara F, Parra G, Guigo R, Bourque G, Tesler G, Pevzner PA, Smit A, Fulton LA, Mardis ER, Wilson RK (2004) Sequence and comparative analysis of the chicken genome provide unique perspectives on vertebrate evolution. Nature 432:695–716

Hilsenhoff WL (1987) An improved biotic index of organic stream pollution. Great Lakes Entomol 20:31–39

Horai H, Arita M, Kanaya S, Nihei Y, Ikeda T, Suwa K, Ojima Y, Tanaka K, Tanaka S, Aoshima K, Oda Y, Kakazu Y, Kusano M, Tohge T, Matsuda F, Sawada Y, Hirai MY, Nakanishi H, Ikeda K, Akimoto N, Maoka T, Takahashi H, Ara T, Sakurai N, Suzuki H, Shibata D, Neumann S, Iida T, Funatsu K, Matsuura F, Soga T, Taguchi R, Saito K, Nishioka T (2010) MassBank: a public repository for sharing mass spectral data for life sciences. J Mass Spectrom 45:703–714

International Human Genome Sequencing Consortium (2004) Finishing the euchromatic sequence of the human genome. Nature 431:931–945

Lisa T, Casale C, Domenech C (1994) Cholinesterase, acid phosphatase, and phospholipase C of *Pseudomonas aeruginosa* under hyperosmotic conditions in a high-phosphate medium. Curr Microbiol 28:71–76

Jones OAH, Cheung VL (2007) An introduction to metabolomics and its potential application in veterinary science. Comp Med 57:436–442

Jones OAH, Sdepanian S, Lofts S, Svendsen C, Spurgeon DJ, Maguire ML, Griffin JL (2014) Metabolomic analysis of soil communities can be used for pollution assessment. Environ Sci Technol 33:61–64

Kanehisa M, Goto S, Kawashima S, Okuno Y, Hattori M (2004) The KEGG resource for deciphering the genome. Nucleic Acids Res 32:D277–D280

Lathe W, Williams J, Mangan M, Karolchik D (2008) Genomic data resources: challenges and promises. Nature Education 1:3

Lester JN, Birkett JW (1999) Microbiology and chemistry for environmental scientists and engineers, 2nd edn. E and FN Spon Ltd, London

Lettinga G (1995) Anaerobic-digestion and waste-water treatment systems. Antonie Van Leeuwenhoek 67:3–28

Lettinga G, Vanvelsen AFM, Hobma SW, Dezeeuw W, Klapwijk A (1980) Use of the upflow sludge blanket (USB) reactor concept for biological wastewater-treatment, especially for anaerobic treatment. Biotechnol Bioeng 22:699–734

Llewellyn CA, Sommer U, Dupont CL, Allen AE, Viant MR (2015) Using community metabolomics as a new approach to discriminate marine microbial particulate organic matter in the western English Channel. Prog Oceanogr 137(Part B):421–433

Lykidis A, Chen C, Tringe SG, McHardy AC, Copeland A, Kyrpides NC, Hugenholtz P, Macarie H, Olmos A, Monroy O, Liu W (2011) Multiple syntrophic interactions in a terephthalate-degrading methanogenic consortium. ISME J 5:122–130

Margulies M, Egholm M, Altman WE, Attiya S, Bader JS, Bemben LA, Berka J, Braverman MS, Chen YJ, Chen ZT, Dewell SB, Du L, Fierro JM, Gomes XV, Godwin BC, He W, Helgesen S, Ho CH, Irzyk GP, Jando SC, Alenquer MLI, Jarvie TP, Jirage KB, Kim JB, Knight JR, Lanza JR, Leamon JH, Lefkowitz SM, Lei M, Li J, Lohman KL, Lu H, Makhijani VB, McDade KE, McKenna MP, Myers EW, Nickerson E, Nobile JR, Plant R, Puc BP, Ronan MT, Roth GT, Sarkis GJ, Simons JF, Simpson JW, Srinivasan M, Tartaro KR, Tomasz A, Vogt KA, Volkmer GA, Wang SH, Wang Y, Weiner MP, Yu PG, Begley RF, Rothberg JM (2005) Genome sequencing in microfabricated high-density picolitre reactors. Nature 437:376–380

McHugh S, O'Reilly C, Mahony T, Colleran E, O'Flaherty V (2003) Anaerobic granular sludge bioreactor technology. Rev Environ Sci Biotechnol 2:225–245

Methe BA, Nelson KE, Eisen JA, Paulsen IT, Nelson W, Heidelberg JF, Wu D, Wu M, Ward N, Beanan MJ, Dodson RJ, Madupu R, Brinkac LM, Daugherty SC, DeBoy RT, Durkin AS, Gwinn M, Kolonay JF, Sullivan SA, Haft DH, Selengut J, Davidsen TM, Zafar N, White O, Tran B, Romero C, Forberger HA, Weidman J, Khouri H, Feldblyum TV, Utterback TR, Van Aken SE, Lovley DR, Fraser CM (2003) Genome of *Geobacter sulfurreducens*: metal reduction in subsurface environments. Science 302:1967–1969

Meyer F, Paarmann D, D'Souza M, Olson R, Glass EM, Kubal M, Paczian T, Rodriguez A, Stevens R, Wilke A, Wilkening J, Edwards RA (2008) The metagenomics RAST server—a public resource for the automatic phylogenetic and functional analysis of metagenomes. BMC Bioinform 9:386

Min BK, Cheng SA, Logan BE (2005) Electricity generation using membrane and salt bridge microbial fuel cells. Water Res 39:1675–1686

Morrison N, Bearden D, Bundy JG, Collette T, Currie F, Davey MP, Haigh NS, Hancock D, Jones OAH, Rochfort S, Sansone S-A, Stys D, Teng Q, Field D, Viant MR (2007) Standard reporting requirements for biological samples in metabolomics experiments: environmental context. Metabolomics 3:203–210

Nadell CD, Xavier JB, Foster KR (2009) The sociobiology of biofilms. FEMS Microbiol Rev 33:206–224

Nicholson JK, Holmes E, Wilson ID (2005) Gut microorganisms, mammalian metabolism and personalized health care. Nat Rev Microbiol 3:431–438

Oh SE, Logan BE (2005) Hydrogen and electricity production from a food processing wastewater using fermentation and microbial fuel cell technologies. Water Res 39:4673–4682

Peršoh D (2015) Plant-associated fungal communities in the light of meta'omics. Fungal Divers 75:1–25

Petrova OE, Sauer K (2009) A novel signalling network essential for regulating *Pseudomonas aeruginosa* biofilm development. PLoS Pathog 5:e1000668

Phung NT, Lee J, Kang KH, Chang IS, Gadd GM, Kim BH (2004) Analysis of microbial diversity in oligotrophic microbial fuel cells using 16S rDNA sequences. FEMS Microbiol Lett 233:77–82

Portais J-C, Delort A-M (2002) Carbohydrate cycling in micro-organisms: what can 13C-NMR tell us? FEMS Microbiol Rev 26:375–402

Qin J, Li R, Raes J, Arumugam M, Burgdorf KS, Manichanh C, Nielsen T, Pons N, Levenez F, Yamada T, Mende DR, Li J, Xu J, Li S, Li D, Cao J, Wang B, Liang H, Zheng H, Xie Y, Tap J, Lepage P, Bertalan M, Batto JM, Hansen T, Le Paslier D, Linneberg A, Nielsen HB, Pelletier E, Renault P, Sicheritz-Ponten T, Turner K, Zhu H, Yu C, Li S, Jian M, Zhou Y, Li Y, Zhang X, Li S, Qin N, Yang H, Wang J, Brunak S, Dore J, Guarner F, Kristiansen K, Pedersen O, Parkhill J, Weissenbach J, Bork P, Ehrlich SD, Wang J (2010) A human gut microbial gene catalogue established by metagenomic sequencing. Nature 464:59–65

Rochfort S, Ezernieks V, Mele P, Kitching M (2015) NMR metabolomics for soil analysis provide complementary, orthogonal data to MIR and traditional soil chemistry approaches—a land use study. Magn Reson Chem 53:719–725

Sales CM, Lee PKH (2015) Resource recovery from wastewater: application of meta-omics to phosphorus and carbon management. Curr Opin Biotechnol 33:260–267

Scullion J, Elliott GN, Huang WE, Goodacre R, Worgan H, Darby R, Bailey MJ, Gwynn-Jones D, Griffith GW, Winson MK, Williams PA, Clegg C, Draper J (2003) Use of earthworm casts to validate FT-IR spectroscopy as a 'sentinel' technology for high-throughput monitoring of global changes in microbial ecology: the 7th international symposium on earthworm ecology, Cardiff, Wales, 2002. Pedobiologia 47:440–446

Snape JR, Maund SJ, Pickford DB, Hutchinson TH (2004) Ecotoxicogenomics: the challenge of integrating genomics into aquatic and terrestrial ecotoxicology. Aquat Toxicol 67:143–154

Sogin ML, Morrison HG, Huber JA, Mark Welch D, Huse SM, Neal PR, Arrieta JM, Herndl GJ (2006) Microbial diversity in the deep sea and the underexplored "rare biosphere". Proc Natl Acad Sci USA 103:12115–12120

Swenson TL, Jenkins S, Bowen BP, Northen TR (2015) Untargeted soil metabolomics methods for analysis of extractable organic matter. Soil Biol Biochem 80:189–198

Timmins ÉM, Howell SA, Alsberg BK, Noble WC, Goodacre R (1998) Rapid differentiation of closely related *Candida* species and strains by pyrolysis mass spectrometry and fourier transform infrared spectroscopy. J Clin Microbiol 36:367–374

Tremaroli V, Workentine M, Weljie AM, Vogel H, Ceri H, Viti C, Tatti E, Zhang P, Hynes A, Turner R, Zannoni D (2009) Metabolomic investigation of the bacterial response to a metal challenge. Appl Environ Microbiol 75:719–728

Trigal C, Garcia-Criado F, Alaez CF (2007) Macroinvertebrate communities of mediterranean ponds (North Iberian Plateau): importance of natural and human-induced variability. Freshw Biol 52:2042–2055

van Gulik WM (2010) Fast sampling for quantitative microbial metabolomics. Curr Opin Biotechnol 21:27–34

Venter JC, Remington K, Heidelberg JF, Halpern AL, Rusch D, Eisen JA, Wu D, Paulsen I, Nelson KE, Nelson W, Fouts DE, Levy S, Knap AH, Lomas MW, Nealson K, White O, Peterson J, Hoffman J, Parsons R, Baden-Tillson H, Pfannkoch C, Rogers YH, Smith HO (2004) Environmental genome shotgun sequencing of the Sargasso Sea. Science 304:58–60

Viant MR (2007) Metabolomics of aquatic organisms: the new 'omics' on the block. Mar Ecol Prog Ser 332:301–306

Vogel TM, Simonet P, Jansson JK, Hirsch PR, Tiedje JM, van Elsas JD, Bailey MJ, Nalin R, Philippot L (2009) TerraGenome: a consortium for the sequencing of a soil metagenome. Nat Rev Microbiol 7:252

Wilkins M (2009) Proteomics data mining. Expert Rev Proteomics 6:599–603

Workentine ML, Harrison JJ, Weljie AM, Tran VA, Stenroos PU, Tremaroli V, Vogel HJ, Ceri H, Turner RJ (2010) Phenotypic and metabolic profiling of colony morphology variants evolved from *Pseudomonas fluorescens* biofilms. Environ Microbiol 12:1565–1577

Zannoni D, Borsetti F, Harrison JJ, Turner RJ (2008) The bacterial response to the chalcogen metalloids Se and Te. Adv Microb Physiol 53:1–72

Chapter 8
Metabolomics: Applications to Food Safety and Quality Research

Farhana R. Pinu

1 Introduction

Food safety is a major concern worldwide, which has gained enormous attention in the last two decades mainly because of the emergence of new food-borne pathogens and other chemical hazards. Additionally, there has been an extreme increase in the food-borne illness incidences along with large-scale outbreaks. Many factors contribute to these growing concerns, including the industrialisation and mass production of agricultural products, the increasing number of imported food products, and changes in food consumption patterns due to consumer lifestyle changes (Motarjemi et al. 2008). Ready to eat and fast foods, especially, have gained great popularity, and thus also resulted in many food-borne illness incidences in recent years. Moreover, the tremendous upsurge in global population is also almost forcing producers towards mass production of agricultural products without giving much attention to quality and safety issues (Motarjemi and Lelieveld 2014). The world has already seen an introduction of a massive amount of genetically modified (GM) crops in the last 40 years, in response to the need to feed over 6.5 billion people. Consumers are doubtful about the safety of consuming GM food products; therefore, scientific attention is required to unravel the facts about the safety of eating such foods (Pinu 2015).

In many third-world countries, food quality and safety are often overlooked. In addition, many food vendors deliberately contaminate food products with unwanted materials (e.g. melamine in powdered milk) to increase profits, and others do so

F.R. Pinu (✉)
The New Zealand Institute for Plant and Food Research Limited,
Private Bag, 92169, Auckland 1142, New Zealand
e-mail: farhana.pinu@plantandfood.co.nz

F.R. Pinu
The Liggins Institute, University of Auckland, Private Bag, 92019, Auckland, New Zealand

D.J. Beale et al. (eds.), *Microbial Metabolomics*,
DOI 10.1007/978-3-319-46326-1_8

unintentionally because of their lack of knowledge of food hygiene and safety (Unnevehr 2015). Therefore, insecurity about raw foods and other food products is on the rise because of the deterioration of quality in different food products that are produced in large amounts and the lack of adequate information about their nutritional properties. However, the good news is that consumers around the world are now more aware about the risks and safety of different food products because of the availability of information via the internet and social media (Pinu 2015). Therefore, there is growing interest and a need from both food producers and consumers in implementing proper management of food safety and quality.

Quality control is a routine activity in different food industries, and can ensure the adequate safety of food, helping to maintain the trust of consumers. However, there is a strong demand for the development of techniques that might help to determine the key chemical, microbiological and nutritional features of a food product. Although many technologies are already available for the routine checking of many food products, these can be inadequate and often cannot determine emerging chemical and microbiological hazards present in foods. There is indeed scope for improvement and this is where metabolomics has a potential in food safety and quality research—mainly by introducing, developing and improving techniques to detect contaminants (both microbiological and chemical) present in food products. In this chapter, the ways in which metabolomics can be used (both targeted and non-targeted approaches) to ensure the safety and quality of raw, processed and GM foods and food products will be discussed. The importance of improvement on different analytical approaches that may lead to the management of toxins and food-borne illnesses caused by different microorganisms will also be highlighted.

2 The Concept of Food Safety and Current Practices

Food safety issues around the world are mainly monitored by the World Health Organisation (WHO) and The Food and Agriculture Organisation (FAO) of the United Nations. The Codex Alimentarius Commission (CAC) is an intergovernmental body that is operated under WHO and FAO. In 1997, CAC defined food safety as the assurance that a food or food product will not cause any harm to the consumers when it is prepared and/or eaten according to its intended use. Therefore, a food will only be considered as safe when it will not cause any long- or short-term illness to the consumer. It is the responsibility of food safety management to make sure that a food is safe before it is sold to its intended consumers. The maintenance of food hygiene is another important parameter during industrial food production that can ensure the safety and quality of food products. However, consumers are also responsible for following the instructions provided on the packaging to ensure the post-sale safety of food. Therefore, the key responsibilities that allow the proper maintenance of food safety follow a chain of action from regulators, to producers and then to consumers (Fig. 1).

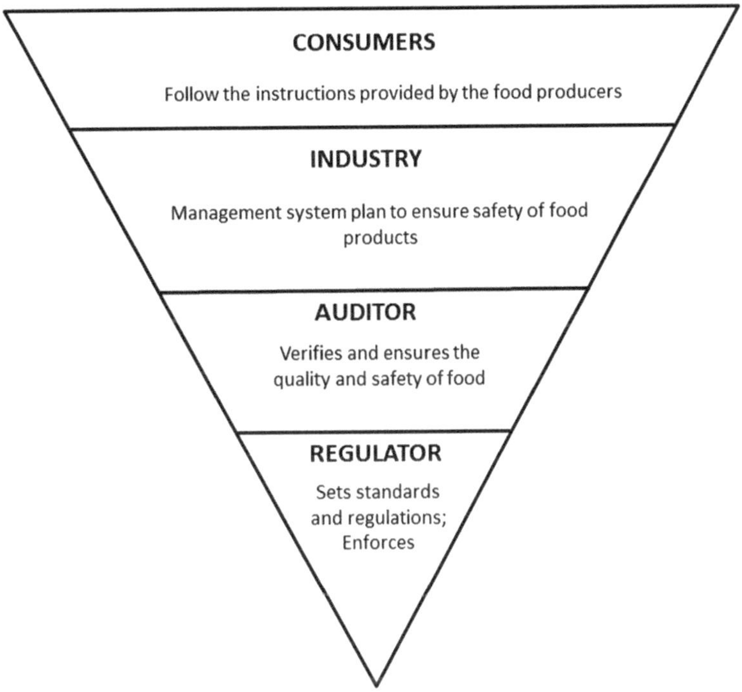

Fig. 1 The food safety regulatory model

The role of regulatory organisations and bodies is of foremost importance (Fig. 1), and relies on the cooperation of different sectors, including government, industry, consumers, and even academia (Motarjemi and Mortimore 2014). While the main task of a government in this area is to foresee all infrastructure and public health services needed for food safety management and to implement and enforce the laws and regulations, the industry is the one that actually ensures the safety of a food product by following all the regulatory protocols. The codes of good practices (CGP) are the first line of defence for any food industry and are a set of principles and measures that have been previously identified, based on past incidences. Generally, CGPs are applicable to most food industries; however, these may vary slightly depending on the type of industries. Hazard Analysis Critical Control Point (HACCP) is the second line of defence, which is also widely implemented in different food industries. There are seven principles under HACCP (Fig. 2) that include identifying the potential hazards or problems, taking measurements and then appropriate precautions to control the situation. Although HACCP is very beneficial in an industrial setting to ensure the safety of the product, industry staff may feel burdened by administrative requirements (Motarjemi 2014). However, the implementation of HACCP rules allows setting up critical control points (CCPs) that can be monitored, and corrective actions that can be undertaken if necessary to avoid any further loss of industrial food production (Fig. 2). Verifications are

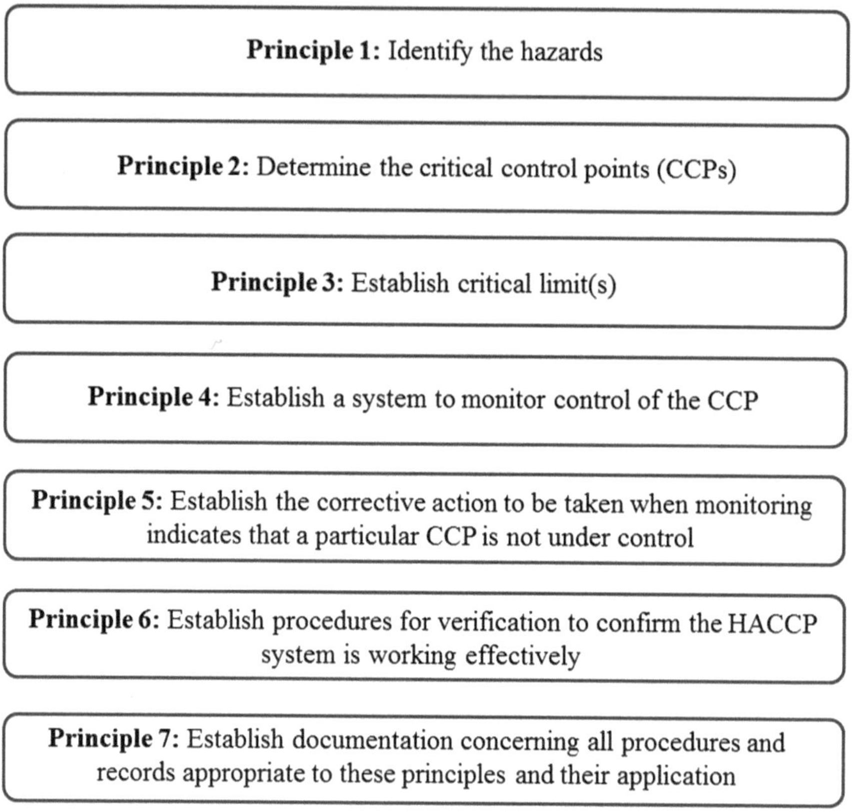

Fig. 2 The seven principles of Hazard Analysis Critical Control Points (HACCP)

carried out to determine the food quality by testing the raw material and the end product, by monitoring the environment before releasing the products. Auditing and consumer complaints handling are also associated with HACCP implementation in an industry.

In addition to regulatory bodies and industries, consumers also play a vital role in maintaining food hygiene and safety. Many consumers prefer to buy inexpensive food materials that do not have appropriate labels and safety information, and consumptions of these foods might pose a health risk (Motarjemi et al. 2001). Moreover, some ignore the instructions provided within the packaging of the food products and some never report defective foods that may cause public health concerns. Therefore, the liability of consumers cannot be overlooked and raising awareness might help to improve current food safety issues. Food safety-related campaigns organised by government and non-governmental organisations and social media can help in raising awareness in consumers.

The role of the scientific community is also important in maintaining food safety and quality. Ongoing research related to food products is always beneficial, bringing new insight into different food pathogens and chemicals that contaminate food and food products. Moreover, toxicological and ecological knowledge of microbial and chemical spoilage provided by scientists allows us to manage the food safety situation in a better way by undertaking control measurements (Motarjemi and Lelieveld 2014).

3 Food Safety and Quality Research: The Main Problems

The current mass production and industrialisation of different foods and food products have been initiated in order to feed the increasing world population. The 'green revolution' has also taken place in the last decades, enabling production of vast amounts of crops mainly using genetically modified (GM) plants. There is still an insecurity about GM food products, as many believe that long-term consumption of these food materials may cause some deleterious effects on human health including allergies and other immunological disorders (Maghari and Ardekani 2011; Krimsky 2015). However, this is still under question as at least two different groups of scientists exist who either consider GM food as safe or harmful (Krimsky 2015). Due to the lack of long-term and consistent research on the effect of GM food on human health, it is still not possible to come to a conclusion and this is why consumers are even more doubtful about the safety of GM products. In addition, climate change is another key factor that has prompted development of new plants or crop varieties that will withstand the gradual changes in environmental conditions. For instance, many efforts have already been undertaken to develop new rice varieties that can withstand considerably unfavourable conditions including heat stress, drought and high salinity (Nokkoul and Wichitparp 2014; Van Oort et al. 2015; Nguyen et al. 2016). Barley is another crop variety that has been used widely as a model to study and develop climate-resilient crops (Dawson et al. 2015). Therefore, the world has observed huge changes in the quality and type of foods and raw materials, which pose both a significant advantage and sometimes a new threat to food safety and quality. Chemical contaminants and xenobiotic molecules, including pesticide residues and organic halogenated compounds, also pose significant potential impacts on both human health and the environment, as these molecules can take a very long time to break down (over 50 years) in the environment. The effects of pesticides and other xenobiotics on aquatic and other environments (e.g. infertility of sea birds) is well documented, and attempts have already been undertaken to reduce the rate of contamination by banning many known chemical contaminants (Walker 1990; Falkowska and Reindl 2015; Gustafson et al. 2015; Pérez et al. 2015). In addition to chemical hazards, food safety is also threatened by microbial exposure. For instance, overall changes in lifestyle have steadily forced us to adapt to comparatively different food habits, and introduced new types of foods (e.g. raw food). As a result, many new food-borne pathogens have emerged or other pathogens re-emerged because of these newly found transmission vehicles. Thus, many food-borne outbreaks which occurred in the last 20 years were caused by bacteria, viruses, and protozoa, and many more pathogens are being introduced via food contamination every year. Therefore, there is an ongoing need for a proper risk assessment and management system to control outbreaks or even a pandemic related to any food-borne pathogen.

Major current challenges can be identified as follows:

I. Emergence of new food pathogens and control of outbreaks
II. Potential effects from genetically modified foods
III. Emerging chemical contaminants and xenobiotics
IV. Adulterations of food materials.

4 Current State of the Art of Food Safety and Quality Research

Food safety is not only a public health issue, but also has serious social and economic consequences. Food-borne illness can cause havoc in any country and sometimes worldwide when it is an epidemic that cause loss of many lives. For instance, food- and water-borne diseases are the main reasons of death of over 2.2 million people annually (WHO 2006), children, immunocompromised and elderly populations are often the most affected. This is not only a burden for many poor countries, but also endangers the international developmental efforts that have been undertaken to combat poverty (Kuchenmüller et al. 2009). These food-borne illnesses can be caused by various agents including pathogenic microbes, heavy metals, chemical contaminants and other toxic materials found in different food sources (e.g. toxins in wild mushrooms) (Anater et al. 2016; Signes-Pastor et al. 2016). Among pathogenic microbes, different strains of *Escherichia coli, Listeria monocytenes, Salmonella* spp. and even some food-borne viruses including norovirus and Hepatitis C have gained particular attention because of outbreaks related to contaminated food consumption in many countries (Marušic 2011). Moreover, many incidents were also reported of contamination of foods due to the presence of heavy metals and chemicals. In particular, the presence of arsenic in rice and other crops in many developing countries and melamine in milk products have gained public attention in recent years (Signes-Pastor et al. 2016). Most previous research on food quality and safety have mainly focused on determining the food composition and stress response of food pathogens, and have helped to add knowledge on what should be present in a particular food and how the growth of pathogenic microbes can be controlled using different techniques.

4.1 Study of Food Composition and Its Importance in Food Quality and Safety Research

To know more about the quality of food and food products, it is important to determine the composition, which also allows us to decide whether a given food is safe to consume, by providing information about any potential hazard. Much research has already been undertaken by focusing mainly on food composition of different crops, fresh food materials, and raw ingredients for other food products. Determination of food composition is not only focused on differentiating between different food products to ensure their safety, but is also intended to gain insight about the originality of some speciality foods and fermented food products, e.g. fruit juices, balsamic vinegars, and wines. One of the main reasons for determining originality is to control the adulteration of food material with unwanted ingredients that should not be present in that specific food. For instance, nuclear magnetic resonance (NMR) profiling of fruit juices has been carried out to determine the country of origin and also to discover if those fruit juices were produced using real juice or concentrate (Spraul et al. 2009a, b; Tomita et al. 2015). In addition, volatile compound analysis of balsamic vinegars through the use of gas chromatography and mass spectrometry (GC-MS) allowed determination of the effects of ageing materials or woods on vinegar quality (Chinnici et al. 2009; Callejón et al. 2010). Many studies have also focused on regional or country-specific food products including fresh vegetables, seafood, fish and ready-to-eat foods. This is mainly true for those countries either with low average incomes or where environmental pollution is a considerably bigger problem than in other developed countries. In countries that lack proper implementation of laws, regulations and auditing systems required for maintaining food safety, food stuffs can become deliberately contaminated with unwanted materials by vendors in order to gain more profit (Pinu 2015). In this situation, food composition studies that help to determine the type and diversity of contaminants are very useful. However, there is still a huge scope for improvements to technologies for rapid detection and identification of those food contaminants. Using these technologies, food auditors will be able to manage situations better by analysing food samples in real time while examining a food processing unit. Such technologies are still very limited currently, and research should focus on developing more user-friendly techniques and detectors using cutting-edge technologies.

4.2 Food Safety: Pathogenic Microbes and Their Stress Responses

One of the most demanding areas of research in food safety is the study of food-borne pathogens, which is mainly because of the many outbreaks related to the consumption of food products contaminated with pathogenic microbes. Food is mostly considered as an ideal growth medium for a wide range of microorganisms, and thus also an ideal vehicle for transmission of food-borne illnesses. Although vast numbers of people from most low-income countries where food hygiene is not maintained properly suffer from food-borne illnesses frequently, few data are available about those illnesses or outbreaks, mainly because of the lack of appropriate reporting systems. Therefore, most of the available data about food-borne outbreaks related to the consumption of contaminated food materials from restaurants to homes are from developed countries, where these cases are well documented. In recent years (2009–2015), a few developed countries, including Germany and the United States, have seen food-borne outbreaks associated with pathogenic *Escherichia coli* strains (O104:H4, O157:H7, O27 and O121) from the consumption of fresh vegetables, sprouted foods, ground beef, chicken salad and chipotle Mexican grill (Raupp 2014; Gelting et al. 2015; Ison et al. 2015; Radosavljevic et al. 2015). Different species of *Salmonella* have also caused food-borne outbreaks in developed countries including the USA and Australia from the consumption of chicken, cucumber, nut butter and raw cashew cheese (OzFoodNet Working 2014; Herman et al. 2015; Laufer et al. 2015; McWhorter et al. 2015). In addition to bacteria, food-borne illnesses associated with viruses, such as norovirus and rotavirus, were also quite common, and caused severe gastroenteritis disorders in many countries (Trivedi et al. 2012; Pacilli et al. 2015).

Food pathogens are very efficient in responding to the environmental changes during the different steps of food processing and they evolve to adapt to these harsh environments, mainly by changing their gene expression or by producing quorum sensing molecules, thus showing potential for causing disease in vulnerable hosts (Humphrey 2004; Begley and Hill 2015; Rul and Monnet 2015). Therefore, many studies have been undertaken to find out the mechanisms of adaptation of pathogenic microbes under various stresses including heat, acid, salt and oxidative stress (Samelis et al. 2001; Humphrey 2004; Koutsoumanis and Sofos 2004; Tiganitas et al. 2009; Shen and Fang 2012; Stackhouse et al. 2012; Alvarez-Ordóñez et al. 2015; O'Leary et al. 2015). For instance, biofilm formation is one of the key adaptation mechanisms that has been observed in many food pathogens, including *Listeria monocytogenes* and *Salmonella enterica* (Rodrigues et al. 2011; O'Leary et al. 2015). Preventive measures include the use of either chemical or physical forces and using bacteriophages that do not allow these food pathogens to form the biofilms, thus diminishing their chances of growing in the food products (Soni and Nannapaneni 2010; Van Houdt and Michiels 2010; Da Silva and De Martinis 2013). Therefore, studies of stress adaptation of food pathogens are very useful in the area of food safety, and generate in-depth knowledge about the molecular mechanisms behind this process. Control measures can be undertaken or developed based on what causes food pathogens to adapt to these adverse situations. Moreover, these may also lead to the development of techniques to detect the microbial contamination in the early stages of growth, thus preventing the infection from being transmitted widely by the consumption of food.

5 Metabolomics and Food Safety and Quality Research

5.1 Food Metabolomics

Metabolomics is one of the most recently introduced 'omics' that aims to analyse small molecules (metabolites) in a given biological sample. Although metabolomics was initially defined as the technical area by which it would be possible to detect and identify all the metabolites produced by a cell or an organism (Fiehn 2002; Bino et al. 2004), this has not been realised, mainly due to the diversity of metabolites. However, metabolomics is already established as a powerful tool for studying the metabolism and physiology of many living organisms, and thus this approach has been applied to a diverse range of research areas, including biomarker and drug discovery, agriculture, nutrition, bioremediation, plant biotechnology and also food science (Anizan et al. 2012; Badilita et al. 2014; Hall and de Maagd 2014; Li et al. 2014; Lima et al. 2014; Booth et al. 2015; El Amrani et al. 2015; Melnik 2015). Metabolomics has been used to study food systems including food ingredients, food processing and food pathogens; it has gained popularity in the last 10 years and numerous studies have already been carried out (Vikram et al. 2004; Badilita et al. 2014; Kusano et al. 2014; Inoue et al. 2015; Le Boucher et al. 2015; Ragone et al. 2015). Therefore, a distinct research area entitled 'food metabolomics' is well established that refers to the application of metabolomics in food system processes, from farm to consumers (Kim et al. 2016). Recently a term 'foodomics' has been introduced within the scientific community that refers to the application of 'omics approaches including genomics, proteomics, transcriptomics and metabolomics in food science (Herrero et al. 2012; Cifuentes and Rutledge 2013; D'Alessandro and Zolla 2013; Ibáñez and Cifuentes 2014; Laghi et al. 2014; Inoue and Toyo'oka 2015). However, food metabolomics deals only with the most downstream product of cell metabolism, metabolites, present in a given food or food system.

The food metabolome is very complex in nature, and also widely variable, depending on the type of food and the raw materials. Therefore, many thousands of metabolites are present in food are highly variable in terms of polarity and molecular weight (Shulaev 2006). The differing concentrations of metabolites in food items pose a major challenge in the development of analytical tools to detect as many metabolites as possible within a single analysis (Kueger et al. 2012). So far, there is no such analytical instrument available; therefore, for the better understanding of the metabolome by analysing as many metabolites as possible, the use of multiple analytical technologies is suggested by many scientists (Hall et al. 2002; Dunn and Ellis 2005; Villas-Bôas et al. 2007; Sumner 2010; Hall and Hardy 2012; Pinu et al. 2014). In addition, one of the major challenges that a food scientist needs to overcome while analysing food is the complex sample matrix. The matrix effect (ME) can create challenges in the detection and quantification of compounds that are present at very low concentrations in different food. ME is also responsible for poor and unreliable data that can affect the reproducibility, repeatability, linearity and accuracy of the methods used by various analytical instruments (Trufelli et al. 2011). To avoid and reduce the ME, a sample clean-up step using Solid Phase Extraction (SPE) or Solid Phase Microextraction (SPME) or liquid extraction is usually necessary before analysis by other methods (Jiang et al. 2012). More efficient chromatographic separation is also suggested by Trufelli et al. (2011), in which two-dimensional separation techniques (both GC and LC) can also be applied (Marriott et al. 2012; Mondello et al. 2012). However, pre-analytical steps can be time-consuming, arduous and often can cause loss of analytes, which is not appropriate for an unbiased profiling approach (Villas-Bôas et al. 2007; Cappiello et al. 2010; Trufelli et al. 2011). Although metabolomics was initially introduced mainly as an unbiased and non-targeted approach, both targeted and non-targeted analyses are performed for any biological samples to better answer the research questions. Therefore, both targeted and non-targeted metabolomics are gaining popularity for the analysis food products.

A significant improvement has also been achieved in metabolomics workflow, including sample preparation, quenching, metabolite extraction and acquisition of data (Fig. 3). In recent years, many sample preparation protocols have been published that allow better detection of metabolites with a wide range of chemical properties (Anizan et al. 2010; Biais et al. 2012; Teo et al. 2013; Brennan 2014; Chan et al. 2014; Le Gall 2015; Rejczak and Tuzimski 2015). In short, an appropriate quenching method is used to stop the ongoing enzymatic activities after the collection of a food or microbial samples. After quenching, metabolites are extracted using a suitable extraction solvent (e.g. chloroform/methanol/water). However, it is noteworthy that metabolite profiles may vary depending on the metabolite extraction protocols; therefore, it is better to use at least a few extraction protocols to obtain a global metabolite profile of any biological sample (Duportet et al. 2012, Jäpelt et al. 2015). Once metabolites are extracted from the sample, they are ready for analysis by an instrument of choice. The acquired data then needs to be explored using different statistical and chemometric approaches (Aggio et al. 2011, 2014; Gowda et al. 2014; Pluskal et al. 2010; Robotti and Marengo 2016; Smith et al. 2006; Xia et al. 2012): including feature detection, alignment, biomarker identification and chemical structure elucidation (Fig. 3).

5.2 *Recent Advancements of Analytical Instruments in Metabolomics*

Due to advancements in different analytical instruments in the last decade, it is now possible to analyse thousands of metabolites from a food sample in a single analysis. The sample preparation and data handling processes have also been improved tremendously, making it easier for scientists to analyse samples in a short time. Moreover, there are many commercial and in-house metabolite databases available, which is also beneficial for the identification and sometime structural elucidation of

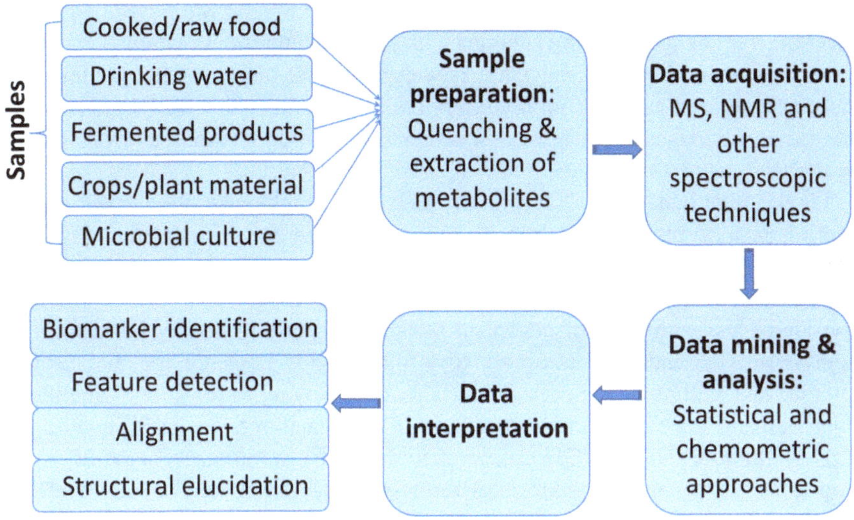

Fig. 3 Workflow for the analysis of the food metabolome. Here, *MS* Mass spectrometry and *NMR* Nuclear Magnetic Resonance spectroscopy

unknown metabolites present in a food. It is well known that powerful detectors are the main factors for the analysis of metabolites. Two technologies, Nuclear Magnetic Resonance (NMR) and Mass Spectrometry (MS) have been employed widely in metabolomics (Shulaev 2006; Villas-Bôas et al. 2007; Dieterle et al. 2011; Herrero et al. 2012; Kueger et al. 2012; Zhang et al. 2012; Balan et al. 2013; Ibáñez et al. 2013; Badilita et al. 2014; Senyuva et al. 2015). However, there are many other instrumental techniques, including Fourier transform infra-red spectroscopy (FTIR), which are available for metabolite profiling of food samples.

NMR has been broadly used for untargeted metabolite profiling of complex mixtures (i.e. fruit juices, wines, spirits, urine and blood) (Ogrinc et al. 2003). NMR spectroscopy is increasingly renowned for its efficacy, non-invasiveness (non-destructive), throughput and linearity (Laghi et al. 2014). Moreover, NMR spectroscopy also provides structural, chemical-kinetics and other information in multidimensional applications (Dieterle et al. 2011). Thus, high resolution NMR spectroscopy along with multivariate data analysis has been used for direct characterisation of fruit juices (Cuny et al. 2008), wine (Lee et al. 2009; Pinu et al. 2014), grape berries (Pereira et al. 2006; Mulas et al. 2011), olive oil (Del Coco et al. 2012; Piccinonna et al. 2016) and beer (Almeida et al. 2006; Rodrigues et al. 2011). To obtain a global metabolite profile of a complex sample, NMR needs to be coupled with another non-targeted analytical approach (e.g. MS). But there are some serious drawbacks to using NMR for metabolome analysis. Different parameters of the sample, e.g. salinity, pH and concentrations of metal ions, can affect the sensitivity of NMR spectrometers and can cause difficulty in bioinformatics-based resonance assignments (Lewis et al. 2012). To avoid these problems, consistency in sample preparation is required. Moreover, an NMR spectrometer cannot detect metabolites with low concentrations; samples may need to be concentrated before analysing, as well as require larger sample volumes. To dissolve the dried samples, high concentration of deuterated solvents are required and these solvents also cause bias in sample preparation and data analysis (Lewis et al. 2012).

On the other hand, the MS technique has gone through remarkable developments and it is now an important instrument for many researchers (El-Aneed et al. 2009). MS is the most extensively used instrumental approach in metabolomics (Villas-Bôas et al. 2005; Dunn 2011; Vestal 2011; Kueger et al. 2012). It is also considered the technique of choice in metabolite profiling mainly because of its high sensitivity and also its ability to profile a wide range of metabolites in a mixture and within a single analysis (Glinski and Weckwerth 2006). Significant improvements have been achieved in terms of mass analysers, including quadrupole (Q), quadrupole ion-trap (QIT), time of flight (ToF), orbitrap, ion-mobility spectrometry (IMS) and Fourier transform ion cyclotron resonance (FTICR). Quadrupole mass analysers are very robust, low cost and simple to use, but they provide lower mass resolution and accuracy than other mass analysers (Villas-Bôas et al. 2005). ToF, FTICR and orbitrap are considered excellent instruments that offer the highest mass resolution of all mass analysers.

MS is mostly used in combination with a few powerful separation techniques to enhance its identification power. Gas chromatography (GC), liquid chromatography (LC) and capillary electrophoresis (CE) are the most common separating techniques used in combination with MS, which allow maximum separation of metabolites in a complex biological sample (Villas-Bôas et al. 2003; Ramautar et al. 2011; Theodoridis et al. 2012). However, direct infusion (DI) is also widely used for metabolite profiling, which is usually referred to as metabolic footprinting or fingerprinting, depending on whether the analysis is of extra- or intracellular metabolites (Villas-Bôas et al. 2005; Han et al. 2009). Because of the development of interfacing systems such as atmospheric pressure ionisation (API), DI-MS can be used to analyse a sample to obtain mass spectra of the metabolites within a few seconds (Villas-Bôas et al. 2005). The requirement for a small amount of sample is the major advantage of using DI-MS. Moreover, no derivatisation is required for this analysis, and more metabolites are detected by DI-MS than by GC-MS, making this technique best suited for high throughput non-targeted metabolite profiling (Mas et al. 2007). However, DI-MS shows poor reproducibility when analysing complex mixtures because of the matrix effect. The identification of metabolites by DI-MS is also very troublesome, and stereoisomers cannot be resolved using this technique (Villas-Bôas et al. 2005; Glinski and Weckwerth 2006; Pope et al. 2007). GC is one of most efficient separation techniques in metabolomics, allowing the separation of hundreds of metabolites within a single analysis and requiring a very small sample volume (1–2 μL). The coupling of GC with MS is a perfect match because metabolites in an inert gas phase can be ionised much more easily at the

MS ion source, making GC-MS the best combination of separation technique with mass spectrometry detection. The instrumentation has been developed considerably during the last 50 years. GC-MS is a highly sensitive analytical platform, which also provides excellent instrument repeatability, around 5 % or below (Villas-Bôas et al. 2005). However, an extra step of sample derivatisation is required for the analysis of semi- and non-volatile metabolites, and GC-MS has been widely used for the last 50 years for the analysis of a wide range of metabolites present in foods (Table 1). LC is another powerful separation technique that allows rapid analysis of small amounts of sample. The main advantage of this separation technique over GC is that no previous derivatisation of the sample is required to analyse the non-volatile compounds (Scalbert et al. 2009). This technique is often coupled with MS and sometimes also with NMR (Shulaev 2006; Zhou et al. 2012). For LC-MS, a wide range of different detectors is used, ranging from ultra-high resolution MS such as FTICR or orbitrap to low-resolution MS such as ion traps and triple quads and hybrid systems. The most recent addition is the ion-mobility TOF-MS system (Lanucara et al. 2014). The development of methods depends on the nature of the metabolites to be analysed. LC-MS has been applied widely in metabolite profiling (both targeted and non-targeted) of complex biological samples (Berg et al. 2012; Theodoridis et al. 2012). CE is an efficient and rapid technique that can separate a wide range of charged metabolites within a single analytical run (Villas-Bôas et al. 2005). CE is often coupled to MS and it is considered a promising analytical instrument in metabolomics (Shulaev 2006). CE-MS has high resolving power and requires very small volumes of samples (1–20 nL). Thus, it has been used for both targeted and non-targeted high throughput analysis of metabolites (Nevedomskaya et al. 2010; Sato and Yanagisawa 2010; Ramautar et al. 2011). The major drawback for CE-MS is its poor sensitivity; thus the detection limit is several magnitudes higher than those of chromatographic methods (Table 1). The introduction of a small volume of sample is inadequate for the detection of many metabolites (Cai and Henion 1995). Moreover, low recovery and irreversible adsorption of analytes onto the capillary wall also can occur in CE. For these reasons, CE-MS is mostly used in metabolomics as a combination of different protocols targeting different groups of metabolites combined with sample preparation steps to concentrate the metabolites in the samples, making it a lower throughput technology in metabolomics and certainly more suitable for targeted analysis.

Fourier transform infrared spectroscopy (FTIR) is another analytical technique that is extensively applied in the food industry because it is rapid, highly automated, reproducible, non-destructive and cost-effective (Bauer et al. 2008; Versari et al. 2010). Nevertheless, FTIR is insensitive for complex liquid samples and water in the samples often increases the background noise, increasing the limit of detection and reducing linearity (Achá et al. 1998). The data obtained from FTIR are also very complex and there few databases available for assisting with identification of metabolites (Berthomieu and Hienerwadel 2009).

Table 1 Comparisons of different analytical instruments used in food safety and quality research

Analytical technique	Advantages	Disadvantages	Application in food safety and quality research
GC-MS	• High chromatographic resolution • Sensitive and robust • Simultaneous analysis of different groups of metabolites • Large linear range, availability of commercial and in-house MS libraries	• Derivatisation is required for non-volatile metabolites • Unable to analyse thermo-labile compounds	Gilbert-López et al. (2010), Anizan et al. (2012), Canellas et al. (2012), Banerjee et al. (2013), Caligiani et al. (2013), Cao et al. (2014), Guo et al. (2014), Duedahl-Olesen et al. (2015a, b), Giri et al. (2015), Grimalt et al. (2015), Janssens et al. (2015), Walorczyk et al. (2015)
LC-MS	• High sensitivity • Derivatisation not usually required • Large sample capacity • Thermo-stable compounds can be analysed	• Average to poor chromatographic resolution • De-salting may be required • Limited commercial libraries • Tough restrictions on LC eluents • Matrix effects	Da Rosa et al. (2012), García-Gómez et al. (2012), Russo et al. (2012), Gottfried and Herebian (2013), Boutsiadou-Theurillat et al. (2014), Fan et al. (2014), Zeng et al. (2014), Arroyo-Manzanares et al. (2015), Juan-Borrás et al. (2015), Khan (2015), Squadrone et al. (2015), Zhang et al. (2015), Han et al. (2016)
CE-MS	• High resolution, small volume of sample required • Rapid analysis • Usually no derivatisation required	• Poor reproducibility • Poor sensitivity • Buffer incompatibility with MS • Difficulty in interfacing with MS • Limited commercial libraries	Font et al. (2008), Lara et al. (2008), Nevedomskaya et al. (2010)

(continued)

Table 1 (continued)

Analytical technique	Advantages	Disadvantages	Application in food safety and quality research
Direct MS	• High resolution depending on the mass analyser • Sensitive • Rapid analysis	• Interference of salts and strong ions • Limited identification power • Poor reproducibility	Ackerman et al. (2009), Self and Wu (2012), Shen et al. (2012), Farré et al. (2013), Self (2013), Doué et al. (2015)
NMR	• Rapid analysis • Non-destructive • Minimal sample preparation • Quantitative	• Low sensitivity • More than one peak per component • Identification is laborious because of complex matrix	Spraul et al. (2009a, b), Balan et al. (2013), Osselaere et al. (2013), Badilita et al. (2014), Lima et al. (2014), Qi et al. (2015), Ragone et al. (2015), Tomita et al. (2015), Wan et al. (2015)

GC-MS Gas Chromatography and Mass Spectrometry; *LC-MS* Liquid Chromatography and Mass Spectrometry; *CE-MS* Capillary Electrophoresis and Mass Spectrometry and *NMR* Nuclear Magnetic Resonance

6 Application of Metabolomics in Food Safety and Quality Research

Although the main aim of metabolomics is to generate hypotheses based on data using an unbiased and non-targeted approach, both targeted and non-targeted methodologies are frequently applied in food safety and quality research (Table 1). It is especially true when metabolomics has been used for the analysis of chemical contaminants including pesticides in different types of food materials (Forsberg et al. 2011; Arroyo-Manzanares et al. 2015; Chatterjee et al. 2015). However, when the main purpose of a study is to determine biomarkers either for microbial spoilage or other sort of contamination, it is better to choose a non-targeted approach that might help in finding novel compounds or metabolic pathways relevant to a food system. The two main areas in food safety where metabolomics has been widely applied are in determining either chemical or microbiological hazards present in a food or food processing system. However, consumers these days are also concerned about food prepared using GM ingredients, as the long-term effects of consumption of such foods are still not adequately known; therefore, a growing research trend is observed in this area.

6.1 Chemical Food Safety

A large number of chemicals have entered our food chain as a result of the wide application of different growth-promoting agents (e.g. clenbuterol in pigs, recombinant growth hormones in fish and steroids in bovine animals), antibiotics and pesticides that help us to produce large amounts of agricultural products. Therefore, both food ingredients and the environment are facing the burden of chemical exposures, many of which are unwanted and pose a threat to human health. Moreover, deliberate contamination of food ingredients with unwanted chemicals (e.g. melamine) also poses a serious risk to consumers. Therefore, these chemicals should be banned or their use should be limited as set by the regulatory authorities. To achieve this, development of high throughput methods that would enable study of these contaminants is urgently needed. Both targeted and non-targeted metabolomics demonstrate enormous potential in detecting and identifying these chemicals in our food. For instance, targeted analysis using a suitable analytical platform (e.g. NMR and MS) is beneficial for the study of already-known chemical contaminants in food. However, it is more problematic when the contaminants are either not known or are unknown breakdown products of a familiar compound. An untargeted metabolomics approach could be a better way to study those novel compounds and also to determine and validate candidate biomarkers that could be used for tackling illegal practices in food production (Dervilly-Pinel et al. 2012).

Although NMR was initially the method of choice for detecting and identifying chemical contaminants in food because of its capacity to quantify and elucidate the structure of molecules (Dieterle et al. 2011), MS is currently more popular due to its sensitivity, and robustness of coupling with various separation systems (e.g. GC, LC and CE). Moreover, tandem MS/MS or MS^n experiments can now elucidate the structure of any unknown metabolites, and thus are useful for the development of multi-residue methods to detect and identify chemical contaminants accurately. Many LC-MS-based methods have been published recently that have been used to study chemical contaminants (Table 1) in different crops, including herbal teas, nutraceuticals, rice and maize, and also in animals (Castro-Puyana and Herrero 2013). For instance, 255 veterinary drug residues and other chemical contaminants were determined within 10 min using a UHPLC-MS/MS method (Zhan et al. 2012). In addition, high resolution MS/MS has been widely applied for the analysis of pesticides, toxins and antibiotics. More than 500 pesticides were screened in fruits and vegetables using an orbitrap tandem MS/MS method (Alder et al. 2011). De Dominicis et al. (2012) also reported an orbitrap-MS/MS-based protocol for analysing pesticides and toxins simultaneously in bakery and other foods. In addition to analysing pesticides and other xenobiotic molecules, attempts have also been made to develop methods using LC-MS to identify other contaminants in food, including melamine in infant formula (Inoue et al. 2015). Many other LC-MS/MS protocols have already been published that analysed a wide range of food and fermented products including fruits, vegetables, wine, baby foods and cereals (Lacina et al. 2012; Pérez-Ortega et al. 2012; Fan et al. 2014).

While LC has been the choice of separation technique for the analysis of contaminants in food materials, GC still remains the mostly used system coupled with high resolution MS for the determination of environmental pollutants in food such as polychlorinated biphenyls (PCBs), polycyclic aromatic hydrocarbons (PAHs) or polybrominated diphenyl ethers (PBDEs) in various samples including fruits, vegetables, cereals, teas, fish muscle, dietary supplements and sheep milk (Mastovska and Wylie 2012; Storelli et al. 2012; Banerjee et al. 2013; Cao et al. 2014; Duedahl-Olesen et al. 2015a, b; Walorczyk et al. 2015). For example, Sapozhnikova and Lehotay (2013) accurately quantified over 80 pollutants (PCBs, PAHs and pesticides) in fish within 9 min using a low pressure GC-MS/MS method. In addition to tandem MS used with a GC system, two-dimensional GC (GCxGC) analysis is also gaining popularity for the analysis of pollutants, which allows determination of these compounds in very small amounts (0.1 μg/kg) (Kalachova et al. 2012, 2013; Giri et al. 2015).

Some applications of food safety in terms of chemical contaminant determination also make use of direct MS analysis either as direct analysis in real time (DART) or as desorption electrospray analysis (DESI). Rapid and high throughput analysis of food samples can be obtained using DART-MS, as it directly analyses samples from the surface, thus decreasing sample preparation time significantly compared with those for the other coupled MS systems (Castro-Puyana et al. 2013). DART-MS combined with high resolution analysers has been applied to analyse pesticides in fruits and grains (Schurek et al. 2008; Farré et al. 2013). Similarly, DESI-MS has been used for the determination of pesticides in fruit peels and vegetables (García-Reyes et al. 2009; Zhang et al. 2009).

6.2 Early Detection of Food Pathogens and Food Spoilage Microorganisms

Food pathogens are one of the major threats for food safety, and large numbers of people around the world suffer from diseases including diarrhoea, dysentery and other forms of food poisoning that are directly caused by various food pathogens, such as *Salmonella* spp., *Shigella* spp., *Listeria monocytogenes, Campylobacter jejuni* and *E. coli*. Moreover, food spoilage microorganisms (e.g. *Pseudomonas* spp., *Acinetobacter* spp., *Botrytis* spp.) are not necessarily pathogenic for people, but can cause severe economic damage because of the spoilage of a wide range of food materials. If there was a way to detect these pathogens or spoilage microbes in the early stages of their growth in food products, it would be possible to reduce dramatically the numbers of food-borne outbreaks and subsequent significant losses of food by undertaking appropriate measurements to control their further growth. Traditional identification and cultural methods for pathogens or food spoilage microbes are time consuming, and therefore it is necessary to develop techniques that would enable the rapid detection of microbes soon after the contamination occurred in a food system (Xu et al. 2010). The metabolomics approach has already shown huge potential for developing analytical methods to detect food pathogens in their early stages of growth in a food product (Li et al. 2011; Beale et al. 2014). In this approach, an experiment is performed using both a contaminated and a non-contaminated food, and the metabolites from both are analysed. Multivariate or discriminant analyses are performed to determine a group of potential candidate biomarkers that can distinguish between the two conditions. Once validated, these biomarkers can be used for the determination of pathogens in a real food sample, or a simpler screening technique (e.g. enzymatic or colorimetric) can be developed to be used routinely by food auditors.

GC-MS has been the choice of analytical instrument for determining pathogenic growth in a food, as this technique is very efficient in determining the volatile organic compounds (VOC) produced by microorganisms. Food spoilage microbes generally produce many VOCs as part of their metabolism, and this causes alteration in the sensory properties of the food (Li et al. 2011). VOC analysis is attractive for food sampling and is also advantageous because of rapid and non-invasive sample preparation. Once collected in the headspace using an appropriate SPME cartridge, VOCs produced by spoilage or pathogenic microorganisms are ready to be analysed by GC-MS. Xu et al. (2010) published a VOC-based GC-MS metabolite profiling approach to determine the potential biomarkers of spoilage of pork by *S. typhimurium,* and identified 16 metabolites that clearly distinguished the naturally spoiled from *S typhimurium*-contaminated pork meats. Moreover, Li et al. (2011) demonstrated the capability of metabolomics in assigning biomarkers to the spoilage microbes, *Botrytis allii* and *Burkholderia cepacia,* and they identified 16 volatile metabolites related to post-harvest onion spoilage. In addition to volatile metabolites, other primary metabolites including sugars and amino acids (dextrose, glycine, tyrosine and histidine) were also characterised as potential biomarkers that could be used for the early detection of *E. coli* O157:H7 and of different strains of *Salmonella* in ground beef and chicken (Cevallos-Cevallos et al. 2011).

Among other MS-based techniques, Matrix-assisted laser desorption/ionisation coupled to time of flight-MS (MALDI-TOF-MS) also has been used to identify different strains of pathogenic microbes in biological samples, including food and beverages (Böhme et al. 2010, 2012; Picariello et al. 2012; Ojima-Kato et al. 2014; Beale et al. 2014; Jadhav et al. 2015). MALDI-TOF-MS is very useful for identifying microbial strains using either whole cells or cell extracts and has been applied to epidemiological studies, biological warfare agents, detection of antibiotic resistance pathogens, detection of water- and food-borne pathogens and enterotoxins (Tsilia et al. 2012). This technique is rapid, sensitive and inexpensive, and thus well received by microbiologists (Singhal et al. 2015). Recently, rapid detection and source tracking of *L. monocytogenes* was carried out using MALDI-TOF-MS in Australian dairy products (Jadhav et al. 2015). *Staphylococcus aureus* strain characterisation was also performed in Italian dairy products using MALDI-TOF-MS fingerprinting (Böhme et al. 2012). However, the main drawback of using this technique is the inability to identify new species, as the identification process depends on the existing spectral database that contains the peptide mass fingerprints of the type strains (Singhal et al. 2015).

6.3 Microbial Food Toxins

In contrast to chemical contaminants, food toxins that pose serious health hazards for consumers are mainly produced naturally by different microorganisms (fungi, algae and others) growing on food substrates or during the food preparation and storage. Many of these toxins are produced as by-products of fungal metabolism (mainly *Penicillium*, *Aspergillus* and *Fusarium*) in contaminated foods including beer, wine, bread, rice and maize, and are thus known as mycotoxins. The determination of mycotoxins is very similar to that of chemical contaminants and mainly carried out by multi-residue analysis (Giacometti et al. 2013; Gottfried and Herebian 2013; Aniołowska and Steininger 2014; Pizzutti et al. 2014; Rodríguez-Carrasco et al. 2015). In addition to mycotoxins, toxins produced by some algae are also of interest for food safety. This type of toxin is usually found in contaminated seafood and fish and may affect people, such as fish and different types of shellfish poisoning (e.g. neurotoxic, paralytic, diarrhoeic and amnesic). It is important to study algal toxins to know about their modes of action and toxic doses in detail, to help to take appropriate precautions to prevent an outbreak. MS-based metabolomics approaches have been developed to detect and quantify several groups of algal toxins, such as ciguatoxins and domoic acid (Yogi et al. 2011; Beach et al. 2014; Stewart and McLeod 2014).

6.4 Study of GM Crops and Food Materials

GM or transgenic crops have gained popularity since the first commercial plantings in 1996. GM technology has allowed us to produce crops and food materials with enhanced nutritional properties, increased yield; and moreover, these crops can be resistant to various pests and diseases, and adverse environments, including drought and salinity (Parrott et al. 2010). A wide range of GM crops are now commercially cultivated around the world, including soybean, maize, cotton, canola, potatoes and tomatoes. Although the new traits in GM crops are mainly from the introduction of small RNAs or some regulatory proteins, it is still unknown if even that small amount of change could cause an overall alteration in other metabolic pathways, and thus adverse modification of cellular downstream products (Delaney 2015). The effects of growing transgenic plants on the environment have also been well documented (Liu et al. 2012; Brookes and Barfoot 2013; Knight et al. 2013), which has in turn raised concerns about the long-term consumption of GM food materials on human and animal health.

As the long-term effects of consuming GM food ingredients are still under question, it is time to do more research on toxicological aspects that could provide assurance to consumers about their safety. A few research projects have already been carried out that addressed the effects of consumption of GM feed ingredients by animal models. For instance, Sheng et al. (2014) evaluated the toxicity and allergenicity of GM rice as expressed in human serum albumin in rats, and Qi et al. (2015) assessed the safety of consumption of similar GM rice by rats using their urine metabolome. Although both sets of authors found significant differences in profiles between GM and non-GM rice-consuming rats, they concluded that GM rice could be considered safe. However, both these studies were performed for a limited period of time (only 90 days), and therefore it is still too soon to conclude if long-term consumption (e.g. 5–10 years) of such GM rice by the rats would have any adverse effects on their health.

Much research has been undertaken in last 10 years to acquire more knowledge regarding the compositional differences between GM and non-GM crops and food materials (Barros et al. 2010; Asiago et al. 2012; Cao et al. 2012; Liu et al. 2012; Clarke et al. 2013; Kusano et al. 2014). This provides valuable insight into the nutritional properties compared with those of conventional crops, and if any unintended chemicals or proteins are present in that given crop that might pose a threat to consumers. Metabolomics (both targeted and non-targeted), especially metabolite profiling, is very useful in this regard, as it allows the generation of data on the comprehensive composition of any GM crop (Rischer and Oksman-Caldentey 2006). For example, Kusano et al. (2014) recently published a study using metabolomics and ionomics approaches to determine the chemical diversity of a soybean lineage representing 35 years of breeding. The authors used different analytical platforms, including CE-TOF-MS, GC-TOF-MS and LC-qTOF-MS, to determine the global metabolite profiles of soybeans, and found that newer varieties are completely different from older ones. However, they found no significant differences in metabolite composition between conventional and GM soybeans, thus indicating that GM soybeans could be safe as food and feed; their findings were in accordance those of with Clarke et al. (2013). Similar studies have been performed for other GM crops (e.g. maize and potato). Using a hierarchical metabolomics approach, it has been shown that metabolite composition is very similar between conventional and GM potato crops (Catchpole et al. 2005). However, it is noteworthy that compositional differences also exist among conventional crops, depending on their origin, environmental conditions, and genetics (Reynolds et al. 2005; Harrigan et al. 2007). The use of 'omics' approaches, including metabolomics, in determining the safety aspect of GM food materials still has scope for improvement, and more research should be undertaken to highlight the toxicological aspects in addition to the compositional studies.

6.5 Food Quality and Traceability

Both targeted and non-targeted metabolomics have been widely applied to determine food quality by analysing any adulteration during food processing, or to confirm authenticity of a food or beverage. Adulteration is usually performed deliberately to gain more profit, using unwanted substances, such as the addition of melamine to milk powder, synthetic dyes in spices, and the use of horsemeat in other meat products (Senyuva et al. 2015). Some of these adulterations might cause serious health issues (Chu and Wang 2013) and some of them are related to economic adulterations to consumers (Everstine et al. 2013). However, in all cases, it is essential to know the overall quality of the food materials and also to determine if the food product is adulterated or not before it can be sold in the market. Authentication of food is a useful step, and attempts have been made to determine potential markers in different food matrices that can differentiate between adulterated and normal food. For example, both targeted and untargeted metabolomics approaches using LC-MS/MS have been used to authenticate Indian citrus fruits, and several metabolites (didymin, rhoifolin, isorhoifolin, neohesperidin, hesperidin, naringin, narirutin, limonin glucoside, and vicenin) were characterised as potential markers (Jandrić et al. 2015). Arias et al. (2016) recently published a metabolomics study using UHPLC–QTof-MS (ESI + mode) to distinguish the control (non-medicated pigs) and pigs treated with ronidazole, dimetridazole and metronidazole where they have identified at least four ionic features that could be used as potential biomarkers of illegal 5-nitroimidazole abuse. GC-MS is another popular analytical platform that has been successfully implemented to determine the quality of various food products. For instance, GC-MS has been used to monitor the kimchi fermentation process (Park et al. 2016). Moreover, fatty acid profiling using GC-MS has also been found to be useful in determining adulteration in flaxseed oil (Sun et al. 2015). Recently, Isotope-ratio MS (IRMS) along with either GC or LC has become another technique that can be used for authentication of foods including wines, essential oils in mandarin, and flavoured strawberry foods (Schipilliti et al. 2010; Guyon et al. 2011;

Schipilliti et al. 2011). In addition to MS-based techniques, other spectroscopic instruments and imaging systems have also been used successfully to authenticate different foods. NMR-based metabolomics was applied to determine adulteration in foods and beverages, including saffron and fruit juice (Balan et al. 2013; Ordoudi et al. 2015). NMR metabolomics approach has also been successfully used to determine the freshness of shucked mussels and a significant increase of many primary metabolites including acetate, lactate, succinate, alanine, branched chain amino acids and trimethylamine was observed during storage of mussels between 0 and 4 °C (Aru et al. 2016). Yang et al. (2016) recently reported the use of both NMR and LC-MS-based metabolomics study of milk adulteration and they identified some metabolites (choline and succinic acid) that can distinguish Holstein milk from that of other cows. Moreover, the use of Raman imaging could also identify the adulteration agents (e.g. melamine and dihydrocyanide) in powdered food (Qin et al. 2014).

One major area where much attention has been paid and which helps to prevent adulteration is the determination of geographic origins of foods and beverages. In many countries, food and condiment preparation is considered an art (e.g. wine-making) and strong regulation already exists to ensure the quality of the product before it can be marketed under protected designation of origin (PDO) or protected geographic indication (PGI). This is especially true for many European food products (balsamic vinegars, olive oils and cheeses). Traditional balsamic vinegars are important product from Italy, produced in many regions; however, Modena is famous for production of balsamic vinegars with PGI certification. GC-MS analysis of volatile compounds was found to be very useful for distinguishing balsamic vinegar samples with PGI (Chinnici et al. 2009; Cirlini et al. 2011). Extra virgin olive oil is another product for which PDO status is very important. Many producers follow fraudulent practices to obtain PDO for low-quality olive oils for which consumers will pay comparatively higher prices. Pizarro et al. (2011) identified volatile markers related to the geographic origin of Spanish extra virgin olive oils using headspace-SPME-GC-MS analysis. GC-MS along with multivariate data analysis has also been used to study the origin and quality of Italian buffalo mozzarella cheese (Pisano et al. 2016). Moreover, different types of NMR experiments (^{1}H, ^{13}C, ^{31}P) have also been successfully employed for quality assessment and authentication of olive oils (Dais and Hatzakis 2013).

7 Conclusion

Metabolomics has already been employed widely in food science, especially to address issues of food safety and quality. Although recent analytical developments allow us to analyse over 1000 metabolites in a single analysis (or using multiple analytical instruments), a big challenge still exists in the handling of large data sets (Skov et al. 2014). This is where all the 'omics' approaches currently have major difficulties and, frequently, the huge amount of data generated cannot be appropriately interpreted biologically because of the inability to analyse the data. Therefore, genuine efforts should be undertaken to improve the data analysis platforms, to improve the overall quality of research in the metabolomics area. In addition, ongoing technical improvement should lead to more cost-effective, user-friendly and high throughput methods that could be used for the analysis of various food matrices. Other aspects that need attention include the creation of more food-based databases to make identification and discrimination of foods much easier. It would be also easier to formulate regulatory actions to maintain the quality and safety of foods and food products. We have already seen a tremendous improvement in food safety and quality research in the last few years, and the application of metabolomics will allow us to look at a holistic overview of the food system, thus improving the overall objectives of ensuring food safety and quality.

Acknowledgements The author would like to thank Dr Damian Martin, Megan Jones and Science and Publishing Office of Plant and Food Research Ltd for their support during the writing of this manuscript.

References

Achá V, Naveau H, Meurens M (1998) Extractive sampling methods to improve the sensitivity of FTIR spectroscopy in analysis of aqueous liquids. Analusis 26:157–163

Ackerman LK, Noonan GO, Begley TH (2009) Assessing direct analysis in real-time-mass spectrometry (DART-MS) for the rapid identification of additives in food packaging. Food Addit Contam Part A 26:1611–1618

Aggio R, Villas-Bôas SG, Ruggiero K (2011) Metab: an R package for high-throughput analysis of metabolomics data generated by GC-MS. Bioinformatics 27:2316–2318

Aggio RBM, Mayor A, Reade S, Probert CSJ, Ruggiero K (2014) Identifying and quantifying metabolites by scoring peaks of GC-MS data. BMC Bioinformatics 15:374

Alder L, Steinborn A, Bergelt S (2011) Suitability of an orbitrap mass spectrometer for the screening of pesticide residues in extracts of fruits and vegetables. J AOAC Int 94:1661–1673

Almeida C, Duarte IF, Barros A, Rodrigues J, Spraul M, Gil AM (2006) Composition of beer by 1H NMR spectroscopy: effects of brewing site and date of production. J Agric Food Chem 54:700–706

Alvarez-Ordóñez A, Broussolle V, Colin P, Nguyen-The C, Prieto M (2015) The adaptive response of bacterial food-borne pathogens in the environment, host and food: implications for food safety. Int J Food Microbiol 213:99–109

Anater A, Manyes L, Meca G, Ferrer E, Luciano FB, Pimpão CT, Font G (2016) Mycotoxins and their consequences in aquaculture: a review. Aquaculture 451:1–10

Aniołowska M, Steininger M (2014) Determination of trichothecenes and zearalenone in different corn (*Zea mays*) cultivars for human consumption in Poland. J Food Compos Anal 33:14–19

Anizan S, Bichon E, Monteau F, Cesbron N, Antignac JP, le Bizec B (2010) A new reliable sample preparation for high throughput focused steroid profiling by gas chromatography-mass spectrometry. J Chromatogr A 1217:6652–6660

Anizan S, Bichon E, Duval T, Monteau F, Cesbron N, Antignac JP, Le Bizec B (2012) Gas chromatography coupled to mass spectrometry-based metabolomic to screen for anabolic practices in cattle: identification of 5α-androst-2-en-17-one as new biomarker of 4-androstenedione misuse. J Mass Spectrom 47:131–140

Arias M, Chevallier OP, Graham SF, Gasull-Gimenez A, Fodey T, Cooper KM, Crooks SRH, Danaher M, Elliott CT (2016) Metabolomics reveals novel biomarkers of illegal 5-nitromimidazole treatment in pigs. Further evidence of drug toxicity uncovered. Food Chem 199:876–884

Arroyo-Manzanares N, Diana Di Mavungu J, Uka V, Malysheva SV, Cary JW, Ehrlich KC, Vanhaecke L, Bhatnagar D, de Saeger S (2015) Use of UHPLC high-resolution Orbitrap mass spectrometry to investigate the genes involved in the production of secondary metabolites in *Aspergillus flavus*. Food Addit Contam Part A 32:1656–1673

Aru V, Pisano MB, Savorani F, Engelsen SB, Cosentino S, Cesare Marincola F (2016) Metabolomics analysis of shucked mussels' freshness. Food Chem 205:58–65

Asiago VM, Hazebroek J, Harp T, Zhong C (2012) Effects of genetics and environment on the metabolome of commercial maize hybrids: a multisite study. J Agric Food Chem 60:11498–11508

Badilita V, Adhikari SS, MacKinnon N, Wallrabe U, Korvink JG (2014) Disposable 1H NMR detectors for food quality monitoring: application to butter and citrus juice discrimination. Chem Biol Microsyst Soc 1710–1712

Balan M, Nicolescu A, Stavarache C, Ciobanu M, Deleanu C (2013) Fast NMR juice identification based on sugars and other plant metabolites from fruits. Rev Roum Chim 58:175–182

Banerjee K, Mujawar S, Utture SC, Dasgupta S, Adsule PG (2013) Optimization of gas chromatography-single quadrupole mass spectrometry conditions for multiresidue analysis of pesticides in grapes in compliance to EU-MRLs. Food Chem 138:600–607

Barros E, Lezar S, Anttonen MJ, van Dijk JP, Röhlig RM, Kok EJ, Engel KH (2010) Comparison of two GM maize varieties with a near-isogenic non-GM variety using transcriptomics, proteomics and metabolomics. Plant Biotechnol J 8:436–451

Bauer R, Nieuwoudt H, Bauer FF, Kossmann J, Koch KR, Esbensen KH (2008) FTIR spectroscopy for grape and wine analysis. Anal Chem 80:1371–1379

Beach DG, Walsh CM, McCarron P (2014) High-throughput quantitative analysis of domoic acid directly from mussel tissue using Laser Ablation Electrospray Ionization—tandem mass spectrometry. Toxicon 92:75–80

Beale DJ, Morrison PD, Palombo EA (2014) Detection of *Listeria* in milk using non-targeted metabolic profiling of *Listeria* monocytogenes: a proof-of-concept application. Food Control 42:343–346

Begley M, Hill C (2015) Stress adaptation in food-borne pathogens. Ann Rev Food Sci Technol 6:191–210

Berg M, Vanaerschot M, Jankevics A, Cuypers B, Breitling R, Dujardin JC (2012) LC-MS metabolomics from study design to data-analysis—using a versatile pathogen as a test case. Comput Struct Biotechnol J 4:e201301002

Berthomieu C, Hienerwadel R (2009) Fourier transform infrared (FTIR) spectroscopy. Photosynth Res 101:157–170

Biais B, Bernillon S, Deborde C, Cabasson C, Rolin D, Tadmor Y, Burger J, Schaffer AA, Moing A (2012) Precautions for harvest, sampling, storage, and transport of crop plant metabolomics samples. Methods Mol Biol 860:51–63

Bino RJ, Hall RD, Fiehn O et al (2004) Potential of metabolomics as a functional genomics tool. Trends Plant Sci 9:418–425

Böhme K, Fernández-No IC, Barros-Velázquez J, Gallardo JM, Calo-Mata P, Cañas B (2010) Species differentiation of seafood spoilage and pathogenic gram-negative bacteria by MALDI-TOF mass fingerprinting. J Proteome Res 9:3169–3183

Böhme K, Morandi S, Cremonesi P, Fernández No IC, Barros-Velázquez J, Castiglioni B, Brasca M, Cañas B, Calo-Mata P (2012) Characterization of *Staphylococcus aureus* strains isolated from Italian dairy products by MALDI-TOF mass fingerprinting. Electrophoresis 33:2355–2364

Booth SC, Weljie AM, Turner RJ (2015) Metabolomics reveals differences of metal toxicity in cultures of *Pseudomonas pseudoalcaligenes* KF707 grown on different carbon. Front Microbiol 6:827

Boutsiadou-Theurillat X, Meier P, Richard C (2014) Development and in-house validation of a rapid LC-MS/MS method for the semi-quantification of eleven mycotoxins in maize samples. Chimia 68:716–720

Brennan L (2014) NMR-based metabolomics: from sample preparation to applications in nutrition research. Prog Nucl Magn Reson Spectrosc 83:42–49

Brookes G, Barfoot P (2013) Key environmental impacts of global genetically modified (GM) crop use 1996-2011. GM Crops Food 4:109–119

Cai J, Henion J (1995) Capillary electrophoresis-mass spectrometry. J Chromatogr A 703:667–692

Caligiani A, Malavasi G, Palla G, Marseglia A, Tognolini M, Bruni R (2013) A simple GC-MS method for the screening of betulinic, corosolic, maslinic, oleanolic and ursolic acid contents in commercial botanicals used as food supplement ingredients. Food Chem 136:735–741

Callejón RM, Torija MJ, Mas A, Morales ML, Troncoso AM (2010) Changes of volatile compounds in wine vinegars during their elaboration in barrels made from different woods. Food Chem 120:561–571

Canellas E, Vera P, Domeño C, Alfaro P, Nerín C (2012) Atmospheric pressure gas chromatography coupled to quadrupole-time of flight mass spectrometry as a powerful tool for identification of non intentionally added substances in acrylic adhesives used in food packaging materials. J Chromatogr A 1235:141–148

Cao S, He X, Xu W, Luo Y, Yuan Y, Liu P, Cao B, Shi H, Huang K (2012) Safety assessment of transgenic *Bacillus thuringiensis* rice T1c-19 in Sprague-Dawley rats from metabonomics and bacterial profile perspectives. IUBMB Life 64:242–250

Cao YZ, Liu YM, Wang N, Ji XX, Yao CC, Li X, Shi LL, Chang QY, Fan CL, Pang GF (2014) Multiresidual determination of 295 pesticides and chemical pollutants in animal fat by gel permeation chromatography (GPC) cleanup coupled with GC-MS/MS, GC-NCI-MS, and LC-MS/MS. In: High-throughput analysis for food safety. Wiley Blackwell, New York, pp 117–166

Cappiello A, Famiglini G, Palma P, Trufelli H (2010) Matrix effects in liquid chromatography-mass spectrometry. J Liq Chromatogr Relat Technol 33:1067–1081

Castro-Puyana M, Herrero M (2013) Metabolomics approaches based on mass spectrometry for food safety, quality and traceability. TrAC Trend Anal Chem 52:74–87

Castro-Puyana M, Mendiola JA, Ibáñez E, Herrero M (2013) MS-based metabolomics approaches for food safety, quality, and traceability. In: Foodomics: advanced mass spectrometry in modern food science and nutrition. Wiley, New York, pp 453–470.

Catchpole GS, Beckmann M, Enot DP et al (2005) Hierarchical metabolomics demonstrates substantial compositional similarity between genetically modified and conventional potato crops. Proc Natl Acad Sci USA 102:14458–14462

Cevallos-Cevallos JM, Danyluk MD, Reyes-De-Corcuera JI (2011) GC-MS based metabolomics for rapid simultaneous detection of *Escherichia coli* O157:H7, *Salmonella typhimurium*, *Salmonella muenchen*, and *Salmonella hartford* in ground beef and chicken. J Food Sci 76: M238–M246

Chan PH, Zhang WL, Lau CH, Cheung CY, Keun HC, Tsim KWK, Lam H (2014) Metabonomic analysis of water extracts from different angelica roots by 1 h-nuclear magnetic resonance spectroscopy. Molecules 19:3460–3470

Chatterjee NS, Banerjee K, Utture S, Kamble N, Rao BM, Panda SK, Mathew S (2015) Assessment of polyaromatic hydrocarbons and pesticide residues in domestic and imported *pangasius* (*Pangasianodon hypophthalmus*) fish in India. J Sci Food Agric 96:2373

Chinnici F, Guerrero ED, Sonni F, Natali N, Marín RN, Riponi C (2009) Gas chromatography-mass spectrometry (GC-MS) characterization of volatile compounds in quality vinegars with protected European geographical indication. J Agric Food Chem 57:4784–4792

Chu CY, Wang CC (2013) Toxicity of melamine: the public health concern. J Environ Sci Health Part C 31:342–386

Cifuentes A, Rutledge D (2013) Foodomics and food analysis in the post-genomics era. TrAC Trend Anal Chem 52:1

Cirlini M, Caligiani A, Palla L, Palla G (2011) HS-SPME/GC-MS and chemometrics for the classification of Balsamic Vinegars of Modena of different maturation and ageing. Food Chem 124:1678–1683

Clarke JD, Alexander DC, Ward DP, Ryals JA, Mitchell MW, Wulff JE, Guo L (2013) Assessment of genetically modified soybean in relation to natural variation in the soybean seed metabolome. Sci Rep 3:3083

Cuny M, Vigneau E, Le Gall G, Colquhoun I, Lees M, Rutledge DN (2008) Fruit juice authentication by 1H NMR spectroscopy in combination with different chemometric tools. Anal Bioanal Chem 390:419–427

D'Alessandro A, Zolla L (2013) Foodomics to investigate meat tenderness. TrAC Trend Anal Chem 52:47–53

Da Rosa JS, Freitas-Silva O, Pacheco S, de Oliveira Godoy RL, de Rezende CM (2012) LC-MS based screening and targeted profiling methods for complex plant: coffee a case study. Curr Drug Metab 13:1244–1250

Da Silva EP, De Martinis ECP (2013) Current knowledge and perspectives on biofilm formation: The case of *Listeria monocytogenes*. Appl Microbiol Biotechnol 97:957–968

Dawson IK, Russell J, Powell W, Steffenson B, Thomas WTB, Waugh R (2015) Barley: a translational model for adaptation to climate change. New Phytol 206:913–931

De Dominicis E, Commissati I, Suman M (2012) Targeted screening of pesticides, veterinary drugs and mycotoxins in bakery ingredients and food commodities by liquid chromatography-high-resolution single-stage Orbitrap mass spectrometry. J Mass Spectrom 47:1232–1241

Delaney B (2015) Safety assessment of foods from genetically modified crops in countries with developing economies. Food Chem Toxicol 86:132–143

Del Coco L, Schena FP, Fanizzi FP (2012) ^{1}H Nuclear magnetic resonance study of olive oils commercially available as Italian products in the United States of America. Nutrients 4:343–355

Dervilly-Pinel G, Courant F, Chéreau S, Royer AL, Boyard-Kieken F, Antignac JP, Monteau F, Le Bizec B (2012) Metabolomics in food analysis: application to the control of forbidden substances. Drug Test Anal 4:59–69

Dais P, Hatzakis E (2013) Quality assessment and authentication of virgin olive oil by NMR spectroscopy: a critical review. Anal Chim Acta 765:1–27

Dieterle F, Riefke B, Schlotterbeck G, Ross A, Senn H, Amberg A (2011) NMR and MS methods for metabonomics. Methods Mol Biol 691:385–415

Doué M, Dervilly-Pinel G, Pouponneau K, Monteau F, le Bizec B (2015) Direct analysis in real time—high resolution mass spectrometry (DART-HRMS): a high throughput strategy for identification and quantification of anabolic steroid esters. Drug Test Anal 7:603–608

Duedahl-Olesen L, Navaratnam MA, Jewula J, Jensen AH (2015a) PAH in some brands of tea and coffee. Polycyclic Aromat Compd 35:74–90

Duedahl-Olesen L, Aaslyng M, Meinert L, Christensen T, Jensen AH, Binderup ML (2015b) Polycyclic aromatic hydrocarbons (PAH) in Danish barbecued meat. Food Control 57:169–176

Dunn WB (2011) Mass spectrometry in systems biology: an introduction. Methods Enzymol 500:15–35

Dunn WB, Ellis DI (2005) Metabolomics: current analytical platforms and methodologies. TrAC Trend Anal Chem 24:285–294

Duportet X, Aggio RBM, Carneiro S, Villas-Bôas SG (2012) The biological interpretation of metabolomic data can be misled by the extraction method used. Metabolomics 8:410–421

El-Aneed A, Cohen A, Banoub J (2009) Mass spectrometry, review of the basics: electrospray, MALDI, and commonly used mass analyzers. Appl Spectrosc Rev 44:210–230

El Amrani A, Dumas AS, Wick LY, Yergeau E, Berthomé R (2015) "Omics" Insights into PAH degradation toward improved green remediation biotechnologies. Environ Sci Technol 49:11281–11291

Everstine K, Spink J, Kennedy S (2013) Economically motivated adulteration (EMA) of food: common characteristics of EMA incidents. J Food Prot 76:723–735

Falkowska L, Reindl AR (2015) Dietary exposure to, and internal organ transfer of, selected halogenated organic compounds in birds eating fish from the Southern Baltic. J Environ Sci Health Part A Toxiic/Hazard Subst Environ Eng 50:1029–1039

Fan S, Zhao P, Yu C, Pan C, Li X (2014) Simultaneous determination of 36 pesticide residues in spinach and cauliflower by LC-MS/MS using multi-walled carbon nanotubes-based dispersive solid-phase clean-up. Food Addit Contam Part A 31:73–82

Farré M, Picó Y, Barceló D (2013) Direct peel monitoring of xenobiotics in fruit by direct analysis in real time coupled to a linear quadrupole ion trap-orbitrap mass spectrometer. Anal Chem 85:2638–2644

Fiehn O (2002) Metabolomics—the link between genotypes and phenotypes. Plant Mol Biol 48:155–171

Font G, Ruiz MJ, Fernández M, Picó Y (2008) Application of capillary electrophoresis-mass spectrometry for determining organic food contaminants and residues. Electrophoresis 29:2059–2078

Forsberg ND, Wilson GR, Anderson KA (2011) Determination of parent and substituted polycyclic aromatic hydrocarbons in high-fat salmon using a modified quechers extraction, dispersive SPE and GC-MS. J Agric Food Chem 59:8108–8116

García-Gómez D, García-Hernández M, Rodríguez-Gonzalo E, Carabias-Martínez R (2012) A fast and reliable method for the quantitative determination of benzimidazoles and metabolites in milk by LC-MS/MS with on-line sample treatment. Anal Bioanal Chem 404:2909–2914

García-Reyes JF, Jackson AU, Molina-Díaz A, Cooks RG (2009) Desorption electrospray ionization mass spectrometry for trace analysis of agrochemicals in food. Anal Chem 81:820–829

Gelting RJ, Baloch MA, Zarate-Bermudez M, Hajmeer MN, Yee JC, Brown T, Yee BJ (2015) A systems analysis of irrigation water quality in an environmental assessment of an *E. coli* O157: H7 outbreak in the United States linked to iceberg lettuce. Agric Water Manag 150:111–118

Giacometti J, Tomljanović AB, Josić D (2013) Application of proteomics and metabolomics for investigation of food toxins. Food Res Int 54:1042–1051

Gilbert López B, Robles-Molina J, García-Reyes JF, Molina-Díaz A (2010) Rapid determination of BTEXS in olives and olive oil by headspace-gas chromatography/mass spectrometry (HS-GC-MS). Talanta 83:391–399

Giri A, Khummueng W, Mercier F, Kondjoyan N, Tournayre P, Meurillon M, Ratel J, Engel E (2015) Relevance of two-dimensional gas chromatography and high resolution olfactometry for the parallel determination of heat-induced toxicants and odorants in cooked food. J Chromatogr A 1388:217–226

Glinski M, Weckwerth W (2006) The role of mass spectrometry in plant systems biology. Mass Spectrom Rev 25:173–214

Gottfried S, Herebian D (2013) Determination of mycotoxins in food matrices using LC-MS/MS compared with high-resolution orbitrap™ MS technology. Curr Anal Chem 9:99–107

Gowda H, Ivanisevic J, Johnson CH, Kurczy ME, Benton HP, Rinehart D, Nguyen T, Ray J, Kuehl J, Arevalo B, Westenskow PD, Wang J, Arkin AP, Deutschbauer AM, Patti GJ, Siuzdak G (2014) Interactive XCMS online: simplifying advanced metabolomic data processing and subsequent statistical analyses. Anal Chem 86:6931–6939

Grimalt S, Harbeck S, Shegunova P, Seghers J, Sejerøe-Olsen B, Emteborg H, Dabrio M (2015) Development of a new cucumber reference material for pesticide residue analysis: feasibility study for material processing, homogeneity and stability assessment. Anal Bioanal Chem 407:3083–3091

Guo C, Huang XY, Yang MJ, Wang S, Ren ST, Li H, Peng XX (2014) GC/MS-based metabolomics approach to identify biomarkers differentiating survivals from death in crucian carps infected by *Edwardsiella tarda*. Fish Shellfish Immunol 39:215–222

Gustafson KD, Belden JB, Bolek MG (2015) The effects of the herbicide atrazine on freshwater snails. Ecotoxicology 24:1183–1197

Guyon F, Gaillard L, Salagoïty MH, Médina B (2011) Intrinsic ratios of glucose, fructose, glycerol and ethanol 13C/12C isotopic ratio determined by HPLC-co-IRMS: toward determining constants for wine authentication. Anal Bioanal Chem 401:1551–1558

Hall R, Beale M, Fiehn O, Hardy N, Sumner L, Bino R (2002) Plant metabolomics: the missing link in functional genomics strategies. Plant Cell 14:1437–1440

Hall RD, Hardy NW (2012) Practical applications of metabolomics in plant biology. Methods Mol Biol 860:1–10

Hall RD, de Maagd RA (2014) Plant metabolomics is not ripe for environmental risk assessment. Trend Biotechnol 32:391–392

Han J, Datla R, Chan S, Borchers CH (2009) Mass spectrometry-based technologies for high-throughput metabolomics. Bioanalysis 1:1665–1684

Han Y, Song L, Zhao P, Li Y, Zou N, Qin Y, Li X, Pan C (2016) Residue determination of glufosinate in plant origin foods using modified Quick Polar Pesticides (QuPPe) method and liquid chromatography coupled with tandem mass spectrometry. Food Chem 197:730–736

Harrigan GG, Stork LG, Riordan SG et al (2007) Impact of genetics and environment on nutritional and metabolite components of maize grain. J Agric Food Chem 55:6177–6185

Herman KM, Hall AJ, Gould LH (2015) Outbreaks attributed to fresh leafy vegetables, United States, 1973–2012. Epidemiol Infect 143:3011–3021

Herrero M, Simõ C, García-Cañas V, Ibáñez E, Cifuentes A (2012) Foodomics: MS-based strategies in modern food science and nutrition. Mass Spectrom Rev 31:49–69

Humphrey T (2004) *Salmonella*, stress responses and food safety. Nat Rev Microbiol 2:504–509

Ibáñez C, García-Cañas V, Valdés A, Simó C (2013) Novel MS-based approaches and applications in food metabolomics. TrAC Trend Anal Chem 52:100–111

Ibáñez E, Cifuentes A (2014) Foodomics: food science and nutrition in the postgenomic era, vol 64. Elsevier, Amsterdam, pp 395–440

Inoue K, Toyo'oka T (2015) Foodomics, vol 68. Elsevier, Amsterdam, pp 653–684

Inoue K, Tanada C, Sakamoto T, Tsutsui H, Akiba T, Min JZ, Todoroki K, Yamano Y, Toyo'Oka T (2015) Metabolomics approach of infant formula for the evaluation of contamination and degradation using hydrophilic interaction liquid chromatography coupled with mass spectrometry. Food Chem 181:318–324

Ison SA, Delannoy S, Bugarel M, Nightingale KK, Webb HE, Renter DG, Nagaraja TG, Loneragan GH, Fach P (2015) Genetic diversity and pathogenic potential of attaching and effacing *Escherichia coli* O26: H11 strains recovered from bovine feces in the United States. Appl Environ Microbiol 81:3671–3678

Jadhav S, Gulati V, Fox EM, Karpe A, Beale DJ, Sevior D, Bhave M, Palombo EA (2015) Rapid identification and source-tracking of *Listeria monocytogenes* using MALDI-TOF mass spectrometry. Int J Food Microbiol 202:1–9

Jandrić Z, Islam M, Singh DK, Cannavan A (2015) Authentication of Indian citrus fruit/fruit juices by untargeted and targeted metabolomics. Food Control (in press)

Janssens G, Mangelinckx S, Courtheyn D, De Kimpe N, Matthijs B, Le Bizec B (2015) Simultaneous detection of androgen and estrogen abuse in breeding animals by gas chromatography-mass spectrometry/combustion/isotope ratio mass spectrometry (GC-MS/C/IRMS) evaluated against alternative methods. J Agric Food Chem 63:7574–7581

Jäpelt KB, Nielsen NJ, Wiese S, Christensen JH (2015) Metabolic fingerprinting of Lactobacillus paracasei: a multi-criteria evaluation of methods for extraction of intracellular metabolites. Anal Bioanal Chem 407:6095

Jiang H, Cao H, Zhang Y, Fast DM (2012) Systematic evaluation of supported liquid extraction in reducing matrix effect and improving extraction efficiency in LC-MS/MS based bioanalysis for 10 model pharmaceutical compounds. J Chromatogr B Anal Technol Biomed Life Sci 891–892:71–80

Juan-Borrás M, Periche A, Domenech E, Escriche I (2015) Routine quality control in honey packaging companies as a key to guarantee consumer safety. The case of the presence of sulfonamides analyzed with LC-MS-MS. Food Control 50:243–249

Kalachova K, Pulkrabova J, Cajka T, Drabova L, Hajslova J (2012) Implementation of comprehensive two-dimensional gas chromatography-time-of-flight mass spectrometry for the simultaneous determination of halogenated contaminants and polycyclic aromatic hydrocarbons in fish. Anal Bioanal Chem 403:2813–2824

Kalachova K, Pulkrabova J, Cajka T, Drabova L, Stupak M, Hajslova J (2013) Gas chromatography-triple quadrupole tandem mass spectrometry: a powerful tool for the (ultra)trace analysis of multiclass environmental contaminants in fish and fish feed. Anal Bioanal Chem 405:7803–7815

Khan MR (2015) Influence of food condiments on the formation of carcinogenic heterocyclic amines in cooked chicken and determination by LC-MS/MS. Food Addit Contam Part A Chem Anal Control Expo Risk Assess 32:307–314

Kim S, Kim J, Yun EJ, Kim KH (2016) Food metabolomics: from farm to human. Curr Opin Biotechnol 37:16–23

Knight JG, Clark A, Mather DW (2013) Potential damage of GM crops to the country image of the producing country. GM Crops Food 4:151–157

Koutsoumanis KP, Sofos JN (2004) Comparative acid stress response of *Listeria monocytogenes*, *Escherichia coli* O157:H7 and *Salmonella Typhimurium* after habituation at different pH conditions. Lett Appl Microbiol 38:321–326

Krimsky S (2015) An illusory consensus behind GMO health assessment. Sci Technol Human Values 40:883–914

Kuchenmüller T, Hird S, Stein C, Kramarz P, Nanda A, Havelaar AH (2009) Estimating the global burden of foodborne diseases—a collaborative effort. 14(18). pii: 19195

Kueger S, Steinhauser D, Willmitzer L, Giavalisco P (2012) High-resolution plant metabolomics: from mass spectral features to metabolites and from whole-cell analysis to subcellular metabolite distributions. Plant J 70:39–50

Kusano M, Baxter I, Fukushima A et al (2014) Assessing metabolomic and chemical diversity of a soybean lineage representing 35 years of breeding. Metabolomics 11:261–270

Lacina O, Zachariasova M, Urbanova J, Vaclavikova M, Cajka T, Hajslova J (2012) Critical assessment of extraction methods for the simultaneous determination of pesticide residues and mycotoxins in fruits, cereals, spices and oil seeds employing ultra-high performance liquid chromatography-tandem mass spectrometry. J Chromatogr A 1262:8–18

Laghi L, Picone G, Capozzi F (2014) Nuclear magnetic resonance for foodomics beyond food analysis. TrAC Trend Anal Chem 59:93–102

Lanucara F, Holman SW, Gray CJ, Eyers CE (2014) The power of ion mobility-mass spectrometry for structural characterization and the study of conformational dynamics. Nat Chem 6:281–294

Lara FJ, García-Campaña AM, Alés-Barrero F, Bosque-Sendra JM (2008) In-line solid-phase extraction preconcentration in capillary electrophoresis-tandem mass spectrometry for the multiresidue detection of quinolones in meat by pressurized liquid extraction. Electrophoresis 29:2117–2125

Laufer AS, Grass J, Holt K, Whichard JM, Griffin PM, Gould LH (2015) Outbreaks of *Salmonella* infections attributed to beef—United States, 1973–2011. Epidemiol Infect 143:2003–2013

Le Boucher C, Courant F, Royer AL, Jeanson S, Lortal S, Dervilly-Pinel G, Thierry A, Le Bizec B (2015) LC–HRMS fingerprinting as an efficient approach to highlight fine differences in cheese metabolome during ripening. Metabolomics 11:1117–1130

Lee JE, Hwang GS, Van Den Berg F, Lee CH, Hong YS (2009) Evidence of vintage effects on grape wines using 1H NMR-based metabolomic study. Anal Chim Acta 648:71–76

Le Gall G (2015) Sample collection and preparation of biofluids and extracts for NMR spectroscopy, vol 1277. Humana Press, New York, pp 15–28

Lewis IA, Shortreed MR, Hegeman AD, Markley JL (2012) Novel NMR and MS approaches to metabolomics. In: Fan T, Higashi RM, Lane AN (eds) The handbook of metabolomics, vol 17. pp 199–230

Li C, Schmidt NE, Gitaitis R (2011) Detection of onion postharvest diseases by analyses of headspace volatiles using a gas sensor array and GC-MS. LWT—Food Sci Technol 44:1019–1025

Li K, Pidatala VR, Shaik R, Datta R, Ramakrishna W (2014) Integrated metabolomic and proteomic approaches dissect the effect of metal-resistant bacteria on maize biomass and copper uptake. Environ Sci Technol 48:1184–1193
Lima MRM, Diaz SO, Lamego I, Grusak MA, Vasconcelos MW, Gil AM (2014) Nuclear magnetic resonance metabolomics of iron deficiency in soybean leaves. J Proteome Res 13:3075–3087
Liu Z, Zhao J, Li Y, Zhang W, Jian G, Peng Y, Qi F (2012) Non-uniform distribution pattern for differentially expressed genes of transgenic rice Huahui 1 at different developmental stages and environments. PLoS One 7:e37078
Maghari BM, Ardekani AM (2011) Genetically modified foods and social concerns. Avicenna J Med Biotechnol 3:109–117
McWhorter AR, Davos D, Chousalkar KK (2015) Pathogenicity of *Salmonella* strains isolated from egg shells and the layer farm environment in Australia. Appl Environ Microbiol 81:405–414
Marriott PJ, Chin ST, Maikhunthod B, Schmarr HG, Bieri S (2012) Multidimensional gas chromatography. TrAC Trends Anal Chem 34:1–20
Marušić A (2011) Food safety and security: what were favourite topics for research in the last decade? J Global Health 1:72–78
Mas S, Villas-Bôas SG, Hansen ME, Åkesson M, Nielsen J (2007) A comparison of direct infusion MS and GC-MS for metabolic footprinting of yeast mutants. Biotechnol Bioeng 96:1014–1022
Mastovska K, Wylie PL (2012) Evaluation of a new column backflushing set-up in the gas chromatographic-tandem mass spectrometric analysis of pesticide residues in dietary supplements. J Chromatogr A 1265:155–164
Melnik BC (2015) Linking diet to acne metabolomics, inflammation, and comedogenesis: an update. Clin Cosmet Investig Dermatol 8:371–388
Mondello L, Zoccali M, Purcaro G, Franchina FA, Sciarrone D, Moret S, Conte L, Tranchida PQ (2012) Determination of saturated-hydrocarbon contamination in baby foods by using on-line liquid-gas chromatography and off-line liquid chromatography-comprehensive gas chromatography combined with mass spectrometry. J Chromatogr A 1259:221–226
Motarjemi Y (2014) Hazard analysis and critical control point system (HACCP). Food safety management: a practical guide for the food industry. Elsevier, Amsterdam, pp 845–872
Motarjemi Y, Mortimore S (2014) Assessment of food safety management systems. In: Motarjemi Y, Lelieveld H (eds) Food safety management: a practical guide for the food industry. Elsevier, Amsterdam, pp 987–1004
Motarjemi Y, Lelieveld H (2014) Food safety management: a practical guide for the food industry. Elsevier, Amsterdam
Motarjemi Y, Vanschothorst M, Käferstein F (2001) Future challenges in global harmonization of food safety legislation. Food Control 12:339–346
Motarjemi Y, Stadler RH, Studer A, Damiano V (2008) Application of the Haccp approach for the management of processing contaminants. Process-induced food toxicants: occurrence, formation, mitigation, and health risks. Wiley, New York, pp 563–620
Mulas G, Galaffu MG, Pretti L, Nieddu G, Mercenaro L, Tonelli R, Anedda R (2011) NMR analysis of seven selections of Vermentino grape berry: metabolites composition and development. J Agric Food Chem 59:793–802
Nevedomskaya E, Ramautar R, Derks R, Westbroek I, Zondag G, Van Der Pluijm I, Deelder AM, Mayboroda OA (2010) CE-MS for metabolic profiling of volume-limited urine samples: application to accelerated aging TTD mice. J Proteome Res 9:4869–4874
Nguyen YTB, Kamoshita A, Dinh VTH, Matsuda H, Kurokura H (2016) Salinity intrusion and rice production in Red River Delta under changing climate conditions. Paddy Water Environ 1–12
Nokkoul R, Wichitparp T (2014) Effect of drought condition on growth, yield and grain quality of upland rice. Am J Agric Biol Sci 9:439–444

Ogrinc N, Košir IJ, Spangenberg JE, Kidrič J (2003) The application of NMR and MS methods for detection of adulteration of wine, fruit juices, and olive oil. A review Anal Bioanal Chem 376:424–430

Ojima-Kato T, Yamamoto N, Suzuki M, Fukunaga T, Tamura H (2014) Discrimination of *Escherichia coli* O157, O26 and O111 from other serovars by MALDI-TOF MS based on the S10-germs method. PLoS One 9:e113458

O'Leary D, McCabe EM, McCusker MP, Martins M, Fanning S, Duffy G (2015) Acid environments affect biofilm formation and gene expression in isolates of *Salmonella enterica Typhimurium* DT104. Int J Food Microbiol 206:7–16

Ordoudi SA, Cagliani LR, Lalou S, Naziri E, Tsimidou MZ, Consonni R (2015) 1H NMR-based metabolomics of saffron reveals markers for its quality deterioration. Food Res Int 70:1–6

Osselaere A, Devreese M, Goossens J, Vandenbroucke V, De Baere S, De Backer P, Croubels S (2013) Toxicokinetic study and absolute oral bioavailability of deoxynivalenol, T-2 toxin and zearalenone in broiler chickens. Food Chem Toxicol 51:350–355

OzFoodNet Working Group (2014) OzFoodNet quarterly report, 1 April to 30 June 2013. Commun Dis Intell Q Rep 38:E376–E382

Pacilli M, Cortese MM, Smith S et al (2015) Outbreak of gastroenteritis in adults due to Rotavirus genotype G12P[8]. Clin Infect Dis 61:e20–e25

Park SE, Yoo SA, Seo SH, Lee KI, Na CS, Son HS (2016) GC–MS based metabolomics approach of Kimchi for the understanding of *Lactobacillus plantarum* fermentation characteristics. LWT—Food Sci Technol 68:313–321

Parrott W, Chassy B, Ligon J, Meyer L, Petrick J, Zhou J, Herman R, Delaney B, Levine M (2010) Application of food and feed safety assessment principles to evaluate transgenic approaches to gene modulation in crops. Food Chem Toxicol 48:1773–1790

Pereira GE, Gaudillere JP, Pieri P, Hilbert G, Maucourt M, Deborde C, Moing A, Rolin D (2006) Microclimate influence on mineral and metabolic profiles of grape berries. J Agric Food Chem 54:6765–6775

Pérez-Ortega P, Gilbert-López B, García-Reyes JF, Ramos-Martos N, Molina-Díaz A (2012) Generic sample treatment method for simultaneous determination of multiclass pesticides and mycotoxins in wines by liquid chromatography-mass spectrometry. J Chromatogr A 1249:32–40

Pérez S, Rial D, Beiras R (2015) Acute toxicity of selected organic pollutants to saltwater (*mysid Siriella armata*) and freshwater (*cladoceran Daphnia magna*) ecotoxicological models. Ecotoxicology 24:1229–1238

Picariello G, Mamone G, Addeo F, Ferranti P (2012) Novel mass spectrometry-based applications of the 'omic': sciences in food technology and biotechnology. Food Technol Biotechnol 50:286–305

Piccinonna S, Ragone R, Stocchero M, del Coco L, de Pascali SA, Schena FP, Fanizzi FP (2016) Robustness of NMR-based metabolomics to generate comparable data sets for olive oil cultivar classification. An inter-laboratory study on Apulian olive oils. Food Chem 199:675–683

Pinu FR (2015) Metabolomics-The new frontier in food safety and quality research. Food Res Int 72:80–81

Pinu FR, Edwards PJB, Jouanneau S, Kilmartin PA, Gardner RC, Villas-Boas SG (2014) Sauvignon blanc metabolomics: grape juice metabolites affecting the development of varietal thiols and other aroma compounds in wines. Metabolomics 10:556–573

Pisano MB, Scano P, Murgia A, Cosentino S, Caboni P (2016) Metabolomics and microbiological profile of Italian mozzarella cheese produced with buffalo and cow milk. Food Chem 192:618–624

Pizarro C, Rodríguez-Tecedor S, Pérez-del-Notario N, González-Sáiz JM (2011) Recognition of volatile compounds as markers in geographical discrimination of Spanish extra virgin olive oils by chemometric analysis of non-specific chromatography volatile profiles. J Chromatogr A 1218:518–523

Pizzutti IR, De Kok A, Scholten J, Righi LW, Cardoso CD, Necchi Rohers G, Da Silva RC (2014) Development, optimization and validation of a multimethod for the determination of 36 mycotoxins in wines by liquid chromatography-tandem mass spectrometry. Talanta 129:352–363

Pluskal T, Castillo S, Villar-Briones A, Orešič M (2010) MZmine 2: modular framework for processing, visualizing, and analyzing mass spectrometry-based molecular profile data. BMC Bioinform 11:395

Pope GA, MacKenzie DA, Defernez M et al (2007) Metabolic footprinting as a tool for discriminating between brewing yeasts. Yeast 24:667–679

Qi X, Chen S, Sheng Y, Guo M, Liu Y, He X, Huang K, Xu W (2015) Safety assessment of genetically modified rice expressing human serum albumin from urine metabonomics and fecal bacterial profile. Food Chem Toxicol 76:1–10

Qin J, Chao K, Kim MS (2014) Authenticating powdered foods and ingredients using a high-throughput Raman chemical imaging method. Am Soc Agric Biol Eng 3:1912–1919

Radosavljevic V, Finke EJ, Belojevic G (2015) *Escherichia coli* O104:H4 outbreak in Germany—Clarification of the origin of the epidemic. Eur J Pub Health 25:125–129

Ragone R, Crupi P, Piccinonna S, Bergamini C, Mazzone F, Fanizzi FP, Schena FP, Antonacci D (2015) Classification and chemometric study of Southern Italy monovarietal wines based on NMR and HPLC-DAD-MS. Food Sci Biotechnol 24:817–826

Ramautar R, Mayboroda OA, Somsen GW, De Jong GJ (2011) CE-MS for metabolomics: developments and applications in the period 2008-2010. Electrophoresis 32:52–65

Raupp J (2014) Social agents and news media as risk amplifiers: a case study on the public debate about the *E. coli* outbreak in Germany 2011. Health Risk Soc 16:565–579

Rejczak T, Tuzimski T (2015) Recent trends in sample preparation and liquid chromatography/mass spectrometry for pesticide residue analysis in food and related matrixes. J AOAC Int 98:1143–1162

Reynolds TL, Nemeth MA, Glenn KC, Ridley WP, Astwood JD (2005) Natural variability of metabolites in maize grain: differences due to genetic background. J Agric Food Chem 53:10061–10067

Rischer H, Oksman-Caldentey KM (2006) Unintended effects in genetically modified crops: revealed by metabolomics? Trend Biotechnol 24:102–104

Robotti E, Marengo E (2016) Chemometric multivariate tools for candidate biomarker identification: LDA, PLS-DA, SIMCA, Ranking-PCA. Methods Mol Biol 1384:237–267

Rodrigues JA, Barros AS, Carvalho B (2011) Probing beer aging chemistry by nuclear magnetic resonance and multivariate analysis. Anal Chim Acta 702:178–187

Rodríguez-Carrasco Y, Fattore M, Albrizio S, Berrada H, Mañes J (2015) Occurrence of *Fusarium* mycotoxins and their dietary intake through beer consumption by the European population. Food Chem 178:149–155

Rul F, Monnet V (2015) How microbes communicate in food: a review of signaling molecules and their impact on food quality. Curr Opin Food Sci 2:100–105

Russo R, Severino V, Mendez A, Lliberia J, Parente A, Chambery A (2012) Detection of buffalo mozzarella adulteration by an ultra-high performance liquid chromatography tandem mass spectrometry methodology. J Mass Spectrom 47:1407–1414

Samelis J, Sofos JN, Kendall PA, Smith GC (2001) Fate of *Escherichia coli* O157:H7, *Salmonella Typhimurium* DT 104, and *Listeria* monocytogenes in fresh meat decontamination fluids at 4 and 10°C. J Food Prot 64:950–957

Sapozhnikova Y, Lehotay SJ (2013) Multi-class, multi-residue analysis of pesticides, polychlorinated biphenyls, polycyclic aromatic hydrocarbons, polybrominated diphenyl ethers and novel flame retardants in fish using fast, low-pressure gas chromatography-tandem mass spectrometry. Anal Chim Acta 758:80–92

Sato S, Yanagisawa S (2010) Capillary electrophoresis-electrospray ionization-mass spectrometry using fused-silica capillaries to profile anionic metabolites. Metabolomics 6:529–540

Scalbert A, Brennan L, Fiehn O, Hankemeier T, Kristal BS, van Ommen B, Pujos-Guillot E, Verheij E, Wishart D, Wopereis S (2009) Mass-spectrometry-based metabolomics: limitations and recommendations for future progress with particular focus on nutrition research. Metabolomics 5:435–458

Schipilliti L, Dugo P, Bonaccorsi I, Mondello L (2011) Headspace-solid phase microextraction coupled to gas chromatography-combustion-isotope ratio mass spectrometer and to enantioselective gas chromatography for strawberry flavoured food quality control. J Chromatogr A 1218:7481–7486

Schipilliti L, Tranchida PQ, Sciarrone D, Russo M, Dugo P, Dugo G, Mondello L (2010) Genuineness assessment of mandarin essential oils employing gas chromatography-combustion-isotope ratio MS (GC-C-IRMS). J Sep Sci 33:617–625

Schurek J, Vaclavik L, Hooijerink H, Lacina O, Poustka J, Sharman M, Caldow M, Nielen MWF, Hajslova J (2008) Control of strobilurin fungicides in wheat using direct analysis in real time accurate time-of-flight and desorption electrospray ionization linear ion trap mass spectrometry. Anal Chem 80:9567–9575

Self RL (2013) Direct analysis in real time-mass spectrometry (DART-MS) for rapid qualitative screening of toxic glycols in glycerin-containing products. J Pharm Biomed Anal 80:155–158

Self RL, Wu WH (2012) Rapid qualitative analysis of phthalates added to food and nutraceutical products by direct analysis in real time/orbitrap mass spectrometry. Food Control 25:13–16

Senyuva HZ, Gökmen V, Sarikaya EA (2015) Future perspectives in Orbitrap™-high-resolution mass spectrometry in food analysis: a review. Food Addit Contam Part A 32:1568–1606

Shen S, Fang FC (2012) Integrated stress responses in Salmonella. Int J Food Microbiol 152:75–81

Shen Y, van Beek TA, Claassen FW, Zuilhof H, Chen B, Nielen MWF (2012) Rapid control of Chinese star anise fruits and teas for neurotoxic anisatin by Direct Analysis in Real Time high resolution mass spectrometry. J Chromatogr A 1259:179–186

Sheng Y, Qi X, Liu Y, Guo M, Chen S, He X, Huang K, Xu W (2014) Subchronic toxicity study in vivo and allergenicity study in vitro for genetically modified rice that expresses pharmaceutical protein (human serum albumin). Food Chem Toxicol 72:242–246

Shulaev V (2006) Metabolomics technology and bioinformatics. Brief Bioinform 7:128–139

Signes-Pastor AJ, Carey M, Meharg AA (2016) Inorganic arsenic in rice-based products for infants and young children. Food Chem 191:128–134

Singhal N, Kumar M, Kanaujia PK, Virdi JS (2015) MALDI-TOF mass spectrometry: an emerging technology for microbial identification and diagnosis. Front Microbiol 6:791

Skov T, Honoré AH, Jensen HM, Næs T, Engelsen SB (2014) Chemometrics in foodomics: handling data structures from multiple analytical platforms. TrAC Trend Anal Chem 60:71–79

Smith CA, Want EJ, O'Maille G, Abagyan R, Siuzdak G (2006) XCMS: processing mass spectrometry data for metabolite profiling using nonlinear peak alignment, matching, and identification. Anal Chem 78:779–787

Soni KA, Nannapaneni R (2010) Removal of *Listeria* monocytogenes biofilms with bacteriophage P100. J Food Prot 73:1519–1524

Spraul M, Schütz B, Humpfer E, Mörtter M, Schäfer H, Koswig S, Rinke P (2009a) Mixture analysis by NMR as applied to fruit juice quality control. Magn Reson Chem 47:S130–S137

Spraul M, Schütz B, Rinke P, Koswig S, Humpfer E, Schäfer H, Mörtter M, Fang F, Marx UC, Minoja A (2009b) NMR-based multi parametric quality control of fruit juices: SGF profiling. Nutrition 1:148–155

Squadrone S, Ciccotelli V, Prearo M, Favaro L, Scanzio T, Foglini C, Abete MC (2015) Perfluorooctane sulfonate (PFOS) and perfluorooctanoic acid (PFOA): emerging contaminants of increasing concern in fish from Lake Varese, Italy. Environ Monit Assess 80:155–158

Stackhouse RR, Faith NG, Kaspar CW, Czuprynski CJ, Wong ACL (2012) Survival and virulence of *Salmonella enterica serovar enteritidis* filaments induced by reduced water activity. Appl Environ Microbiol 78:2213–2220

Stewart I, McLeod C (2014) The laboratory mouse in routine food safety testing for marine algal biotoxins and harmful algal bloom toxin research: past, present and future. J AOAC Int 97:356–372

Storelli MM, Scarano C, Spanu C, De Santis EPL, Busco VP, Storelli A, Marcotrigiano GO (2012) Levels and congener profiles of polychlorinated dibenzo-p-dioxins (PCDDs), polychlorinated dibenzofurans (PCDFs) and polychlorinated biphenyls (PCBs) in sheep milk from an industrialised area of Sardinia, Italy. Food Chem Toxicol 50:1413–1417

Sumner LW (2010) Recent advances in plant metabolomics and greener pastures. F1000 Biology Reports 2

Sun X, Zhang L, Li P, Xu B, Ma F, Zhang Q, Zhang W (2015) Fatty acid profiles based adulteration detection for flaxseed oil by gas chromatography mass spectrometry. LWT—Food Sci Technol 63:430–436

Teo CC, Chong WPK, Ho YS (2013) Development and application of microwave-assisted extraction technique in biological sample preparation for small molecule analysis. Metabolomics 9:1109–1128

Theodoridis G, Gika H, Franceschi P, Caputi L, Arapitsas P, Scholz M, Masuero D, Wehrens R, Vrhovsek U, Mattivi F (2012) LC-MS based global metabolite profiling of grapes: solvent extraction protocol optimisation. Metabolomics 8:175–185

Tiganitas A, Zeaki N, Gounadaki AS, Drosinos EH, Skandamis PN (2009) Study of the effect of lethal and sublethal pH and aw stresses on the inactivation or growth of *Listeria monocytogenes* and *Salmonella typhimurium*. Int J Food Microbiol 134:104–112

Tomita S, Nemoto T, Matsuo Y, Shoji T, Tanaka F, Nakagawa H, Ono H, Kikuchi J, Ohnishi-Kameyama M, Sekiyama Y (2015) A NMR-based, non-targeted multistep metabolic profiling revealed l-rhamnitol as a metabolite that characterised apples from different geographic origins. Food Chem 174:163–172

Trivedi TK, DeSalvo T, Lee L, Palumbo A, Moll M, Curns A, Hall AJ, Patel M, Parashar UD, Lopman BA (2012) Hospitalizations and mortality associated with norovirus outbreaks in nursing homes, 2009-2010. J Am Med Assoc 308:1668–1675

Trufelli H, Palma P, Famiglini G, Cappiello A (2011) An overview of matrix effects in liquid chromatography-mass spectrometry. Mass Spectrom Rev 30:491–509

Tsilia V, Devreese B, De IB, Mesuere B, Rajkovic A, Uyttendaele M, Van De Wiele T, Heyndrickx M (2012) Application of MALDI-TOF mass spectrometry for the detection of enterotoxins produced by pathogenic strains of the *Bacillus cereus* group. Anal Bioanal Chem 404:1691–1702

Unnevehr L (2015) Food safety in developing countries: moving beyond exports. Glob Food Secur 4:24–29

Van Houdt R, Michiels CW (2010) Biofilm formation and the food industry, a focus on the bacterial outer surface. J Appl Microbiol 109:1117–1131

Van Oort PAJ, de Vries ME, Yoshida H, Saito K (2015) Improved climate risk simulations for rice in arid environments. PLoS One 10:e0118114

Versari A, Parpinello GP, Scazzina F, Rio DD (2010) Prediction of total antioxidant capacity of red wine by Fourier transform infrared spectroscopy. Food Control 21:786–789

Vestal ML (2011) The future of biological mass spectrometry. J Am Soc Mass Spectrom 22:953–959

Vikram A, Prithiviraj B, Hamzehzarghani H, Kushalappa AC (2004) Volatile metabolite profiling to discriminate diseases of McIntosh apple inoculated with fungal pathogens. J Sci Food Agric 84:1333–1340

Villas-Bôas SG, Koulman A, Lane GA (2007) Analytical methods from the perspective of method standardization. In: Nielsen J, Jewett MC (eds) vol 18, pp 11–52

Villas-Bôas SG, Delicado DG, Åkesson M, Nielsen J (2003) Simultaneous analysis of amino and nonamino organic acids as methyl chloroformate derivatives using gas chromatography-mass spectrometry. Anal Biochem 322:134–138

Villas-Bôas SG, Mas S, Åkesson M, Smedsgaard J, Nielsen J (2005) Mass spectrometry in metabolome analysis. Mass Spectrom Rev 24:613–646

Walker CH (1990) Persistent pollutants in fish-eating sea birds—bioaccumulation, metabolism and effects. Aquat Toxicol 17:293–324

Walorczyk S, Kopeć I, Szpyrka E (2015) Pesticide residue determination by gas chromatography-tandem mass spectrometry as applied to food safety assessment on the example of some fruiting vegetables. Food Anal Methods 9:1155–1172

Wan Q, Wu G, He Q, Tang H, Wang Y (2015) The toxicity of acute exposure to T-2 toxin evaluated by the metabonomics technique. Mol Biosyst 11:882–891

World Health Organization (2006) WHO consultation to develop a strategy to estimate the global burden of foodborne diseases. World Health Organization, Geneva, p vii. Available from http://www.who.int/foodsafety/publications/foodborne_disease/fbd_2006.pdf

Xia J, Mandal R, Sinelnikov IV, Broadhurst D, Wishart DS (2012) MetaboAnalyst 2.0-a comprehensive server for metabolomic data analysis. Nucleic Acids Res 40:W127–W133

Xu Y, Cheung W, Winder CL, Goodacre R (2010) VOC-based metabolic profiling for food spoilage detection with the application to detecting *Salmonella typhimurium*-contaminated pork. Anal Bioanal Chem 397:2439–2449

Yang Y, Zheng N, Zhao X, Zhang Y, Han R, Yang J, Zhao S, Li S, Guo T, Zang C, Wang J (2016) Metabolomic biomarkers identify differences in milk produced by Holstein cows and other minor dairy animals. J Proteomics 136:174–182

Yogi K, Oshiro N, Inafuku Y, Hirama M, Yasumoto T (2011) Detailed LC-MS/MS analysis of ciguatoxins revealing distinct regional and species characteristics in fish and causative alga from the pacific. Anal Chem 83:8886–8891

Zeng W, Hazebroek J, Beatty M, Hayes K, Ponte C, Maxwell C, Zhong CX (2014) Analytical method evaluation and discovery of variation within maize varieties in the context of food safety: transcript profiling and metabolomics. J Agric Food Chem 62:2997–3009

Zhan J, Yu XJ, Zhong YY, Zhang ZT, Cui XM, Peng JF, Feng R, Liu XT, Zhu Y (2012) Generic and rapid determination of veterinary drug residues and other contaminants in raw milk by ultra performance liquid chromatography-tandem mass spectrometry. J Chromatogr B Anal Technol Biomed Life Sci 906:48–57

Zhang A, Sun H, Wang P, Han Y, Wang X (2012) Modern analytical techniques in metabolomics analysis. Analyst 137:293–300

Zhang D, Park JA, Kim DS, Kim SK, Shin SJ, Shim JH, Shin SC, Kim JS, Abd El-Aty AM, Shin HC (2015) A simple extraction method for the simultaneous detection of tetramisole and diethylcarbamazine in milk, eggs, and porcine muscle using gradient liquid chromatography tandem mass spectrometry. Food Chem 192:299–305

Zhang XZ, Li CJ, Chen SS, Li XJ, Han H, Ma XD (2009) Direct determination of atrazine residue on Chinese cabbage leaf using desorption electrospray ionization-tandem mass spectrometry and its application for diagnosing atrazine drift phytotoxicity. J AOAC Int 92:1587–1592

Zhou B, Xiao JF, Tuli L, Ressom HW (2012) LC-MS-based metabolomics. Mol Bio Syst 8:470–481

Chapter 9
Microbial Metabolomics in Biomass Waste Management

Avinash V. Karpe, David J. Beale, Ian H. Harding and Enzo A. Palombo

1 Introduction

Plant biomass has a complex structure that consists of cellulose and hemicellulose surrounded by lignin (Fig. 9.1a) together with trace amounts of protein and fatty acids (Eriksson et al. 1990; Hori et al. 1985; Sánchez 2009). Cellulose is the most abundant and most studied polysaccharide and it forms the basic structure of cell walls of all plants and algae. Cellulose is predominantly found within the primary layer of the plant cell wall and its presence decreases in the secondary and tertiary cell wall layers. It occurs as an aggregated, lateral, microfibril network that forms highly complex mesh like structures. To date, cellulose has only been partially understood using advanced technologies such as nuclear magnetic resonance (NMR) spectroscopy, X-ray crystallography and electron microscopy (O'Sullivan 1997; Fernandes et al. 2011; Harris et al. 2012).

Cellulose is an almost linear molecule composed of β-D-glucopyranose, linked by β-1, 4-polyanhydroglucose, with cellobiose as the smallest repetitive unit (Fig. 9.1b), and is thus a β-glucan (Kumar et al. 2008; O'Neill et al. 2004; Stone 1958). Hemicelluloses form the secondary structural component of the lignocellulose complex (Fig. 9.1c). They are composed of hetero-structures of acelyted and other derived monosaccharides, mostly pentoses. Various hemicelluloses have been reported to be directly or indirectly associated with cellulose. They include sugars such as xylans, xyloglucans, mannans, pectins, homogalacturonans, rhamnogalacturonans, arabinans and galactans.

A.V. Karpe (✉) · I.H. Harding · E.A. Palombo
Department of Chemistry and Biotechnology, Swinburne University of Technology, PO Box 218, Hawthorn, Melbourne, VIC 3122, Australia
e-mail: akarpe@swin.edu.au

A.V. Karpe · D.J. Beale
Land and Water, Commonwealth Scientific and Industrial Research Organization (CSIRO), PO Box 2583, Brisbane, QLD 4002, Australia

D.J. Beale et al. (eds.), *Microbial Metabolomics*,
DOI 10.1007/978-3-319-46326-1_9

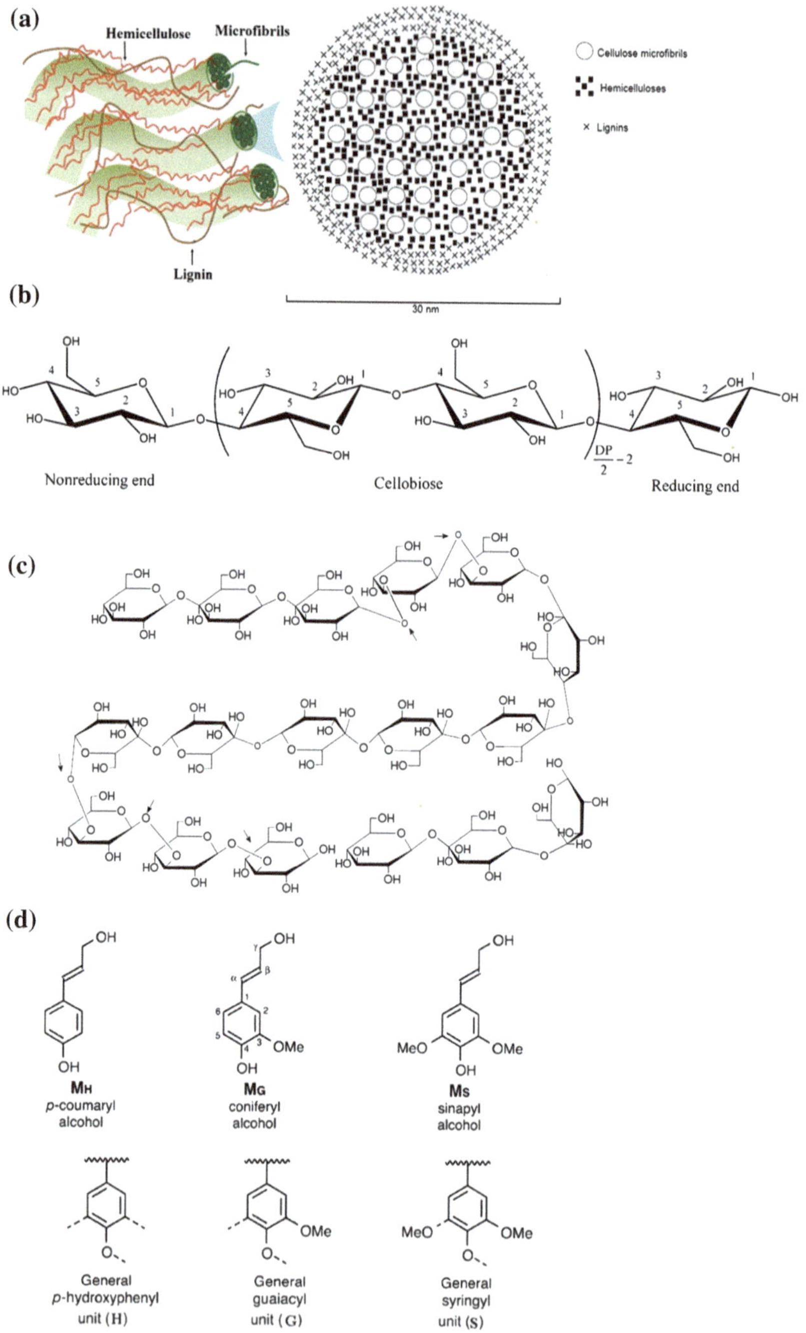

Fig. 9.1 **a–d** Complex structure of Plant biomass. **a** Schematic characterization of plant biomass structure comprising lignocellulose complex. **b** Fragment of cellulose with the reducing, non-reducing and cellobiose component of cellulose. **c** Representative example of hemicellulose (β-glucans). **d** Structures of (i) Hydroxycinnamyl alcohols-p-coumaryl alcohol, coniferyl alcohol and sinapyl alcohol and (ii) derived structural units H, G and S

Lignin is the second most abundant biopolymer. In most plants, especially in higher plants, it functions as the prominent component of xylem. Due to its hydrophobic nature, lignin aids in water transport throughout the plant system. This property also makes lignin tolerant towards biodegradation (Boerjan et al. 2003). Lignin biosynthesis and deposition occurs in the secondary cell wall after the cell has completed its growth. During primary growth, cellulose and hemicellulose are deposited on the cell wall, followed by sequential deposition of lignin in the secondary walls. However, most lignin deposition takes place in the next stage, which is followed by cell death (Baucher et al. 1998). Lignins are made up of hydroxycinnamyl alcohols, i.e, p-coumaryl alcohol, coniferyl alcohol and sinapyl alcohol (Fig. 9.1d) and their methoxy derivatives (Boerjan et al. 2003; Higuchi 2006; Vanholme et al. 2010). The main methoxy products of these three hydroxy cinnamyl alcohols, which form the structural units of lignin polymers, are known as general p-hydroxyphenyl units (H), general guaiacyl units (G) and general syringyl unit (S) (Baucher et al. 1998; Boerjan et al. 2003; Higuchi 2006; Ralph et al. 2004; Vanholme et al. 2010). However, due to the very complex nature of the lignin polymer, they have not been successfully isolated and characterised from natural sources (Vanholme et al. 2010). Vascular plants have greater amounts of G and S units, while grasses and other monocots are rich in H units (Baucher et al. 1998; Boerjan et al. 2003; Yang et al. 2010a).

The complex structure and sheer volume of biomass makes it a major waste management challenge. According to recent estimates from the "*Food and Agriculture Organisation of the United Nations*", the global agricultural output during the year 2011 was 5.6×10^9 metric tonnes (FAO 2015). A majority of this agricultural production was from cereals such as wheat, rice, barley, sorghum and maize among others, comprising up to about 4.3×10^9 metric tonnes. During the same year, biomass waste generation from all agricultural activities was estimated at about 1.06×10^9 metric tonnes. It has long been observed that vast quantities of biomass generated from agricultural activities is burned (ca. 7.43×10^8 metric tonnes in 2012) (FAO 2015).

1.1 Biomass Waste Management

Biomass waste management comprises various approaches including the conversion of biomass to commercially/industrially useful products. The process involves pre-treatment designed to break several linkages and bonds between cellulose, hemicellulose and lignin; thus exposing considerable amounts of hemicellulose and cellulose to subsequent enzyme/microbial based degradation (Sarkar et al. 2012). There are several physical, chemical and biological methods of biomass pre-treatment, which are used according to the biomass type, source and type of product recovery required. No single pre-treatment by itself is adequate for complete biomass conversion. However, a mixture of two or more pre-treatments (e.g., physical–chemical, physical–biological and chemical–biological) has been observed to generate higher biodegradation conversion in a range of biomass matrices (Khuong et al. 2014; Kovacs et al. 2009; Lu et al. 2013; Maeda et al. 2011; Pribowo et al. 2012).

Physical treatment methods are generally applied to increase the overall surface area of biomass, thus exposing higher amounts of cellulose and hemicellulose residues. This leads to a decrease in recalcitrance, enabling subsequent chemical, enzymatic or microbial treatment to be more effective due to the availability of a higher proportion of the substrate. There are various types of physical pre-treatments reported and they can be broadly categorised as milling (Karimi et al. 2013), irradiation (Budarin et al. 2010; Ma et al. 2014) sonication (Karimi et al. 2013; Rehman et al. 2013), extrusion (de Melo et al. 2014) and hydrothermal treatment (Karpe et al. 2014).

Physicochemical methods use a combination of physical disruption and chemical treatment techniques to decompose/degrade the lignocellulose complex. Standalone physical methods such as milling and pyrolysis require high amounts of input energy to break down the lignocellulose complex sufficiently for degradation to occur. To achieve enhanced degradation, physicochemical techniques utilise physical disruption attributes (temperature and pressure) along with the chemical (aqueous, acids and alkaline) treatments to either separate biomass components or alter the recalcitrant structure to make it more degradable. Hydrothermal treatment (HT) (Ando et al. 2000; Goto et al. 2004; Karpe et al. 2014), steam explosion (SE) (Duarte et al. 2012; Shevchenko et al. 2000) and ammonia fibre explosion (AFEX) (Sarkar et al. 2012) are some of the most common physicochemical methods used at laboratory and industrial scales.

In a steam explosion (SE) experiment where 3 % H_2SO_4 was applied at 120 °C, the release of about 14–36 % pentoses and 18–27 % hexoses (Shevchenko et al. 2000) and up to 90 % hemicellulose removal was reported (Karimi et al. 2013). Similarly, an application of 2 % NaOH in combination with ultrasonic treatment for 20 min at 50 °C has been reported to remove about 81 % hemicellulose and 91 % lignin from sugarcane bagasse. Follow-up bacterial treatment with *Cellulomonas flavigena* (MTCC 7450) resulted in about 91 % glucose yield (Velmurugan and Muthukumar 2012).

Organosolvent treatments use organic solvents in either standalone (e.g. 100 % methanol) mode or in combination (e.g. methanol: acetonitrile: acetone in a *v/v* ratio of 40:20:40) to degrade biomass. A *Pinus radiata* substrate treated with an acetone-water mixture at 195 °C for 5 min resulted in an ethanol yield of approximately 99.5 %. The disadvantages of this treatment method are its high costs and volatility of non-polar molecules which are flammable and explosive at elevated treatment temperatures (Haghighi et al. 2013).

Biological pre-treatments rely on microbial cells such as bacteria and fungi to degrade biomass. Fungi such as *Trichoderma* spp., *Aspergillus* spp., *Phanerochaete* spp., *Trametes* spp., *Phlebia* spp., *Postia* spp. and *Laetiporus* spp. are widely used in biomass treatment (Alvira et al. 2010; Karpe et al. 2015a; Liu et al. 2012; Zuroff et al. 2012). The process is cost-effective compared to chemical and physical pre-treatments, however the overall effectiveness of standalone biological pre-treatment is still low without physical or chemical pre-treatments. In combination, the pre-treatment process of milling and enzyme (extracted from *Acremonium* spp.) has been found to be highly effective and has reported to yield as much as 90 % hexose and 77 % pentose degradation (Sarkar et al. 2012). Khuong et al. (2014) reported that the subsequent applications of 5 % NaOH and the fungus *Phlebia* sp. MG-60 resulted in an ethanol yield of approximately 66 %.

Thermophilic bacteria such as *Clostridium thermocellum, C. thermohydrosulfuricum* and *Thermoanaerobacter ethanolicus* have high rates of bioethanol production with an increased tolerance towards ethanol of up to 5 % (Georgieva et al. 2007) and 8 %, respectively (Rani and Seenayya 1999). Additionally, in an independent experiment, it was observed that a combination of *T. brockii* β-glucosidase and *C. thermocellum* cellulosome in ammonia soaked rice straw for 7 days at 60 °C resulted in the conversion of about 91 % glucan to glucose (Waeonukul et al. 2012).

1.2 Biomass Degradation/Conversion Methods

Submerged fermentation/shake flask fermentation (SmF) is an aqueous phase fermentation process where the medium-to-substrate ratio is generally high (in excess of 10:1). The technique is well-established for wine and related alcohol production, and has been industrially applied in Western countries since the late 1940s for enhanced penicillin production from *Penicillium chrysogenum*. The basic methods have now been greatly optimised due to the requirement for scaling up (>3000–5000 litres) for agricultural use. Such optimization includes supplementation of oxygen, heat exchange, agitation and foaming and culture load prevention (Humphrey 1998). However, even with such a long history, and with this more recent optimization, SmF methods are not fully adequate for either enhanced production of lignocellulolytic enzyme or effective degradation of the lignocellulose complex—as evidenced in several experimental reports (Hideno et al. 2011; Juhász et al. 2005; Merino et al. 2007).

Solid State Fermentation (SSF) is a microbial bioprocessing method conducted at very low levels of free water (Hölker and Lenz, 2005a). The technique has been traditionally used for a long time in bread making, and for at least 3000 years for food processing in Asian countries. For example, the Koji process for food fermentation utilises *Aspergillus oryzae*. Production of sake is another example of SSF and utilises *Trichoderma* spp. Both of these applications are well-known in Japanese and Chinese culture. In Europe, SSF is also used to make traditional French blue cheese (Couto and Sanromán 2006; Hölker and Lenz 2005b).

More recently, SSF has been developed for a number of industrial bioprocesses such as biomass conversion (Bak et al. 2009; Brethauer and Studer 2014; Cheng and Liu 2012). Degradation and bioconversion of the lignocellulose complex from various biomass sources has increased the production of industrial metabolites, biofuels and secondary metabolites which can be used, for example, as medicinal compounds. The use of biomass addresses issues related to managing the considerable amounts of waste material generated during numerous agricultural and allied processes. SSF employs a mixture of different organisms and/or enzymes in a single step to generate considerably higher bioconversion of biomass as compared to SmF methodologies (Brijwani et al. 2010; Kausar et al. 2010; Lee 1997; Sarkar et al. 2012).

Recently, Cheng and Liu (2012) have shown the effects of pre-treatment in hydrogen production from milled cornstalk. SSF meditated by *T. reesei* Rut-30 was applied to this substrate, followed by sludge seeding at 35 and 55 °C. It was observed that within 4 days of seeding, the 6-day pre-treated substrate generated about 200 mL H_2 when incubated at 55 °C. Additionally, considerable amounts of low molecular weight fatty acids and ethanol were produced. In a similar fashion, rice straw pre-treated with *Phanerochaete chrysosporium* displayed considerable lignin peroxidase (LiP) and manganese peroxidase (MnP) activities. It was observed that after 30 days of pre-treatment, these enzymes had degraded about 33 % of the lignin. Follow-up enzymatic degradation resulted in the degradation of about 17 % of the glucan and 3 % of the xylan (Bak et al. 2009).

Consolidated bioprocessing (CBP) (Fig. 9.2) is a relatively new methodology for improving biomass conversion to products of commercial interest (Brethauer

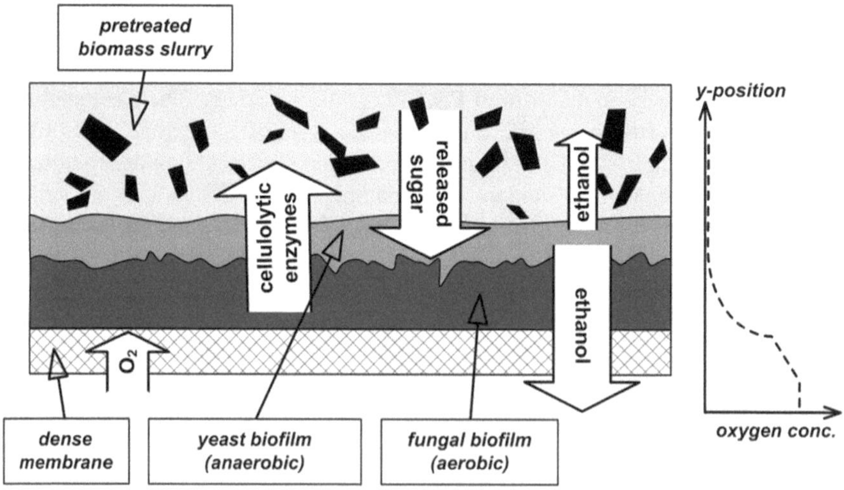

Fig. 9.2 A conceptual design of a multi-species biofilm membrane (MBM) reactor for CBP to generate ethanol from pre-treated wheat straw. Diagram taken from Brethauer and Studer 2014; published by The Royal Society of Chemistry (Wiethölter et al. 2003) on behalf of the Centre National de la Recherche Scientifique (CNRS) and the RSC

and Studer 2014; Lynd et al. 2005). CBP combines the individual processes of biomass hydrolysis and subsequent fermentation to generate products such as ethanol in a single step. It eliminates the time-consuming separate biological pre-treatment process. It also eliminates the need for separate fermentations of different sugars, such as tetroses or pentoses, which cannot be fermented by general industrial fermenters such as *Saccharomyces cerevisiae* (Lynd et al. 2005).

CBP relies on the development of cellulolytic organisms via strategies such as metabolic engineering or genetic engineering of native species. The first strategy has been reported to generate about 0.47 g ethanol/g hexose using the thermophilic bacterium, *Geobacillus thermoglucosidasius. Clostridium thermocellum* and *C. cellulolyticum*, two thermophilic biomass degrading bacteria, have been reported to generate about 50 g/L ethanol on pre-treated cellulose. The second strategy involves recombinant DNA technology to mediate heterologous expression of cellulolytic enzymes in naturally occurring biomass degrading microorganisms. These strategies have been successful at the laboratory scale; however, success under industrial conditions has not yet been reported (Olson et al. 2012). In a fungal (mushroom)-based CBP experiment, Kamei et al. (2014) described the use of spent sawdust waste from *Lentinula edodes* cultivation as a substrate for white-rot fungus, *Phlebia* spp. MG 60. They observed about 45 % ethanol production from the mushroom sawdust waste in about 400 h due to increased saccharification.

Symbiotic fermentation may be carried out as part of CBP. The strategy is to develop a symbiotic consortium of different microorganisms, or their lignocellulolytic enzymes, in either a single batch or continuous batch fermentation. It has been shown previously that fungi such as *Trichoderma* spp. and *Aspergillus* spp. have the ability to co-culture during cellulose degradation (Brijwani et al. 2010; Gupte and Madamwar 1997). A similar approach has been investigated in lignocellulose degradation by termites (Brune 2014) and lignin degradation by co-culture of wood-rot fungi (Qi-he et al. 2011). Mixed microbiota taken from biogas digested sludge inoculated onto wheat stalk was able to produce 37 mL/g of H_2 after 200 h of incubation. In addition, the concentration of volatile fatty acids, such as acetate, propionate and butyrate, also increased considerably. Furthermore, about 75 % of the cellulose was solubilized during this process (Chu et al. 2011). Additionally, Cheng and Liu (2012) reported that corn stalk substrate pre-treated with *T. reesei* RUT C30 by SSF and then seeded with winery sludge was able to generate H_2 gas up to 200 mL within 6 days.

An experiment displaying the synergistic properties of cellulolytic enzymes was recently reported. Crude, filtered and dialysed fungal enzyme samples obtained from SSF were mixed to an already operating SmF biodegradation sample. The resulting symbiosis of *Trichoderma* and *Aspergillus* enzymes caused degradation of cellulose and hemicellulose, as evidenced from a conversion of up to 64 % biomass to reducing sugars. Under anaerobic conditions, this enzyme mix in combination with *S. cerevisiae* was able to generate up to 43 g/L ethanol (about 84 % of the theoretical maximum) within 7 h (Pirota et al. 2014).

2 Microbial Process Metabolomics and Its Application to Biomass

The study of metabolomics has the potential to provide biochemical information related to fungal biomass degradation that increases our understanding of the microbial processes and characterises the various mechanisms utilised. Previously, metabolomic techniques have been applied to investigate bacterial processes related to preventative health (Bi et al. 2013; Marcinowska et al. 2011), environmental pollution (Beale et al. 2013), food (Beale et al. 2014) and fungal metabolism on various substrates including benzoic acid and Chardonnay grape berries (Hong et al. 2012; Matsuzaki et al. 2008). However, within the context of fungal-mediated biomass degradation, its application has been limited compared to other sectors. Analysis of the metabolic flux, however, has been shown to have the capability to enable a good understanding of the nature, time-dependence and substrate-based limitations in regard to the metabolism of fungal cells. Metabolomics also has the potential to assist in extricating the correlation between cell phenotypes and their metabolic patterns and stoichiometry (Meijer et al. 2009). Metabolic flux studies have previously been applied in toxicology and medicine (Maier et al. 2009; Niklas et al. 2010; Saunders et al. 2015), plant proteomes (Nelson et al. 2014) and microbial respiratory systems (Driouch et al. 2012; Pedersen et al. 1999). Numerous methods are available that monitor the flux (change) of metabolite production/consumption over time (i.e. metabolic flux analysis). The majority of these methods involve the use of heavy isotopes due to their superior tracking abilities. In commonly utilised analyses, such as nuclear NMR and gas chromatography-mass spectrometry (GC-MS), metabolite tracking is performed using isotopes such as ^{1}H, ^{2}H, ^{13}C, ^{14}N, ^{15}N and ^{18}O. While each isotope has its advantages and shortcomings, ^{13}C has been used by most researchers (Nelson et al. 2014), probably because of its wide-spread use in general chemistry. ^{2}H is a much more abundant element in nature, comprising about half of the peptide atomic population (Yang et al. 2010b). It is much more economical compared to other isotopes, has a comparatively higher invasive rate and a slower decay rate (Kim et al. 2012). Complete ^{2}H labelling is not feasible, however, due to the limited tolerance of multicellular organisms, thereby resulting in partial labelling, ranging from 8 to 30 % (Yang et al. 2010b; Kim et al. 2012).

2.1 Classification of Metabolomic Approaches

Various fungi such as those from the divisions of Ascomycota and Basidiomycota have been reported as effective biomass degraders. They generate an array of enzymes including endo- and exo-glucanases, β-glucosidase, xylanases, arabinofuranosidases and pectinases (Brink and Vries 2011; González-Centeno et al. 2010). The resultant degradation caused by these enzymes generates useful industrial and medicinal biomolecules such as ethanol, flavonoids, phenolic compounds, anthocyanins and hydroxybenzoic acid (Arvanitoyannis et al. 2006; Sánchez 2009; Strong and Burgess 2008). Additionally, fungi such as *Penicillium* spp. and *Phanerochaete* spp. can be used for lignin mineralization during the degradation process (Arora and Sharma 2010; Rodríguez et al. 1994).

The application of metabolomics in environmental and non-clinical areas is still in a nascent stage as compared to other 'omic' based approaches such as genomics, transcriptomics and proteomics. Unlike the other approaches, metabolomics deals with the complete metabolic behaviour of an organism by profiling the substrates and their metabolic products (e.g. saccharides, lipids, amino acids, organic acids, vitamins and phenolics) thereby providing a direct functional output of the biochemical behavioural pattern of the living system under the study (Fiehn 2002). Largely, due to this, in recent years, metabolomics has arguably been the most rapidly emerging field in biosciences (Roessner and Bowne 2009). However, due to the highly complex physiology, biochemistry and behaviour of fungal (or eukaryotic) cells, metabolomic studies of these organisms has presented considerable difficulties as compared to prokaryotic systems (Niklas et al. 2010). The metabolic behaviour of the fungus, therefore, needs to be elucidated by multiple approaches in either an individual or combined fashion. These approaches include untargeted and targeted metabolic profiling, metabolic fingerprinting and metabolic flux analysis.

2.2 *Metabolic Profiling: Targeted and Untargeted Approaches*

Metabolic profiling involves qualitative and quantitative analysis of the complete range of metabolites involved in an organism's metabolism (Roessner and Bowne 2009). This approach, therefore, may be considered as a precursor to either metabolic fingerprinting or metabolic flux analysis (Karpe et al. 2015b; Roessner and Bowne 2009). Because of its wide scope of application, metabolic profiling has been widely used in the study of environmental (Beale et al. 2013; Jones et al. 2014, 2015), bioprocess (Zha et al. 2014; Karpe et al. 2015c) and other industrial microbiology and treatment sectors (Ding et al. 2012; Matsuzaki et al. 2008; Beale et al. 2016). Besides these areas, it has also been used in clinical research (Ng et al. 2012; Kim et al. 2012; Kouremenos et al. 2010), food processing (Beale et al. 2014) and plant–microbial interactions (both parasitic and commensal) (Draper et al. 2011; Hong et al. 2012). Overall, the field of metabolic profiling is vast; however, this chapter will only be considering the metabolic profile of bioprocessing, specifically related to fungal/bacterial biomass conversions.

Microbial biomass processing is an emerging domain application with respect to most other fields in which metabolic profiling approaches have been used. One of the earlier reports that considered the potential of metabolomics was the perspective article by Dixon et al. (2006) detailing the application of metabolomics approaches in agriculture. While the Dixon et al. report considered issues such as data acquisition, data accuracy and a number of gaps in metabolic pathway information, it is apparent that the same issues have relevance in biomass related metabolomics. For example, in biomass metabolomics, the analyst has to deal with differential concentrations of larger substrates such as oligo- and polysaccharides, peptones and smaller, but varied, metabolites such as sugars, amino acids, fatty acids and aromatics. One of the reported studies considering an untargeted metabolomic approach in biomass degradation was the rumen bacterial mediated degradation of birch wood xylan and apple pectin (Villas-Bôas et al. 2006). The bacterial species *Clostridium proteoclasticum*, *Butyrivibrio fibrisolvens* and *Streptococcus bovis* were used. The biomass degradation quantified by GC-MS indicated the presence of about 40 different sugars, sugar acids, fatty acids and acids of lignin and tannin degradation including gallic acid, coumaric acid and quinic acid (among others). In our research, winery biomass was degraded using ascomycete fungi such as *T. harzianum*, *A. niger*, *P. chrysogenum*, *P. citrinum* and *S. cerevisiae* (as a negative control). An untargeted approach using GC-MS based metabolomic analysis was used which yielded 233 metabolite peaks. These consisted of biomass degraded mono- and oligosaccharides, sugar alcohols, sugar acids, fatty acids, amino acids, organic acids and aromatic acids, although the major metabolites present in the biomass degrading fungal profile were pentoses, hexoses and to a lesser degree fatty acids and amino acids. The results were in agreement with considerable enzyme activities of cellulases, xylanases and β-glucosidase in the biomass degrading fungi (Karpe et al. 2015c). In further research, the winery biomass was degraded by a symbiotic mixture of all the above mentioned fungi, except *S. cerevisiae*; and the degradation level increased considerably, indicated primarily by the higher enzyme activities, especially xylanase. The enhanced enzyme activity affected overall metabolic output where the higher metabolite concentration consisted of fatty acids such as oleanolic acid, glyceric acid, heptadecanoic acid and tetradecanoic acid. Additionally, the mixed fungal degradation was also able to yield some industrially useful secondary metabolites such as gallic acid, inositol, β-sitosterol and mannitol in higher concentrations (Karpe et al. 2015a).

By comparison to the above untargeted approach, targeted metabolomics deals with quantitative analysis of a limited number of the metabolites which represent the complete metabolome output of an organism. On a superficial level, targeted metabolomics resembles metabolic fingerprinting, however they are significantly different. While, the fingerprinting approach classifies different metabolic (functional) groups and identifies signature metabolic patterns to each organism (qualitatively or quantitatively), the targeted approach selectively screens any non-targeted metabolites and quantifies only targeted metabolites. The presence of targeted metabolites or metabolic groups in an organism's metabolome is therefore a prerequisite for targeted metabolomics (Shulaev 2006). The quantitative approach combined with selectivity increases accuracy and sensitivity of the analysis (for those specific metabolites). The main issues associated with wider use of a targeted approach are the current incomplete knowledge of organism specific metabolic pathways, the identification of intermediate metabolites, and their structural elucidation and availability of metabolites in pure form that can be used as a reference standard and, in general, an increased level of complexity. A targeted approach, therefore, has been used less compared to both non-targeted and fingerprinting metabolomic approaches. It has been argued that targeted approaches are largely limited to either genetic behavioural studies or clinical applications, as indicated in a recent review by Putri et al. (2013).

In spite of these limitations, targeted metabolomics methods are becoming more prominent in studies that encompass multiple 'omics' approaches. One recent study on targeted metabolomics involved the quantification of sugars, sugar phosphates, organic and amino acids within the *S. cerevisiae* metabolome (Kato et al. 2012). On a GC-MS system, around 99 metabolites were targeted using mass-to-charge (m/z) channel based screening of a mass spectrum range of 89–299 m/z units. Based on the data collected, the samples were then quantitatively re-analysed on an LC-QQQ-MS. The combined approach was able to detect metabolites concentrations of as low as to 1.4 femtomole in the yeast metabolome to a very high accuracy (R^2 = 99.94 %) (Kato et al. 2012). Another reported work involved the quantification of the *Clostridium thermocellum* metabolic profile. The thermophilic bacterium was grown on cellobiose and its metabolic output was analysed by LC-QQQ-MS/MS (using ^{13}C-isotopes). About 45 metabolites including amino acids and organic acids were quantified, some present in concentrations as low as 0.8 ng/mL of detection limit and 2.7 ng/mL of quantification level (Cui et al. 2013).

The targeted approach has also been applied in the assessment of metabolites that may be responsible for development of fungal tolerance in horticulture crops. A good example is the polyphenol assessment of wine-grape varieties displaying tolerance towards powdery and downy mildew; the fungal infections are caused by *Uncinula necator* and *Plasmopara viticola*, respectively. The analysis of secondary metabolites belonging to phenolic acids, dihydrochalcones, stilbenes, flavanols and flavanols was performed by the combination of two different UPLC-QQQ-MS systems. About 55 metabolites including gallic and coumaric acids, resveratrol, procyanidine, cathechins and naringenin were quantified with results ranging from 30 ng/mL to 203 μg/mL (Ehrhardt et al. 2014).

An interesting application of combined (targeted and non-targeted) metabolomic approaches was recently demonstrated by Godelmann et al. (2013). This group performed untargeted and targeted metabolic profiling using GC-TOF-MS and ^{1}H-NMR, respectively, to differentiate the grape variety, its vintage and geographical distribution. Specifically, 16 metabolites were quantified which included organic acids such as lactate, citrate, malate, succinate, acetate, fumarate and tartrate; alcohols such as methanol, ethanol, 3-methylbutanedioland glycerol; and secondary metabolites such as shikimate, caftarate and acetone. The concentrations ranged from 79.3 mg/L for shikimate to 120.2 g/L for ethanol. Chemometric statistical analysis was able to classify the geographic regions and vintage year to about 90 and 97 % accuracy, respectively. The experiment provides not only quality parameters for wine grapes, but also indicates the potential of applying specific fungal/bacterial species to extract targeted metabolites on a commercial scale from those substrates.

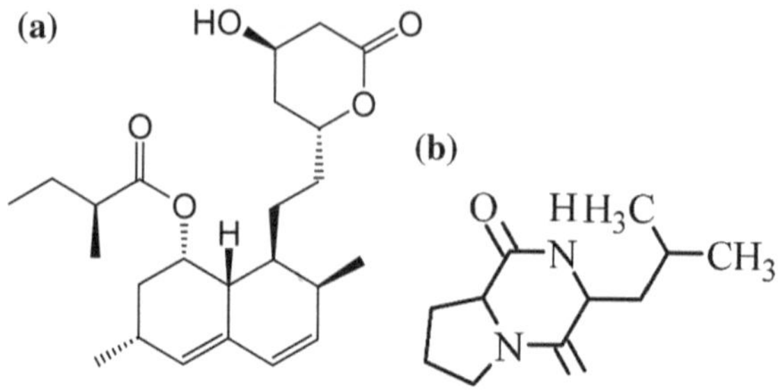

Fig. 9.3 The general structure of chief antibacterial metabolites secreted by *P. communes* fungus. The structures represent **a** Lovastatin and **b** 3-isobutyl hexahydropyrrolo[1, 2-a]pyrazine-1, 4-dione (α-Ketoisovaleryl-Leu-Pro lactam)

A metabolomics approach can also be applied to prevent or minimise process inhibition, either competitive or product, as shown in recent research using GC-MS-based analysis of yeast (Zha et al. 2014). It was observed that biomass pre-treatment products such as hydroxymethyl furfural (HMF) and furfural prolonged the log-phase of yeast growth, thereby causing considerable growth inhibition. Similarly, other metabolites such as aldehydes, phenolics, vanillin and sorbic acid negatively affected the growth rate. *S. cerevisiae* was found to counter the effects of HMF and furfural by converting these molecules to their respective acids. It was also noted that amino acids, when initially present in considerable amounts, compensated for the inhibitory and toxic effects of various molecules during fermentation (Zha et al. 2014).

In related research, the satins and anti-pathogenic metabolites from environmental *Penicillium commune* were analysed by a combination of analytical techniques such as thin layer chromatography (TLC), Fourier transformed infrared spectroscopy (FTIR), direct insertion probe/electron ionisation/ion trap detection (DIP/EI/ITD) mass spectrometry and GC-MS. The multiple techniques identified lovastatin (Fig. 9.3a) and 3-isobutyl hexahydropyrrolo[1, 2-a]pyrazine-1, 4-dione (or α-Ketoisovaleryl-Leu-Pro lactam) (Fig. 9.3b) (Diblasi et al. 2015). These metabolites or the fractions containing these metabolites were shown to be effective antibacterial agents; the lovastatin containing fraction was shown to be effective against *Pseudomonas aeruginosa* and the fraction containing 3-isobutyl hexahydropyrrolo[1, 2-a]pyrazine-1, 4-dione was observed to be inhibitory towards *Staphylococcus aureus*.

2.3 Metabolic Fingerprinting

Metabolic fingerprinting can be defined as the quantification of selective metabolites from a larger metabolic output of an organism at a particular given period and grouping them together for potential further assessment. As this approach considers the measurement of a subset of metabolites and is focused on profiling metabolites associated with the organism(s) under experimental conditions, it provides more accurate tracking of metabolic behaviour of the organism relative to its surroundings (Kosmides et al. 2013). Currently, most of the methods used for metabolic fingerprinting are NMR based, with some studies using GC-MS and LC-MS (Roessner and Bowne 2009).

As previously stated, there has been limited metabolomics-based research related to biomass conversion compared to clinical applications. Nevertheless, a number of useful studies, especially in the area of metabolic fingerprinting, have been recently described. For example, the supramolecular effects of pre-treatment and successive microbial degradation of rice straw using NMR were described by Ogura et al. (2013). This biomass was pre-treated by ZrO_2 ball milling for 6 h followed by metabolic analysis using 2D ^{13}C-^{1}H heteronuclear correlation spectra (HETCOR) in solid-state NMR and thermogravimetric analyses. The pre-treated biomass was supplemented with paddy soil and the products were analysed by ^{1}H-NMR spectroscopy. The presence of syringyl, guaiacyl, hydroxyphenyl and coumarate derivatives suggest considerable lignin linkage breakdown. Similarly, sugars and sugar acids resulting from degradation of cellulose and hemicellulose were detected. Deformation of cellulose was noticed from the shift in position of the C4 peak (NMR spectroscopy) of crystalline cellulose and the increasing intensity of amorphous cellulose. With respect to other pre-treatments, the activation energy of ball milling reduced considerably from 166 to 96 kJ/mol. This suggests an increase in amorphous cellulose. Successive microbial degradation of this biomass resulted in further generation of pentoses, hexoses and disaccharides (from 3 to 4 ppm) and short chain fatty acids (SCFAs), especially butyrate (up to 2.2 ppm). Principal component analysis (PCA) of the data suggested the termination of degradation at 21 days (Ogura et al. 2013). In an extension to this experiment, bacterial cellulose generated by *Gluconacetobacter xylinus* was degraded by anaerobic sludge and the products were analysed by solid state-, solution state- and gas state- NMR. While solid-state NMR showed decreasing crystallanity of cellulose over 84 h, solution-state NMR indicated that the major metabolite output consisted of SCFAs and ethanol. These were generated during the early phase of fermentation and most were re-utilised after 60 h of fermentation. Similarly, CO_2 and CH_4 were generated by oxidation and reduction of acetate, as detected by gas-state NMR (Yamazawa et al. 2013).

Another method recently used for metabolic fingerprinting was FTIR. The experiment applying this method of metabolome discrimination used various rye grass varieties of *Lolium perenne* as animal feed. Metabolites generated in rumen over 24 h were periodically analysed by FTIR to build up a preliminary metabolite fingerprint for the fermentation process of each grass variety. The major metabolic groups discriminated by FTIR based multivariate analyses indicated the presence of numerous sugars, amides, fatty acids and alkanes in cow rumen. Production of ammonia and volatile fatty acids (VFAs) was the primary indicator of biomass fermentation. VFAs such as acetate, propionate, butyrate, pentanoate, hexanoate and heptanoate were observed at different concentrations at various sampling times.

Although the FTIR approach was unable to evaluate individual metabolites, it could separate the overall metabolite spectra into several groups providing a defined analysis pathway. The authors suggested using a more specific metabolic flux analysis to obtain a specific fingerprint related to fermentation of rye grass biomass by rumen microbiota. The study also indicated the likelihood of increasing protein utilisation by increasing the availability of soluble carbohydrates to rumen microbiota. The authors also indicated the use of FTIR as a preliminary non-discriminatory technique to establish a rapid understanding of metabolite interactions, which can then be studied in detail by a secondary flux analysis (Kingston-Smith et al. 2013). The study, albeit a preliminary investigation, indicated that the metabolic fingerprint and flux pattern in the fermentation of plant biomass by rumen microbiota were specific to the type of plant source.

Another widely used method for metabolite fingerprinting is liquid chromatography with electrospray ionisation-mass spectrometry (LC-MS). Due to protonation/de-protonation during ionic adduct formation, the analyte is observed as a molecular ion. The mass-to-charge ratio (m/z) calculation by ESI is then used to predict the putative metabolite in numerous types of samples (Draper et al. 2013). Secretome studies of *Trichoderma hamatum* GD12 in relation with Arabidopsis in the presence of other fungi showed the presence of cellulose binding-motif (CBM). Further analysis by combined ^{1}H-NMR and direct infusion-ESI-MS (DI-ESI-MS) showed the presence of 30 metabolites of at least fivefold change, based on unbiased positive and negative ionisation mode discrimination (Studholme et al. 2013). Similarly, in a recent study of *Aspergillus nidulans* mediated degradation of *Quercus suber* cork, the secretome analysis showed ca. 28 active protein species, of which ca. 12 % comprised plant cell wall degradation proteins. These were further classified as β-glucosidases, arabinosidase, alcohol oxidase and oxidoreductases. The metabolic output of this secretome was analysed by ultra-high-performance liquid chromatography coupled to electrospray ionisation-high resolution mass spectrometry (UHPLC-ESI-HRMS). The organic extracts that resulted from *A. nidulans* mediated degradation displayed the presence of signature metabolites such as low molecular weight aromatics (e.g. ferrulates, veratraldehyde and cinnamate), about five β-aryl ether linked metabolites and a pinoresinol group compound (Martins et al. 2014). Secretome analysis of *Fusarium* spp. Q7-31T by ESI coupled with tandem mass spectrometry (LC-MS/MS) indicated the presence of ca. 10 glycoside hydrolase enzymes out of a total of 28, most of them responsible for producing smaller glucan metabolites by hydrolysing biomass substrates (Tian et al. 2015). A further downstream metabolic effect on the sugar release due to *Fusarium* spp. based biomass degradation was analysed by FTIR metabolite fingerprinting. The results showed 13 % decrease in glucose and about 11 % increase in uronic acids for *Fusarium circinatuum* mediated rotting of *Pinus pinaster* biomass. Additionally, the content of methyl esterified polysaccharide in pectins also increased following post-*Fusarium* culturing (Vivas et al. 2014). It has been observed that fingerprinting of biomass degradation has been mostly focused towards proteome/secretome analysis, as indicated from some recent research (Kosmides et al. 2013; Martins et al. 2014; Studholme et al. 2013; Tian et al. 2015;

Vivas et al. 2014; Cattaneo et al. 2014). Additional elucidation of the metabolic output of these secretomes still needs to be applied in the context of biomass conversion, as currently is applied in plant–fungal parasitic or symbiotic and microbiome relations (Draper et al. 2011; Kingston-Smith et al. 2013).

2.4 *Metabolic Flux (Fluxomics)*

Metabolic profiling can be applied to obtain metabolic flux information on metabolite production/consumption rates (fluxomics), in which metabolic engineering tools such as "induction of expression" can be utilised. These techniques not only decrease the overall inhibition, but also aid in developing systems which generate desired products. This has been reported recently where a *Zymomonas mobilis* ethanolic pathway system was expressed in *Sphingomonas* spp. A1 to convert alginate to ethanol (Takeda et al. 2011).

Due to the complex nature of biomass composition and the allosteric nature of most lignocellulolytic enzymes, standard biochemical tests have proven less effective compared to metabolomic approaches for deciphering biomass conversion. Metabolic profiling indicates signature metabolites and critical pathway points which can be exploited to not only improve the biomass degradation process, but also to convert this biomass to the products of industrial and medicinal interests.

In recent developments, flux studies of *A. niger* and *P. chrysogenum*-mediated SSF of grape biomass waste during an 8 day period were performed. The untargeted metabolic flux analysis of *A. niger* was studied with the aid of $^{2}H_2O$. The process not only helped to identify the key points in the *A. niger* metabolism of lignocellulose, but was also able identify the onset of inhibition caused by specific by-products and competitive inhibitors. Overall, 37 unique metabolites were characterised during this flux study, spread across 17 pathways including fatty acid, amino acid and chitin synthesis. It was observed that although the fungal growth rate was constantly increasing over this period, enzyme activities were inhibited from day 5 onwards. Metabolic pathway analysis also showed the potential of applying metabolic engineering in order generate biofuels such as 2-butanol. This could be achieved by termination of the amino acid pathway using enzymes such as 2-keto-acid decarboxylases followed by alcohol dehydrogenases (Karpe et al. 2015b). Similarly, an untargeted metabolic flux analysis of *P. chrysogenum* indicated the presence of 94 signature metabolites spread over 28 different pathways, including xylitol synthesis, lignin and tannin degradation and pentose degradation. The analysis indicated an onset of autolysis from day 6. Interestingly, the pathway analysis also displayed pentose metabolism, possibly leading to ethanol generation via xylose biosynthesis pathway, a property absent in other biomass degrading ascomycetes investigated under similar conditions. Additionally, the fungus also displayed tannin and lignin mineralization, leading to the generation of syringate over 3–5 day SSF. The antioxidant metabolites syringate or gallate (amongst others) have been identified to possess medicinal properties (Alonso et al. 2002, 2004).

Some of the earlier metabolism flux related research was performed on the galactoglycome of *T. reesei* during its growth on lactose. The work showed that lactose, when used as a carbon source, acted an initial inducer for galactokinase and aldose reductase XYL1. These enzymes further form the oligosaccharides which serve as inducers for cellulase formation (Hartl et al. 2007; Seiboth et al. 2007). A more downstream quantitative metabolic flux analysis of this fungus used sophorose as an initial inducer for cellulase formation and activity. The induction did not affect any balance between intracellular and extracellular proteins, the latter of which consist of cellulases and hemicellulases, a critical factor for industrial application of this fungus (Rautio et al. 2006). A similar, but a more amino acid focussed experiment, was conducted to optimize the growth medium composition for the basidiomycete *Pleurotus sapidus* by using ^{13}C-glucose based metabolic flux. It was observed that the isotopic glucose contributed to fungal amino acid synthesis, ranging from 22 (asparagine/aspartate) to 92 % (alanine). In addition, the threonine aldolase enzyme based mechanism to generate glycine from threonine was observed to be inactive or absent in this fungus. Additionally, glucose was utilised at a rate of 0.24 mmol g/hour, with about 35 % of its output going to the pentose phosphate pathway (Fraatz et al. 2014).

Similar approaches have been reviewed by Kubicek (2013) and asserts the necessity of multi 'Omics' approach for understanding cellulolytic properties of *T. reesei* for industrial applications. It also indicated a limited number of approaches used in metabolome studies of *T. reesei* and other fungal enzyme outputs during biomass degradation as compared to genomics, transcriptomics and proteomics.

3 Metabolic Engineering Approaches

As traditional manufacturing sectors transition (e.g. automotive manufacturing), we find ourselves on the verge of the fourth industrial revolution catalysed by a wave of disruptive technological advances and the creation of lean start-ups. The fourth industrial revolution comprises technological and service based advancements such as the 'internet of things', to 3-D manufacturing and synthetic biology/biological engineering (Maynard, 2015). Biological engineering in particular is relevant to metabolomics and omics-based platforms. It is an emerging interdisciplinary field representing the convergence of diverse domains, such as biotechnology, evolutionary biology, molecular biology, systems biology, physics, chemistry, genetic engineering and informatics (Wagner and Alper 2016). It is widely acknowledged that there is already a number of established metabolic engineering and synthetic biology tools available that develop high-performance cell factories (Carbonell et al. 2014). However, relative to the potential opportunities, the number of successful applications has been limited due to the complexity of exploring the metabolic space efficiently. As an outcome of such work, biological engineering has the potential to generate value from traditional wastes and natural systems biology processes that can benefit society within a range of sectors. For example, biological engineering can provide novel in silico design and fabrication materials, the construction of new biological parts, devices, systems and machines, as well as the re-design of existing, natural biological systems for useful purposes (Wagner and Alper 2016; Maynard 2015).

The application of metabolomics to biological engineering will become increasingly important for the production of chemicals of value from agricultural wastes, which are either high-value fine chemicals (e.g. medicinal products) or bulk low-value commodities (e.g. fuels) (Ellis and Goodacre 2012). The integration of metabolomics and fluxomics at each stage of biological engineering projects will enable metabolic pathways with a desired outcome to become the focus of bioengineering design, where the genetic/metabolic optimisation of those pathways will result in more efficient conversion of low-cost starting materials into highly desirable products (Ellis and Goodacre 2012; Chen 2015). For example, this emerging field that has already been identified as a sustainable solution for producing an alternative source of Omega-3 fatty acids that is not reliant on wild fish harvests or aquaculture; work has already been performed to create recombinant sources of Omega-3 fatty acids in a canola and yeast (Adarme-Vega et al. 2012). Bioprocessing of polyols using lactic acid bacteria has been investigated as an alternative to current industrial food production approaches (Ortiz et al. 2013). The major challenge for all bioengineering studies is to balance the flux in the pathway of interest in order to obtain high yields and maintain productivity of the target microorganism (Yuan et al. 2013).

3.1 Metabolic Engineering to Develop Fungal Bioprocessing Abilities

Although lignocellulosic biomass is found in abundance, it is relatively inexpensive and has been used in the past as feedstock. It is also considered a renewable source for bioethanol production. As previously stated, lignocellulosic biomasses are complex mixtures of different fermentable sugars, as well as inhibitors that impact the performance of the target microorganism (Rumbold et al. 2009). However, it should be noted that for all its potential benefits, producing bioethanol from biomass is still not economically viable when compared to other substrates (such as corn starch and cane juice) (Hasunuma and Kondo, 2012). This challenge may have been overcome by Hasunuma and Kondo (2012) through consolidated bioprocessing (CBP). CBP combines enzyme production of engineered microorganisms, saccharification and fermentation in a single step (Hasunuma and Kondo 2012). However, the overproduction and misfolding of heterologous and endogenous proteins can trigger cellular stress, increasing the metabolic burden and retarding growth (Hasunuma et al. 2015). Furthermore, exposure of various lignocellulose-derived inhibitors disrupts the efficiency of hexose and pentose co-fermentation (Hasunuma et al. 2015; Wang et al. 2014). To investigate this further, Wang et al. (2014)

measured the metabolic response of a xylose-fermenting *S. cerevisiae* 424A (LNH-ST) and its parental strain 4124 with and without three typical inhibitors (furfural, acetic acid and phenol). It was concluded through metabolomics that the 424A (LNH-ST) strain had a lower capacity to buffer redox changes caused by inhibitors compared to strain 4124. Furthermore, a lower ethanol yield in glucose and xylose co-fermentation was observed in strain 424A when compared to glucose fermentation in strain 424A (LNH-ST). Lastly, xylose utilisation of strain 424A (LNH-ST) was more significantly reduced by inhibitors than glucose utilisation. This suggests that xylose catabolism and energy supply, rather than xylose uptake, were the limiting steps in xylose utilisation in the presence of inhibitors (Wang et al. 2014).

A systems biology approach including disruptome screening, transcriptomics and metabolomics has been recently exploited to gain insight into the molecular and genetic traits involved in tolerance and adaptation to fermentation inhibitors using direct and inverse metabolic engineering (Fig. 9.4) (Hasunuma and Kondo 2012). It is noteworthy to mention that both direct and inverse metabolic engineering approaches tend to reach the same outcome, principally because they use both a rational design and evolutionary design strategy applied to native strains, albeit in different combinations. A rational design strategy involves selecting and modifying preselected pathways for specific outcomes. Conversely, an evolutionary design strategy involves harnessing the natural evolutionary processes that resulted in the expression of the desired pathway (i.e. tolerance, production, enhanced growth). Irrespective of the approach, metabolomics (and the other omics based platforms) play a prominent role in achieving the desired outcomes.

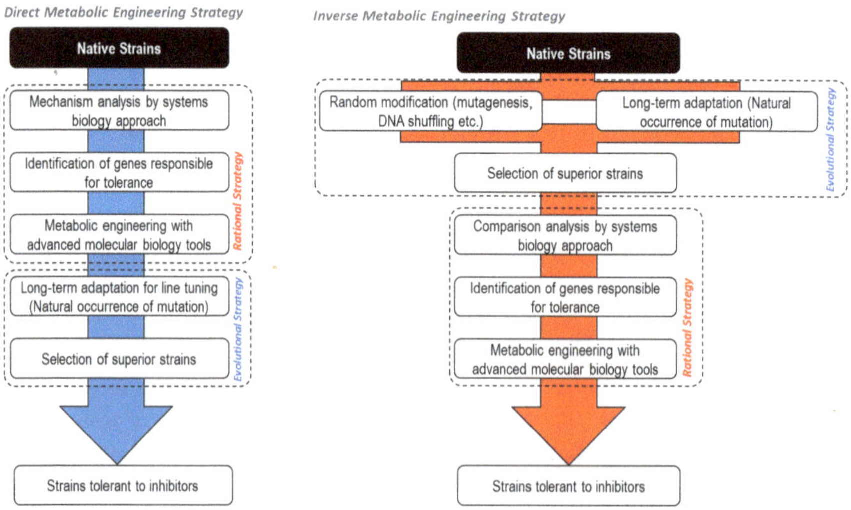

Fig. 9.4 Design strategy of rational and evolutionary direct and inverse metabolic engineering experiments. Adapted from Hasunuma and Kondo (2012)

Papini et al. (2012) described a rational metabolic engineering strategy that was used to express the fungal genes of the phosphoketolase pathway in *S. cerevisiae* and the effects of the expression of this recombinant route in yeast using intracellular flux analysis based on ^{13}C labelling. Several bacterial species and filamentous fungi utilise the phosphoketolase pathway for glucose dissimilation as an alternative to the Embden-Meyerhof-Parnas pathway. However, through the use of fluxomics, Papini et al. (2012) observed yeast using the recombinant pathway and it was observed that the utilisation of the phosphoketolase pathway did not interfere with glucose assimilation through the Embden-Meyerhof-Parnas pathway. Furthermore, the expression of the recombinant pathway contributed to an increase in the acetyl coenzyme supply, therefore enabling future potential metabolic engineering strategies to be considered that utilise acetyl CoA as precursor for the biosynthesis of industrially relevant and value-added compounds (Papini et al. 2012). Huang et al. (2015) applied biological evolutionary engineering to activate the glycerol utilisation pathway for fumaric acid production. In this study, an evolved strain G80 was selected that was tolerant to high concentrations of crude glycerol to produce fumaric acid.

While *A. niger* is an effective organism in the biodegradation of biomass and has been widely applied for the production of high value-added products (Karpe et al. 2015b; Krull et al. 2010; Meijer et al. 2009), the morphology of this fungus, the transport phenomena and the related metabolic activity result in unreliable yield within industrial bioprocessing facilities (Krull et al. 2010). As such, further characterization of the molecular and cell biology of *A. niger*, in particular characterising interactions between the growth conditions, cell morphology, spore-hyphae-interactions and value-added product formation, is of great importance for ensuring a sustainable bioprocessing industry (Krull et al. 2010). Conversely, Rumbold et al. (2009) used a substrate-oriented approach as opposed to a product-oriented approach towards the selection of a microbial production host that avoids the need for extensive metabolic engineering. In this work, six microorganisms were evaluated and the performance of two bacteria (*Escherichia coli* and *Corynebacterium glutamicum*), two yeasts (*S. cerevisiae* and *Pichia stipitis*) and two fungi (*A. niger* and *T. reesei*) were assessed for their ability to degrade lignocellulosic hydrolysates. The study was particularly focused on their ability to utilise monosaccharides, characterising the resistance against lignocellulosic hydrolysates inhibitors and assess growth and utilisation rates on different feedstock (with feedstocks thermally pre-treated under mild acidic conditions followed by enzymatic hydrolysis and non-enzymatic treatments). Rumbold et al. (2009) concluded that carbon source versatility and inhibitor resistance were the major discriminators of the microorganisms tested, with *P. stipitis* and *A. niger* found to the moist effective. *C. glutamicum* and *S. cerevisiae* were shown to be the least adapted to renewable feedstock. These results highlight the complexity of substrate-orientated approaches that remove the over engineering of biological systems and utilising natural metabolic pathways present in substrate matched microorganisms.

As stated previously, *P. chrysogenum* has been applied to the industrial-scale process of penicillin production (Ding et al. 2012). Metabolic engineering methodologies have stimulated new efforts toward optimising fungal production strains of *P. chrysogenum*. Meinert et al. (2013) devised a protocol for fungal metabolomics, intended to inform metabolic engineering of industrial processes using *P. chrysogenum*. In this work, metabolites were quantified directly from biomass extracts using automated sampling and fast filtration methodologies and the use of a ^{13}C tracer (Meinert et al. 2013). The use of ^{13}C tracers is advantageous when investigating complex media containing multiple substrates (Antoniewicz 2013).

Some of the most productive metabolic engineering strategies involve genetic modifications that cause severe metabolic burden to the host cell (Williams et al. 2015). Growth-limiting genetic modifications can be more effective if they are 'switched on' after a population growth phase has been completed. Williams et al. (2015) demonstrated this concept through the engineered dynamic regulation of a synthetic quorum sensing circuit in *S. cerevisiae*, whereby the circuit autonomously triggers gene expression at a high population density through the shikimate pathway for the production of p-hydroxybenzoic acid (PHBA) (Williams et al. 2015). Lastly, while experimental work is needed to confirm desired or undesired metabolic expression of studied microorganisms, computational methods are increasingly being used to extract direct genotype to phenotype information that can be used to predict metabolic function (Udatha et al. 2013).

4 Future Research and Conclusions

Considerable advances have been made in biomass treatment processes in recent years. However, metabolic studies regarding fungal-mediated processes still remain in a nascent phase. Unlike bacteria, fungal cellular systems (including yeasts) are highly complex and are not yet fully understood. This complexity is further increased by various interactions within the fungi (commensal, symbiotic or parasitic) and their substrates (timber, general agriculture and food processing wastes). In spite of the nascent nature of fungal metabolomics, especially with respect to bioprocessing, rapid advances have been recently observed. These developments have improved not only our basic understanding of biomass degrading fungi, but have also provided the key points to improve their metabolism to generate metabolites of commercial interest. Additionally, some of the very recent works have demonstrated the application of these improvements in yeasts to enhance their efficiency in production of various metabolites including food grade alcohols, enzymes and pharmaceutical metabolites.

With continuous developments of different databases such as the Kyoto Encyclopaedia of Genes and Genomes (KEGG), the Joint Genome Initiative (JGI), the Chemical Entities of Biological Interest (ChEBI) and the Human Metabolome Database (HMDB) (amongst others), metabolome studies in the near future are expected to show an exponential growth, as observed with genomic research in the previous decade. A more complete development in metabolomics is also evident from the emergence of a 'multi-omics' approach (a combination of genomics, transcriptomics, proteomics and metabolomics). Another recent approach is the construction of entire novel fungi specifically designed for targeted metabolite production. Although this method is still in its infancy, even from a bacterial system development perspective, a complete 'multi-omics' approach will aid researchers to achieve this development in fungal systems. These approaches are expected to not only close a number of gaps in our current understanding of fungal processes, but will provide an elaborate view towards improving biomass conversion to obtain greater efficiency in metabolite yield.

References

Adarme-Vega TC, Lim DKY, Timmins M, Vernen F, Li Y, Schenk PM (2012) Microalgal biofactories: a promising approach towards sustainable omega-3 fatty acid production. Microb Cell Fact 11:96

Alonso ÁM, Castro R, Rodríguez MC, Guillén DA, Barroso CG (2004) Study of the antioxidant power of brandies and vinegars derived from Sherry wines and correlation with their content in polyphenols. Food Res Int 37:715–721

Alonso ÁM, Guillén DA, Barroso CG, Puertas B, García A (2002) Determination of antioxidant activity of wine byproducts and its correlation with polyphenolic content. J Agr Food Chem 50:5832–5836

Alvira P, Tomás-Pejó E, Ballesteros M, Negro MJ (2010) Pretreatment technologies for an efficient bioethanol production process based on enzymatic hydrolysis: a review. Bioresour Technol 101:4851–4861

Ando H, Sakaki T, Kokusho T, Shibata M, Uemura Y, Hatate Y (2000) Decomposition behavior of plant biomass in hot-compressed water. Ind Eng Chem Res 39:3688–3693

Antoniewicz MR (2013) Using multiple tracers for 13C metabolic flux analysis. In: Alper SH (ed) Systems metabolic engineering: methods and protocols. Humana Press, Totowa

Arora DS, Sharma RK (2010) Ligninolytic fungal laccases and their biotechnological applications. Appl Biochem Biotechn 160:1760–1788

Arvanitoyannis IS, Ladas D, Mavromatis A (2006) Potential uses and applications of treated wine waste: a review. Int J Food Sci Tech 41:475–487

Bak JS, Ko JK, Choi IG, Park YC, Seo J-H, Kim KH (2009) Fungal pretreatment of lignocellulose by *Phanerochaete chrysosporium* to produce ethanol from rice straw. Biotechnol Bioeng 104:471–482

Baucher M, Monties B, Montagu MV, Boerjan W (1998) Biosynthesis and genetic engineering of lignin. Crit Rev Plant Sci 17:125–197

Beale D, Barratt R, Marlow D, Dunn M, Palombo E, Morrison P, Key C (2013) Application of metabolomics to understanding biofilms in water distribution systems: a pilot study. Biofouling 29:283–294

Beale D, Karpe A, Mcleod J, Gondalia S, Muster T, Othman M, Palombo E, Joshi D (2016) An 'omics' approach towards the understanding of laboratory scale anaerobic digesters treating municipal wastewater activated sludge. Water Res 88:346–357

Beale DJ, Morrison PD, Palombo EA (2014) Detection of *Listeria* in milk using non-targeted metabolic profiling of *Listeria monocytogenes*: a proof-of-concept application. Food Control 42:343–346

Bi H, Krausz K, Manna S, Li F, Johnson C, Gonzalez F (2013) Optimization of harvesting, extraction, and analytical protocols for UPLC-ESI-MS-based metabolomic analysis of adherent mammalian cancer cells. Anal Bioanal Chem 405:5279–5289

Boerjan W, Ralph J, Baucher M (2003) Lignin biosynthesis. Annu Rev Plant Biol 54:519–546

Brethauer S, Studer MH (2014) Consolidated bioprocessing of lignocellulose by a microbial consortium. Energy Environ Sci 7:1446–1453

Brijwani K, Oberoi HS, Vadlani PV (2010) Production of a cellulolytic enzyme system in mixed-culture solid-state fermentation of soybean hulls supplemented with wheat bran. Process Biochem 45:120–128

Brink J, Vries RP (2011) Fungal enzyme sets for plant polysaccharide degradation. Appl Microbiol Biot 91:1477–1492

Brune A (2014) Symbiotic digestion of lignocellulose in termite guts. Nat Rev Microbiol 12:168–180

Budarin VL, Clark JH, Lanigan BA, Shuttleworth P, Macquarrie DJ (2010) Microwave assisted decomposition of cellulose: a new thermochemical route for biomass exploitation. Bioresour Technol 101:3776–3779

Carbonell P, Parutto P, Baudier C, Junot C, Faulon J-L (2014) Retropath: automated pipeline for embedded metabolic circuits. ACS Synth Biol 3:565–577

Cattaneo C, Spertino S, Boatti L, Icardi S, Cavaletto M (2014) Protein fingerprinting in the choice of cellulase cocktails for the conversion of lignocellulosic biomass. Anal Method 6:4046–4055

Chen R (2015) The sweet branch of metabolic engineering: cherry-picking the low-hanging sugary fruits. Microb Cell Fact 14:197

Cheng X-Y, Liu C-Z (2012) Fungal pretreatment enhances hydrogen production via thermophilic fermentation of cornstalk. Appl Energ 91:1–6

Chu Y, Wei Y, Yuan X, Shi X (2011) Bioconversion of wheat stalk to hydrogen by dark fermentation: effect of different mixed microflora on hydrogen yield and cellulose solubilisation. Bioresour Technol 102:3805–3809

Couto SR, Sanromán MÁ (2006) Application of solid-state fermentation to food industry—a review. J Food Eng 76:291–302

Cui J, Zhang J, Zhu X, Bai F, Feng Y, Guan W, Cui Q (2013) Separation and quantification of water-soluble cellular metabolites in *Clostridium thermocellum* using liquid chromatography-isotope dilution tandem mass spectrometry. Anal Lett 46:2767–2786

de Melo MMR, Silvestre AJD, Silva CM (2014) Supercritical fluid extraction of vegetable matrices: applications, trends and future perspectives of a convincing green technology. J Supercrit Fluids 92:115–176

Diblasi L, Arrighi F, Silva J, Bardón A, Cartagena E (2015) *Penicillium commune* metabolic profile as a promising source of antipathogenic natural products. Nat Prod Res 29:1–7

Ding M-Z, Lu H, Cheng J-S, Chen Y, Jiang J, Qiao B, Li B-Z, Yuan Y-J (2012) Comparative metabolomic study of *Penicillium chrysogenum* during pilot and industrial penicillin fermentations. Appl Biochem Biotechnol 168:1223–1238

Dixon RA, Gang DR, Charlton AJ, Fiehn O, Kuiper HA, Reynolds TL, Tjeerdema RS, Jeffery EH, German JB, Ridley WP (2006) Applications of metabolomics in agriculture. J Agr Food Chem 54:8984–8994

Draper J, Lloyd A, Goodacre R, Beckmann M (2013) Flow infusion electrospray ionisation mass spectrometry for high throughput, non-targeted metabolite fingerprinting: a review. Metabolomics 9:4–29

Draper J, Rasmussen S, Zubair H (2011) Metabolite analysis and metabolomics in the study of biotrophic interactions between plants and microbes. In: Annual plant reviews, vol 43. Wiley-Blackwell, New York

Driouch H, Melzer G, Wittmann C (2012) Integration of in vivo and in silico metabolic fluxes for improvement of recombinant protein production. Metab Eng 14:47–58

Duarte G, Moreira L, Jaramillo P, Filho E (2012) Biomass-derived inhibitors of holocellulases. Bioenerg Res 5:768–777

Ehrhardt C, Arapitsas P, Stefanini M, Flick G, Mattivi F (2014) Analysis of the phenolic composition of fungus-resistant grape varieties cultivated in Italy and Germany using UHPLC-MS/MS. J Mass Spectrom 49:860–869

Ellis DI, Goodacre R (2012) Metabolomics-assisted synthetic biology. Curr Opin Biotechnol 23:22–28

Eriksson KL, Blanchette RA, Ander P (1990) Microbial and enzymatic degradation of wood and wood components. Springer, Berlin

FAO (2015) FAOSTAT domain [Online]. Rome: The Food and Agriculture Organization of the United Nations. Available: http://faostat3.fao.org/faostat-gateway/go/to/download/Q/QD/E [accessed]

Fernandes AN, Thomas LH, Altaner CM, Callow P, Forsyth VT, Apperley DC, Kennedy CJ, Jarvis MC (2011) Nanostructure of cellulose microfibrils in spruce wood. Proc Natl Acad Sci USA 108:E1195–E1203

Fiehn O (2002) Metabolomics—the link between genotypes and phenotypes. Plant Mol Biol 48:155–171

Fraatz MA, Naeve S, Hausherr V, Zorn H, Blank LM (2014) A minimal growth medium for the basidiomycete *Pleurotus sapidus* for metabolic flux analysis. Fungal Biol Biotechnol 1:9

Georgieva TI, Mikkelsen MJ, Ahring BKR (2007) High ethanol tolerance of the thermophilic anaerobic ethanol producer *Thermoanaerobacter* BG1L1. Cent Eur J Biol 2:364–377

Godelmann R, Fang F, Humpfer E, Schütz B, Bansbach M, Schäfer H, Spraul M (2013) Targeted and nontargeted wine analysis by 1 h nmr spectroscopy combined with multivariate statistical analysis. Differentiation of important parameters: grape variety, geographical origin, year of vintage. J Agr Food Chem 61:5610–5619

González-Centeno M, Rosselló C, Simal S, Garau M, López F, Femenia A (2010) Physico-chemical properties of cell wall materials obtained from ten grape varieties and their byproducts: grape pomaces and stems. LWT-Food Sci Technol 43:1580–1586

Goto M, Obuchi R, Hirose T, Sakaki T, Shibata M (2004) Hydrothermal conversion of municipal organic waste into resources. Bioresour Technol 93:279–284

Gupte A, Madamwar D (1997) Solid state fermentation of lignocellulosic waste for cellulase and β-glucosidase production by cocultivation of *Aspergillus ellipticus* and *Aspergillus fumigatus*. Biotechnol. Progr. 13:166–169

Haghighi Mood S, Hossein Golfeshan A, Tabatabaei M, Salehi Jouzani G, Najafi GH, Gholami M, Ardjmand M (2013) Lignocellulosic biomass to bioethanol, a comprehensive review with a focus on pretreatment. Renew Sustain Energy Rev 27:77–93

Harris DM, Corbin K, Wang T, Gutierrez R, Bertolo AL, Petti C, Smilgies D-M, Estevez JM, Bonetta D, Urbanowicz BR, Ehrhardt DW, Somerville CR, Rose JKC, Hong M, Debolt S (2012) Cellulose microfibril crystallinity is reduced by mutating C-terminal transmembrane region residues CESA1A903 V and CESA3T942I of cellulose synthase. Proc Natl Acad Sci USA 109:4098–4103

Hartl L, Kubicek CP, Seiboth B (2007) Induction of the gal pathway and cellulase genes involves no transcriptional inducer function of the galactokinase in *Hypocrea jecorina*. J Biol Chem 282:18654–18659

Hasunuma T, Ishii J, Kondo A (2015) Rational design and evolutional fine tuning of *Saccharomyces cerevisiae* for biomass breakdown. Curr Opin Chem Biol 29:1–9

Hasunuma T, Kondo A (2012) Development of yeast cell factories for consolidated bioprocessing of lignocellulose to bioethanol through cell surface engineering. Biotechnol Adv 30:1207–1218

Hideno A, Inoue H, Tsukahara K, Yano S, Fang X, Endo T, Sawayama S (2011) Production and characterization of cellulases and hemicellulases by *Acremonium cellulolyticus* using rice straw subjected to various pretreatments as the carbon source. Enzyme Microb Technol 48:162–168

Higuchi T (2006) Look back over the studies of lignin biochemistry. J Wood Sci 52:2–8

Hölker U, Lenz J (2005a) Solid-state fermentation—are there any biotechnological advantages? Curr Opi Microbiol 8:301–306

Hölker U, Lenz J (2005b) Solid-state fermentation—are there any biotechnological advantages? Curr Opin Microbiol 8:301–306

Hong Y-S, Martinez A, Liger-Belair G, Jeandet P, Nuzillard J-M, Cilindre C (2012) Metabolomics reveals simultaneous influences of plant defence system and fungal growth in *Botrytis cinerea*-infected *Vitis vinifera* cv. Chardonnay berries. J Exp Bot 63:5773–5785

Hori H, Elbein AD (1985) Biosynthesis of plant cell wall polysaccharides. In: Higuchi T (ed) Biosynthesis and biodegradation of wood components. Academic Press, Orlando

Huang D, Wang R, Du W, Wang G, Xia M (2015) Activation of glycerol metabolic pathway by evolutionary engineering of *Rhizopus oryzae* to strengthen the fumaric acid biosynthesis from crude glycerol. Bioresour Technol 196:263–272

Humphrey A (1998) Shake flask to fermentor: what have we learned? Biotechnol Prog 14:3–7

Jones OA, Dias DA, Callahan DL, Kouremenos KA, Beale DJ, Roessner U (2015) The use of metabolomics in the study of metals in biological systems. Metallomics 7:29–38

Jones OAH, Sdepanian S, Lofts S, Svendsen C, Spurgeon DJ, Maguire ML, Griffin JL (2014) Metabolomic analysis of soil communities can be used for pollution assessment. Environ Toxicol Chem 33:61–64

Juhász T, Szengyel Z, Réczey K, Siika-Aho M, Viikari L (2005) Characterization of cellulases and hemicellulases produced by Trichoderma reesei on various carbon sources. Process Biochem 40:3519–3525

Kamei I, Nitta T, Nagano Y, Yamaguchi M, Yamasaki Y, Meguro S (2014) Evaluation of spent mushroom waste from Lentinula edodes cultivation for consolidated bioprocessing fermentation by *Phlebia* sp. MG-60. Int Biodet Biodegr 94:57–62

Karimi K, Shafiei M, Kumar R (2013) Progress in physical and chemical pretreatment of lignocellulosic biomass. In: Gupta VK, Tuohy MG (eds) Biofuel technologies. Springer, Berlin

Karpe A, Beale D, Morrison P, Harding I, Palombo E (2015a) Untargeted metabolic profiling of *Vitis vinifera* during fungal degradation. FEMS Microbiol Lett 362. Article fnv060

Karpe AV, Beale DJ, Godhani N, Morrison PD, Harding IH, Palombo EA (2015b) Untargeted metabolic profiling of winery-derived biomass waste degradation by *Aspergillus niger*. J Chem Technol Biotechnol. doi:10.1002/jctb.4749

Karpe AV, Beale DJ, Morrison PD, Harding IH Palombo EA (2015c) Untargeted metabolic profiling of *Vitis vinifera* during fungal degradation. FEMS Microbiol Lett 362, fnv060

Karpe AV, Harding IH, Palombo EA (2014) Comparative degradation of hydrothermal pretreated winery grape wastes by various fungi. Ind Crop Prod 59:228–233

Kato H, Izumi Y, Hasunuma T, Matsuda F, Kondo A (2012) Widely targeted metabolic profiling analysis of yeast central metabolites. J Biosci Bioeng 113:665–673

Kausar H, Sariah M, Mohd Saud H, Zahangir Alam M, Razi Ismail M (2010) Development of compatible lignocellulolytic fungal consortium for rapid composting of rice straw. Int Biodeter Biodegr 64:594–600

Khuong LD, Kondo R, de Leon R, Kim Anh T, Shimizu K, Kamei I (2014) Bioethanol production from alkaline-pretreated sugarcane bagasse by consolidated bioprocessing using *Phlebia* sp. MG-60. Int Biodet Biodegr 88:62–68

Kim T-Y, Wang D, Kim AK, Lau E, Lin AJ, Liem DA, Zhang J, Zong NC, Lam MPY, Ping P (2012) Metabolic labeling reveals proteome dynamics of mouse mitochondria. Mol Cell Proteomics 11:1586–1594

Kingston-Smith AH, Davies TE, Stevens PR, Mur LA (2013) Comparative metabolite fingerprinting of the rumen system during colonisation of three forage grass (*Lolium perenne L.*) varieties. PLoS ONE 8:e82801

Kosmides AK, Kamisoglu K, Calvano SE, Corbett SA, Androulakis IP (2013) Metabolomic fingerprinting: challenges and opportunities. Criti Rev Biomed Eng 41:205–221

Kouremenos KA, Pitt J, Marriott PJ (2010) Metabolic profiling of infant urine using comprehensive two-dimensional gas chromatography: application to the diagnosis of organic acidurias and biomarker discovery. J Chromatogr A 1217:104–111

Kovacs K, Macrelli S, Szakacs G, Zacchi G (2009) Enzymatic hydrolysis of steam-pretreated lignocellulosic materials with *Trichoderma atroviride* enzymes produced in-house. Biotechnol Biofuels 2:14

Krull R, Cordes C, Horn H, Kampen I, Kwade A, Neu TR, Nörtemann B (2010) Morphology of filamentous fungi: linking cellular biology to process engineering using *Aspergillus niger*. Adv Biochem Eng Biotechnol 121:1–21

Kubicek CP (2013) Systems biological approaches towards understanding cellulase production by *Trichoderma reesei*. J Biotechnol 163:133–142

Kumar R, Singh S, Singh O (2008) Bioconversion of lignocellulosic biomass: biochemical and molecular perspectives. J Ind Microbiol Biotechnol 35:377–391

Lee J (1997) Biological conversion of lignocellulosic biomass to ethanol. J Biotechnol 56:1–24

Liu H-Q, Feng Y, Zhao D-Q, Jiang J-X (2012) Evaluation of cellulases produced from four fungi cultured on furfural residues and microcrystalline cellulose. Biodegradation 23:465–472

Lu J, Weerasiri RR, Liu Y, Wang W, Ji S, Lee I (2013) Enzyme production by the mixed fungal culture with nano-shear pretreated biomass and lignocellulose hydrolysis. Biotechnol Bioeng 110:2123–2130

Lynd LR, Zyl WHV, McBride JE, Laser M (2005) Consolidated bioprocessing of cellulosic biomass: an update. Curr Opin Biotech 16:577–583

Ma X, Zheng X, Zhang M, Yang X, Chen L, Huang L, Cao S (2014) Electron beam irradiation of bamboo chips: degradation of cellulose and hemicelluloses. Cellulose 21:3865–3870

Maeda RN, Serpa VI, Rocha VAL, Mesquita RAA, Anna LMMS, de Castro AM, Driemeier CE, Pereira N Jr, Polikarpov I (2011) Enzymatic hydrolysis of pretreated sugar cane bagasse using *Penicillium funiculosum* and *Trichoderma harzianum* cellulases. Process Biochem 46:1196–1201

Maier K, Hofmann U, Bauer A, Niebel A, Vacun G, Reuss M, Mauch K (2009) Quantification of statin effects on hepatic cholesterol synthesis by transient 13C-flux analysis. Metab Eng 11:292–309

Marcinowska R, Trygg J, Wolf-Watz H, Mortiz T, Surowiec I (2011) Optimization of a sample preparation method for the metabolomic analysis of clinically relevant bacteria. J Microbiol Methods 87:24–31

Martins I, Garcia H, Varela A, Núñez O, Planchon S, Galceran MT, Renaut J, Rebelo LPN, Silva Pereira C (2014) Investigating *Aspergillus nidulans* secretome during colonisation of cork cell walls. J Prot 98:175–188

Matsuzaki F, Shimizu M, Wariishi H (2008) Proteomic and metabolomic analyses of the white-rot fungus *Phanerochaete chrysosporium* exposed to exogenous benzoic acid. J Proteome Res 7:2342–2350

Maynard AD (2015) Navigating the fourth industrial revolution. Nat Nanotechnol 10:1005–1006

Meijer S, Otero J, Olivares R, Andersen MR, Olsson L, Nielsen J (2009) Overexpression of isocitrate lyase—glyoxylate bypass influence on metabolism in *Aspergillus niger*. Metab Eng 11:107–116

Meinert S, Rapp S, Schmitz K, Noack S, Kornfeld G, Hardiman T (2013) Quantitative quenching evaluation and direct intracellular metabolite analysis in *Penicillium chrysogenum*. Anal Biochem 438:47–52

Merino S, Cherry J (2007) Progress and challenges in enzyme development for biomass utilization. In: Olsson L (ed) Biofuels. Springer, Berlin

Nelson CJ, Li L, Millar AH (2014) Quantitative analysis of protein turnover in plants. Proteomics 14:579–592

Ng JSY, Ryan U, Trengove RD, Maker GL (2012) Development of an untargeted metabolomics method for the analysis of human faecal samples using *Cryptosporidium*-infected samples. Mol Biochem Parasitol 185:145–150

Niklas J, Schneider K, Heinzle E (2010) Metabolic flux analysis in eukaryotes. Curr Opin Biotechnol 21:63–69

O'Neill MA, Ishii T, Albersheim P, Darvill AG (2004) Rhamnogalacturanan II: structure and function of a borate cross-linked cell wall pectic polysaccharide. Annu Rev Plant Biol 55:109–139

O'Sullivan A (1997) Cellulose: the structure slowly unravels. Cellulose 4:173–207

Ogura T, Date Y, Kikuchi J (2013) Differences in cellulosic supramolecular structure of compositionally similar rice straw affect biomass metabolism by paddy soil microbiota. PLoS ONE 8:e66919

Olson DG, McBride JE, Joe Shaw A, Lynd LR (2012) Recent progress in consolidated bioprocessing. Curr Opin Biotechnol 23:396–405

Ortiz ME, Bleckwedel J, Raya RR, Mozzi F (2013) Biotechnological and in situ food production of polyols by lactic acid bacteria. Appl Microbiol Biotechnol 97:4713–4726

Papini M, Nookaew I, Siewers V, Nielsen J (2012) Physiological characterization of recombinant Saccharomyces *cerevisiae* expressing the *Aspergillus nidulans* phosphoketolase pathway: validation of activity through 13C-based metabolic flux analysis. Appl Microbiol Biotechnol 95:1001–1010

Pedersen H, Carlsen M, Nielsen J (1999) Identification of enzymes and quantification of metabolic fluxes in the wild type and in a recombinant *Aspergillus oryzae* strain. Appl Environ Microb 65:11–19

Pirota RPB, Delabona P, Farinas C (2014) Simplification of the biomass to ethanol conversion process by using the whole medium of filamentous fungi cultivated under solid-state fermentation. BioEnerg Res 7:744–752

Pribowo A, Arantes V, Saddler JN (2012) The adsorption and enzyme activity profiles of specific *Trichoderma reesei* cellulase/xylanase components when hydrolyzing steam pretreated corn stover. Enzyme Microb Technol 50:195–203

Putri SP, Nakayama Y, Matsuda F, Uchikata T, Kobayashi S, Matsubara A, Fukusaki E (2013) Current metabolomics: practical applications. J Biosci Bioeng 115:579–589

Qi-He C, Krügener S, Hirth T, Rupp S, Zibek S (2011) Co-cultured production of lignin-modifying enzymes with *white-rot fungi*. Appl Biochem Biotechnol 165:700–718

Ralph J, Lundquist K, Brunow G, Lu F, Kim H, Schatz PF, Marita JM, Hatfield RD, Ralph SA, Christensen JH, Boerjan W (2004) Lignins: natural polymers from oxidative coupling of 4-hydroxyphenyl- propanoids. Phytochem Rev 3:29–60

Rani KS, Seenayya G (1999) High ethanol tolerance of new isolates of *Clostridium thermocellum* strains SS21 and SS22. World J Microbiol Biotechnol 15:173–178

Rautio J, Smit B, Wiebe M, Penttila M, Saloheimo M (2006) Transcriptional monitoring of steady state and effects of anaerobic phases in chemostat cultures of the filamentous fungus *Trichoderma reesei*. BMC Genom 7:247

Rehman MSU, Kim I, Chisti Y, Han J-I (2013) Use of ultrasound in the production of bioethanol from lignocellulosic biomass. EEST Part A: Energy Sci Res 30:1391–1410

Rodríguez A, Carnicero A, Perestelo F, De La Fuente G, Milstein O, Falcón MA (1994) Effect of *Penicillium chrysogenum* on lignin transformation. Appl Environ Microbiol 60:2971–2976

Roessner U, Bowne J (2009) What is metabolomics all about? Biotechniques 46:363

Rumbold K, van Buijsen HJJ, Overkamp KM, van Groenestijn JW, Punt PJ, van der Werf MJ (2009) Microbial production host selection for converting second-generation feedstocks into bioproducts. Microb Cell Fact 8:64

Sánchez C (2009) Lignocellulosic residues: biodegradation and bioconversion by fungi. Biotechnol Adv 27:185–194

Sarkar N, Ghosh SK, Bannerjee S, Aikat K (2012) Bioethanol production from agricultural wastes: an overview. Renew Energy 37:19–27

Saunders EC, Souza DP, Chambers JM, Ng M, Pyke J, McConville MJ (2015) Use of 13C stable isotope labelling for pathway and metabolic flux analysis in *Leishmania* parasites. In: Peacock C (ed) Parasite genomics protocols. Springer, New York

Seiboth B, Gamauf C, Pail M, Hartl L, Kubicek CP (2007) The D-xylose reductase of *Hypocrea jecorina* is the major aldose reductase in pentose and D-galactose catabolism and necessary for β-galactosidase and cellulase induction by lactose. Mol Microbiol 66:890–900

Shevchenko SM, Chang K, Robinson J, Saddler JN (2000) Optimization of monosaccharide recovery by post-hydrolysis of the water-soluble hemicellulose component after steam explosion of softwood chips. Bioresour Technol 72:207–211

Shulaev V (2006) Metabolomics technology and bioinformatics. Brief Bioinform 7:128–139

Stone BA (1958) Biochemistry of cellulose and related polysaccharides. Nature 182:687–690

Strong PJ, Burgess JE (2008) Treatment methods for wine-related and distillery wastewaters: a review. Bioremediat J 12:70–87

Studholme DJ, Harris B, le Cocq K, Winsbury R, Perera V, Ryder L, Beale M, Ward J, Thornton CR, Grant M (2013) Investigating the beneficial traits of *Trichoderma hamatum* GD12 for sustainable agriculture—insights from genomics. Front Plant Sci 4:258

Takeda H, Yoneyama F, Kawai S, Hashimoto W, Murata K (2011) Bioethanol production from marine biomass alginate by metabolically engineered bacteria. Energy Environ Sci 4:2575–2581

Tian F, Xie Z-L, Zhao L-Z, Guo J, Han X-B, Xie L-F, Wang Y, Chang X-Y (2015) Comparative secretome analysis of *Fusarium* sp. Q7-31T during liquid fermentation using oat straw as a carbon source. Ann Microbiol 65:1–10

Udatha DBRKG, Rasmussen S, Sicheritz-Pontén T, Panagiotou G (2013) Targeted metabolic engineering guided by computational analysis of single-nucleotide polymorphisms (SNPs). In: Alper SH (ed) Systems metabolic engineering: methods and protocols. Humana Press, Totowa

Vanholme R, Demedts B, Morreel K, Ralph J, Boerjan W (2010) Lignin biosynthesis and structure. Plant Physiol 153:895–905

Velmurugan R, Muthukumar K (2012) Sono-assisted enzymatic saccharification of sugarcane bagasse for bioethanol production. Biochem Eng J 63:1–9

Villas-Bôas SG, Noel S, Lane GA, Attwood G, Cookson A (2006) Extracellular metabolomics: a metabolic footprinting approach to assess fiber degradation in complex media. Anal Biochem 349:297–305

Vivas M, Nunes C, Coimbra M, Solla A (2014) Maternal effects and carbohydrate changes of *Pinus pinaster* after inoculation with *Fusarium circinatum*. Trees 28:373–379

Waeonukul R, Kosugi A, Tachaapaikoon C, Pason P, Ratanakhanokchai K, Prawitwong P, Deng L, Saito M, Mori Y (2012) Efficient saccharification of ammonia soaked rice straw by combination of *Clostridium thermocellum* cellulosome and *Thermoanaerobacter brockii* β-glucosidase. Bioresour Technol 107:352–357

Wagner JM, Alper HS (2016) Synthetic biology and molecular genetics in non-conventional yeasts: current tools and future advances. Fungal Genet Biol 89:126–136

Wang X, Jin M, Balan V, Jones AD, Li X, Li BZ, Dale BE, Yuan YJ (2014) Comparative metabolic profiling revealed limitations in xylose-fermenting yeast during co-fermentation of glucose and xylose in the presence of inhibitors. Biotechnol Bioeng 111:152–164

Wiethölter N, Graeßner B, Mierau M, Willats WGT, Knox JP, Moerschbacher BM (2003) Isolation and characterisation of the homogalacturonan from type II cell walls of the commelinoid monocot wheat using HF-solvolysis. Carbohydr Res 338:423–431

Williams TC, Averesch NJH, Winter G, Plan MR, Vickers CE, Nielsen LK, Krömer JO (2015) Quorum-sensing linked RNA interference for dynamic metabolic pathway control in *Saccharomyces cerevisiae*. Metab Eng 29:124–134

Yamazawa A, Iikura T, Shino A, Date Y, Kikuchi J (2013) Solid-, solution-, and gas-state NMR monitoring of 13C-cellulose degradation in an anaerobic microbial ecosystem. Molecules 18:9021–9033

Yang Q, Wu S, Lou R, Lv G (2010a) Analysis of wheat straw lignin by thermogravimetry and pyrolysis-gas chromatography/mass spectrometry. J Anal Appl Pyrol 87:65–69

Yang X-Y, Chen W-P, Rendahl AK, Hegeman AD, Gray WM, Cohen JD (2010b) Measuring the turnover rates of *Arabidopsis* proteins using deuterium oxide: an auxin signaling case study. Plant J 63:680–695

Yuan Y, Du J, Zhao H (2013) Customized optimization of metabolic pathways by combinatorial transcriptional engineering. In: Alper SH (ed) Systems metabolic engineering: methods and protocols. Humana Press, Totowa

Zha Y, Westerhuis JA, Muilwijk B, Overkamp KM, Nijmeijer BM, Coulier L, Smilde AK, Punt PJ (2014) Identifying inhibitory compounds in lignocellulosic biomass hydrolysates using an exometabolomics approach. BMC Biotechnol 14:22

Zuroff T, Curtis W (2012) Developing symbiotic consortia for lignocellulosic biofuel production. Appl Microbiol Biotechnol 93:1–13

Chapter 10
Beyond Metabolomics: A Review of Multi-Omics-Based Approaches

David J. Beale, Avinash V. Karpe and Warish Ahmed

1 Introduction

'Omics'-based techniques comprise a suite of tools and approaches, each with their own specific protocols and frameworks for setting minimum data and reporting standards (Field et al. 2008; Morrison et al. 2006; Sansone et al. 2007; Sumner et al. 2007; Orchard and Kerrien 2010). This suite of techniques primarily comprises metagenomics, transcriptomics, proteomics and metabolomics. There are also a number of other specialized 'omic'-based approaches that are included under the broader 'omics' banner, such as lipidomics, fluxomics (metabolic flux analysis), toxicogenomics, nutrigenomics and foodomics. However, these additional 'omics' approaches are not considered within the context of this chapter, as they are subcategories of the aforementioned suite of techniques. In addition, for completeness, it should be noted that within the context of this chapter, (meta)genomics has been defined as the analysis of genetic material recovered from an organism or environmental samples (Handelsman 2004); (meta)transcriptomics is the analysis of RNA molecules, including messenger RNA (mRNA), Ribosomal RNA (rRNA), transfer RNA (tRNA) and other non-coding RNA produced by an organism or a population of organisms (Pascault et al. 2015); (meta)proteomics is the analysis of proteins produced by an organism or population of organisms, and their function (Douterelo et al. 2014); and lastly, metabolomics is the analysis of the small chemical compounds produced and consumed by an organism or a population of organisms (Beale et al. 2016a).

D.J. Beale (✉) · A.V. Karpe · W. Ahmed
Land and Water, Commonwealth Scientific and Industrial Research Organisation (CSIRO), Ecosciences Precinct, 41 Boggo Road, Dutton Park, QLD 4102, Australia
e-mail: david.beale@csiro.au

A.V. Karpe
Department of Chemistry and Biotechnology, Faculty of Science, Engineering and Technology, Swinburne University of Technology, P.O. Box 218 Hawthorn, VIC 3122, Australia

D.J. Beale et al. (eds.), *Microbial Metabolomics*,
DOI 10.1007/978-3-319-46326-1_10

The application of these 'omics' techniques in isolation has been demonstrated to be beneficial in the understanding and characterization of numerous biological systems within the environment, engineered and industrial systems and treatment processes, in addition to providing insight into human health and clinical investigations (Cohen et al. 2015). It is when these 'omics'-based techniques are applied in combination, however, that their real power can be utilized. These techniques enable researchers to identify and characterize the entire microbial community present at greater depth (i.e. metagenomics) and, in combination with the other techniques studied in parallel, further enable researchers to identify what that community is doing, in terms of gene expression (i.e. transcriptomics), protein production (i.e. proteomics) and the community metabolism (i.e. metabolomics) (Turnbaugh and Gordon 2008). Figure 1 illustrates the chronological order (i.e. the magnitude of information that can be obtained) of these principal 'omics' based techniques, which inform researchers on the microbial potential, functionality, and activity starting at the metagenome through to the metabolome. This multiple 'omics'-based approach has been coined "*multi-omics*".

The objective of this chapter is to highlight examples in the literature that go beyond metabolomics only studies, giving rise to a greater depth of systems biology research that is multi-omics. While multi-omics research is not new, it is a growing area of research that has increased in popularity amongst the scientific community in recent years. Furthermore, the underlying aim of this review is not to provide a detailed chronology of multi-omics or a detailed guide on how to integrate multi-omics datasets; this has been the focus of recent reviews published

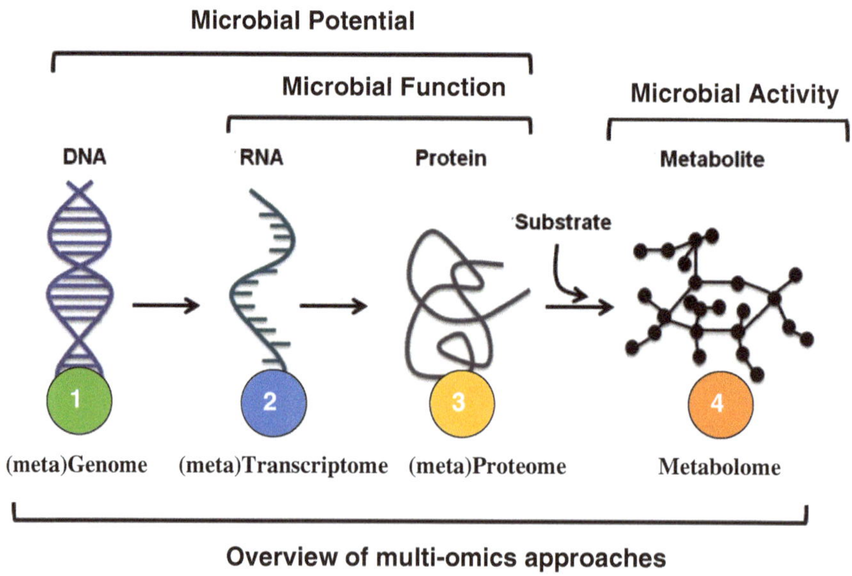

Fig. 1 Chronical order of 'omics' based techniques commonly used in multi-omics studies. Adapted from Abram (2015)

(Barh et al. 2013; Blanchet and Smolinska 2016; Fondi and Liò 2015; Kohl et al. 2014; Schneider and Orchard 2011; Zhang et al. 2010). Instead, the aim of this chapter is to provide an overview of the different approaches used in multi-omics-based research and the tools used to integrate multi-omics data, using examples from the literature from a range of environmental, industrial and biomedical applications to highlight the value of extending beyond metabolomics only research.

2 The Multi-Omic Data Analysis Challenge

Multi-omics based techniques are inherently data-rich studies. For example, the human genome comprises ca. 20,000–25,000 protein coding genes (Pertea and Salzberg 2010) and the human metabolome is estimated to comprise over 40,000 metabolites (Forsythe and Wishart 2009). As such, the challenge in all 'omic-based investigations', which is only compounded further when applying a multi-omics approach, is the handling of these large and complex datasets (Kohl et al. 2014; Röling et al. 2010). Röling et al. (2010) characterized the data processing and quantitative comprehension of multi-omic information as a bottleneck in the overall workflow, requiring input and interpretation of 'systems biologists' and microbiologists. In addition, the authors of this review propose that it needs further input from the bioanalytical chemist/biochemist/biostatistician to first evaluate the quality and validity of the study (experimental) design, as well as the quality of the data acquired from the instrument, before even attempting to integrate and synthesize findings (indeed the old adage of poor data in equals poor data out applies). In any case, assuming the data obtained are of high quality and are valid [following each 'omics' specific protocol and frameworks for minimum data and reporting standards as highlighted by the various societies (Field et al. 2008; Sansone et al. 2007; Sumner et al. 2007)], there are a number of approaches to analyzing and interpreting multi-omics data, namely: post-data analysis integration and integrated data analysis techniques.

In a post-data analysis approach, datasets are analyzed in isolation of each other and key features are *networked* in a post analysis exercise through the synthesis of significant features at joint nodes in the overall model metabolic pathway. This approach has been used in previous studies that focused on characterizing and assessing biological wastewater treatment systems (Beale et al. 2016b), the microbial resistance of marine sediments after an oil spill (Kimes et al. 2013) and characterizing permafrost (Hultman et al. 2015). In contrast, an integrated multi-omics approach employs specialized tools to merge datasets prior to undertaking any data analysis and interpretation (Kuo et al. 2013), thus enabling similarities of each omic approach to be statistically derived, as opposed to relying on human interpretation. The principal differences between the post-data analysis and integrated analysis approaches are graphically presented in Fig. 2.

In addition to post-data analysis integration and integrated data analysis techniques, a third model-based integration approach has been identified. However, a

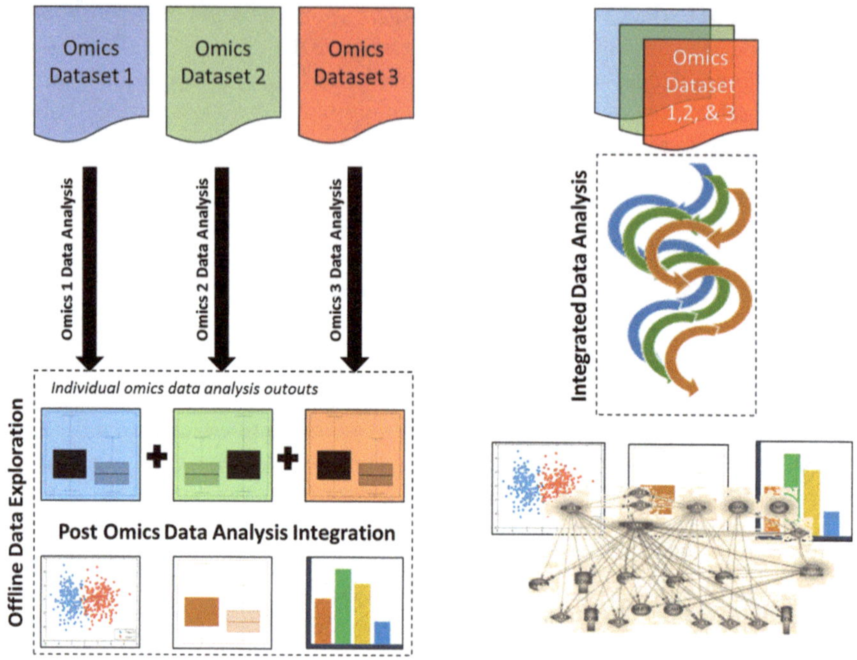

Fig. 2 Principal differences between the post-data analysis and integrated analysis approaches

model-based approach is considered unobtainable, in a practical sense, as stated by Kamburov et al. (2011). Model-based integration methods rely on a well-defined understanding of the system being investigated in order to compare new experimental findings against modelled predictions. That's not to say that a complete multi-omics model-based integration approach is not obtainable as suggested. There are already examples in the literature of its use. For example, Noecker et al. (2016) used a model-based approach to integrate taxonomic and metabolomics data to predict the effects of community ecology on metabolite concentrations and evaluated these predictions with measured metabolomic profiles from the vaginal microbiome. It was concluded that predicted species composition correlated with identified putative metabolic mechanisms underlying these predictions (Noecker et al. 2016). However, it is noteworthy to mention that a model-based integration approach was achieved primarily because of the vaginal microbiome investigated had already been well defined and studied beforehand, utilizing previously published data and publically available datasets (Erickson et al. 2012; Srinivasan et al. 2015; Theriot et al. 2014; Jansson et al. 2009; Jozefczuk et al. 2010). As such, the real challenge for model-based integration approaches is not that they are principally unobtainable, but that not all the systems studied are yet fully characterized as per the example presented by Noecker et al. (2016). Therefore, until a systems biology approach is taken for all studied systems (i.e. a multi-omics approach in order to obtain baseline data), a model-based integration approach is limited to only those systems that are already well defined.

Regardless of the multi-omics approach being applied, there are numerous tools and approaches that assist researchers to integrate omics-based datasets. As illustrated in Table 1, there are tools that have been developed that are context specific (i.e. targeted towards the integration of omics data from specific animal models, medical and clinical studies and selected plant species). There are also tools that are unspecified in terms of the studied system but provide users a set of statistical tools for data normalization and transformation, and a range of chemometric models for interpreting data once integrated.

3 Application of Multi-Omics

There are many examples in the literature of multi-omic studies, with various levels of integration. Some studies comprise simple levels of integrations (i.e. combing two different -omics datasets) through to more comprehensive and computationally demanding studies (i.e. integration of multiple omics datasets). Typically, a two-omics integration study combines either metagenomics or transcriptomics data with proteomics (Sunagar et al. 2016; Wildburger et al. 2015) or metabolomics datasets (Tokimatsu et al. 2005b; Garcia-Alcalde et al. 2011; Kamburov et al. 2011) or combines proteomics and metabolomics datasets (Xu et al. 2015). These studies demonstrate how predicted functional metabolism (metagenomics) or gene expression (transcriptomics) relate to actual protein and metabolite expression and, by extension, provide a means to self-validate findings through cross-referencing experimental findings (Fondi and Liò 2015). The following section provides some examples of multi-omics studies applied to various fields of research, all of which provided an extension beyond a metabolomics only study (or other singular omics-based approaches) and enabled further depth of analysis that would otherwise not be achieved.

3.1 Environmental Contaminants

Marine subsurface environments comprise an abundant and diverse microbial community (El-Serehy et al. 2016; Yanagawa et al. 2014). These communities have the ability to bio-transform and mineralize numerous contaminants (Kimes et al. 2013). The application of omics-based research has been used in the past to characterize and understand sediment and marine environment dynamics. Chariton et al. (2014) used metagenomics to investigate benthic invertebrate diversity in exposed sediments with elevated concentrations of triclosan (antibacterial and antifungal agent). However, a single omics-based approach only provides limited information. As is this case, a metagenomics approach will only provide

Table 1 Summary of multi-omic tools

Software tool	Omics integrated	Domain	Functionality	License	References
3Omics	– Transcriptomics – Proteomics – Metabolomics	Medical (human)	– Correlation network analysis – Co-expression analysis – Phenotype generation – KEGG/HumanCyc pathway enrichment – GO enrichment	Open	Kuo et al. (2013)
Biofomics	– Transcriptomics – Proteomics – Metabolomics	Biofilms	– Experiment library – Data depository	Open	Lourenço et al. (2012)
Escher	– Genomics – Proteomics – Metabolomics	Unspecified	– Web application for visualizing data on biological pathways – Rapidly design new pathway maps based on user data and genome-scale models – Visualize data related to genes or proteins on the associated reactions and pathways – Identify trends in common genomic data types	Open	King et al. (2015)
GIM3E (gene inactivation moderated by metabolism, metabolomics and expression)	– Transcriptomics – Metabolomics	Unspecified	– Establishes metabolite use requirements with metabolomics data – Model-paired transcriptomics data to find experimentally supported solutions – Calculates the turnover (production/consumption) flux of metabolites	Open (Python and requires a COBRApy 0.2.x.)	Machado and Herrgård (2014)
IMPaLA (integrated molecular pathway level analysis)	– Transcriptomics or proteomics – Metabolomics	Medical and clinical	– Enrichment analysis – Pathway analysis	Academic only	Kamburov et al. (2011)
Ingenuity pathway analysis	– Metagenomics – Transcriptomics – Proteomics – Metabolomics	Medical (human) and clinical	– Metabolic pathway analysis – Network visualization – Data integration – Upstream regulator analysis – Mechanistic networks – Causal network analysis – Downstream effects analysis	Commercial	Krämer et al. (2013)

(continued)

Table 1 (continued)

Software tool	Omics integrated	Domain	Functionality	License	References
INMEX (integrative meta-analysis of expression data)	– Transcriptomics – Metabolomics	Medical and clinical	– Meta and integrative analysis of data – Pathway analysis	Open	Xia et al. (2013)
IOMA (integrative omics-metabolic analysis)	– Proteomics – Metabolomics	Unspecified	– Integrates proteomic and metabolomic data to predict flux distributions	Open	Yizhak et al. (2010)
KaPPA-view	– Transcriptomics – Metabolomics	Plants	– Integrates transcriptomics and metabolomic data to map pathways	Open	Tokimatsu et al. (2005a)
MADMAX (management and analysis database for multiple ~ omics experiments)	– Metagenomics – Transcriptomics – Metabolomics	Plants, medical and clinical	– Integrates omics data – Statistical analysis and pathway mapping	Open	Lin et al. (2011)
MapMan	– Metagenomics – Transcriptomics – Metabolomics	Plants (developed for use with Arabidopsis. Includes more species)	– Compare data across these two species – KEGG classification – Classification into KOG clusters – Mapping expression responses	Open	Thimm et al. (2004), Usadel et al. (2005)
MarVis-Pathway (marker visualization pathway)	– Metagenomics – Transcriptomics – Metabolomics	Unspecified	– Toolbox for interactive ranking, filtering, combination, clustering, visualization and functional analysis of data sets containing intensity-based profile vectors	Academic only	Kaever et al. (2015)
MassTrix	– Transcriptomics – Metabolomics	Unspecified	– Integration of data – Generation of colored pathway maps KEGG data analysis	Open	Wagele et al. (2012)
MetScape 2	– Transcriptomics – Metabolomics	Medical and clinical	– Integrates data from KEGG and EHMN databases	Open	Karnovsky et al. (2012)
mixOmics (R package)	– Metagenomics – Transcriptomics – Proteomics – Metabolomics	Unspecified	– Integration of data – Chemometric analysis (similarity/difference)	Open	Günther et al. (2014)
Cytoscape with MODAM and Cytoscape with OmicsAnalyzer	– Transcriptomics – Proteomics – Metabolomics – Fluxomics	Unspecified	– Multi-Omic Data Miner and OmicsAnalyzer were designed as an accessible and handy Cytoscape plugin that facilitates omics analysis – Compile all biologically-relevant information regarding the model system through web link association – Map the network components with multi-omics data – Model omics data	Open	Enjalbert et al. (2011), Xia et al. (2010)

(continued)

Table 1 (continued)

Software tool	Omics integrated	Domain	Functionality	License	References
PaintOmics	– Transcriptomics – Metabolomics	100 top species of different biological kingdoms	– Integration and visualization of transcriptomics and metabolomics data	Open	Garcia-Alcalde et al. (2011)
PathVisio 3	– Transcriptomics – Proteomics – Metabolomics	Unspecified	– Integration of omics data – Visualize omics data based on common data nodes and interactions in the pathway	Open (Apache)	Kutmon et al. (2015)
ProMeTra	– Transcriptomics – Metabolomics	Medical and Clinical	– Interactive visualizations of metabolite concentrations together with transcript measurements mapped on the pathways and GenomeMaps	Open	Neuweger et al. (2009)
SIMCA	– Metagenomics – Transcriptomics – Proteomics – Metabolomics	Unspecified	– Integration of data – Chemometric analysis (similarity/difference)	Commercial	Wheelock and Wheelock (2013)
VANTED (visualization and analysis of networks with related experimental data)	– Metagenomics – Transcriptomics – Proteomics – Metabolomics	Unspecified	– Comparison of multiple omics data sets – Visualization of metabolic maps – Correlation networks analysis	Open	Junker et al. (2006)
VitisNet	– Metagenomics – Transcriptomics – Proteomics – Metabolomics	Grapes	– Integration of data – Visualization of connectivity	Open	Grimplet et al. (2009)

information of the microbial diversity present (and by extension, a theoretical analysis of their functionality). Investigating beyond metagenomics would provide a greater depth to the analysis by measuring the actual function of the studied population using proteomics or metabolomics. Hook et al. (2014) investigated contaminated sediments using a similar approach, however, they used transcriptomics and metabolomics in order to better understand the modes of toxic action within contaminated ecosystems. The inclusion of metabolomics in such a study enabled the assessment of changes in the biochemical profiles of microbial communities living in contaminated sites (Jones et al. 2014; Llewellyn et al. 2015). In the study by Hook et al. (2014), the function of transcripts with altered abundance in *Melita plumulosa* (an epibenthic amphipod) following whole-sediment exposure to a series of common environmental contaminants was investigated. Contaminants included in the study comprised porewater ammonia, bifenthrin and fipronil (pesticides), diesel and crude oil (petroleum products) and metals (Cu, Ni and Zn). Subsequent data integration and hierarchical cluster analysis demonstrated grouped transcriptome and metabolome expression profiles by contaminant class. Many of the transcriptional changes observed were consistent with patterns previously described in other crustaceans (Osborn and Hook 2013).

Furthermore, following the Deepwater Horizon (DWH) oil spill in the Gulf of Mexico, researchers used metagenomic analysis and metabolomic profiling of deep-sea sediment samples (Kimes et al. 2013). Post-data analysis integration of the two datasets identified the presence of aerobic microbial communities and their associated functional genes among all the samples collected, whereas, a greater number of Deltaproteobacteria and anaerobic functional genes were found in the sediments closest to the point of oil contamination. Metabolic profiling revealed a greater number of putative metabolites in sediments surrounding the contamination site relative to background sites. These putative metabolites were identified as a series of benzylsuccinates (with carbon chain lengths from 5 to 10), suggesting that increased exposure to hydrocarbons enriched the Deltaproteobacteria, which are known to be capable of anaerobic hydrocarbon metabolism. Lastly, through a combined multi-omics approach, it was surmised that the sediment samples collected at the site of contamination comprised an active indigenous microbial community capable of metabolising aromatic hydrocarbons in deep-sea sediments of the Gulf of Mexico (Kimes et al. 2013).

Hultman et al. (2015) undertook a similar study investigating the microbial metabolism of permafrost. They used several ‘omics approaches, combined with post-data analysis in order to determine the phylogenetic composition of microbial communities of intact permafrost, the seasonally thawed active layer and thermokarst bog (surfaces of marshy hollows). The multi-omics strategy revealed good correlation of process rates for methanogenesis (the dominant process), in addition to providing insights into novel survival strategies for potentially active microbes in permafrost (Hultman et al. 2015).

The advancement of omics-based techniques and their integration have contributed towards marine biologists and molecular biologists pushing the boundaries of our understanding of marine molecular biology (Thakur et al. 2008). This is best evident in a global holistic approach to the study of marine ecosystems presented by Karsenti et al. (2011). In this study, the application of multi-omics was extended by including additional meta-data, linking biogeography with ecology, genetics and morphology datasets to provide a global perspective to marine systems.

3.2 *Food and Nutrition*

Metagenomics-based characterization of microbial communities provides a very promising and powerful approach to food safety testing, with research to date undertaken focusing on foodborne pathogen biology (Stasiewicz et al. 2015). However, transcriptomics, proteomics and metabolomics approaches have also been demonstrated to have the potential for food safety applications (Valdés et al. 2013; Jadhav et al. 2014, 2015; Beale et al. 2014b). To date, transcriptomics, proteomics and metabolomics have become the three most commonly used techniques in food and nutrition-based 'omics' research (Kato et al. 2011).

Takahashi et al. (2014) combined gene expression profiles using DNA microarrays with proteomic and metabolomics data in order to assess the anti-obesity effects of coffee in mice. The underlying premise was that coffee consumption may reduce the risks of developing obesity and diabetes. As such, gene expression, protein and metabolite profiles in the livers of C57BL/6J mice fed a high-fat diet containing three types of coffee (caffeinated, decaffeinated and green unroasted coffee) were investigated. The data were integrated using an in-house software tool and visualized in KEGG pathways (summarized in Fig. 3). Takahashi et al. (2014) concluded that the alterations within and around the urea cycle were found to be highly consistent between transcripts and metabolites, suggesting that expression of the genes related to the urea cycle were downregulated by a high-fat diet and up-regulated by coffee consumption. These findings were also consistent when integrated post-data analysis of each omics dataset.

For nutrition-based research, to further advance the application of multi-omics research and disseminate current findings more broadly for other researchers to use, a database of nutrition-focused omics (genomics and DNA microarray data) has been created (Nutrigenomics Database) (Saito et al. 2005). Similar approaches have been applied to other areas of research. For example, VitisNet was created in order analyze omics-based data relating to common wine grapes enabling the integration of large datasets, streamlining biological functional processing and improving the understanding of dynamic processes in systems biology experiments (Grimplet et al. 2009). PaintOmics has similar functionality but is targeted towards a defined list of plants (Garcia-Alcalde et al. 2011).

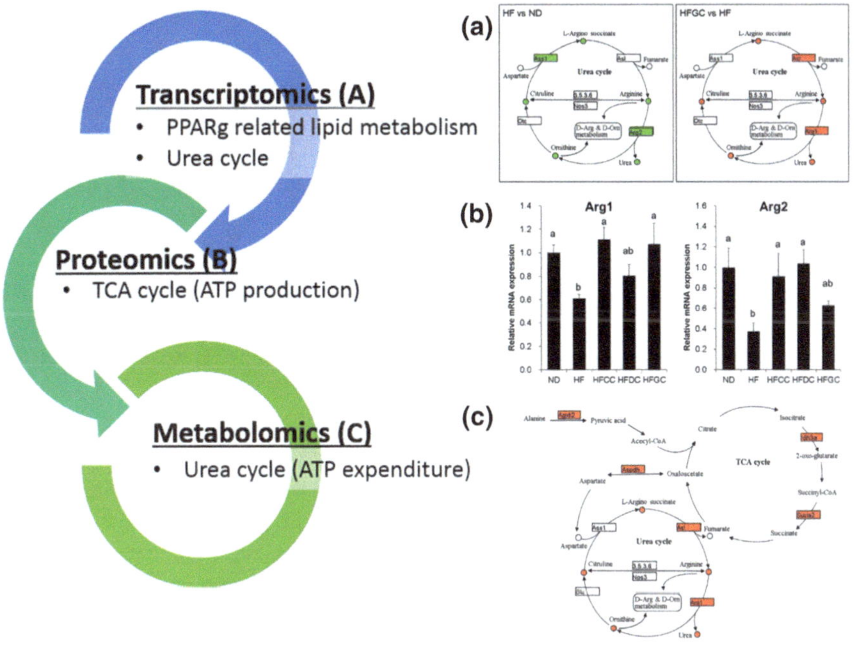

Fig. 3 Integrated analysis of transcriptomics, metabolomics and proteomics of liver tissue samples collected from mice fed a high-fat diet containing high levels of coffee. Adapted from Takahashi et al. (2014)

3.3 *Water and Wastewater Systems*

The application of metagenomics to characterize microbial populations in wastewater treatments systems (Talbot et al. 2008; Vanwonterghem et al. 2014; Pap et al. 2015; Kovács et al. 2015; Ma et al. 2014; Albertsen et al. 2012) and pipes (Vincke et al. 2001; Santo Domingo et al. 2011; Neria-González et al. 2006) is not new. However, to the authors' knowledge, the application of multi-omics approaches is limited. Gomez-Alvarez et al. (2012) analyzed the whole-metagenome to determine the microbial composition and functional genes associated with biomass harvested from sections of a corroded wastewater pipe. Taxonomic and functional analysis demonstrated that approximately 90 % of the total diversity was associated with the phyla *Actinobacteria*, *Bacteroidetes*, *Firmicutes* and *Proteobacteria*. Furthermore, the biofilm was found to have an abundance of sulphur-oxidizing bacteria and sulphate-reducing bacteria. Combined with transcriptomics, Gomez-Alvarez et al. (2012) also demonstrated an enrichment of genes associated with heavy metal resistance, virulence (protein secretion systems) and stress response in the biofilm analyzed.

Mohan et al. (2014) analyzed hydraulic fracturing source water and wastewater produced during fracking using metagenomic and metabolomic techniques.

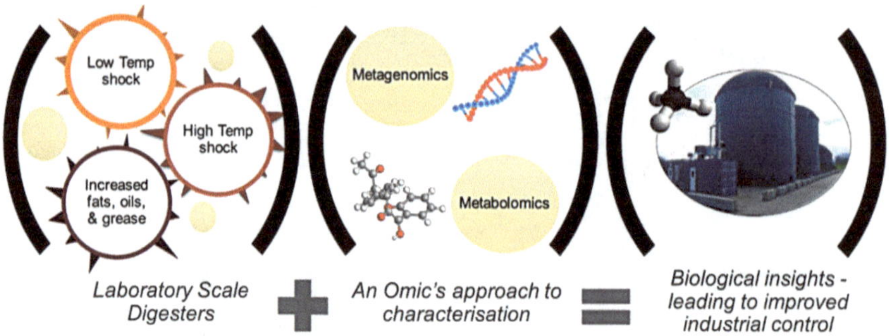

Fig. 4 Post-data analysis approach applied to a multi-omics study investigating operational shocks within laboratory scale digesters treating municipal waste. Adapted from Beale et al. (2016b)

The post-data analysis of the multi-omic datasets revealed a relative increase in genes responsible for carbohydrate metabolism, respiration, sporulation and dormancy, iron acquisition and metabolism, stress response and sulphur metabolism in the produced water samples from a diverse microbial community (Mohan et al. 2014). These results suggest that microbial communities in potable water have an increased genetic ability to handle stress, which has significant implications for infrastructure management, such as biofilm control and combating microbial influenced corrosion.

Another multi-omics study conducted by Beale et al. (2016b) investigated operational shocks in laboratory scale Anaerobic Digesters (AD) treating municipal wastewater (Fig. 4). The laboratory scale digesters were operated to simulate potential shocks to the AD process experienced at a 350 ML/day wastewater treatment plant. The shocks included high (42 °C) and low (32 °C) temperature (either side of the optimum temperature of 37 °C) and a 20 % loading of fats, oil and grease (20 % w:v). These variables were explored at two sludge retention times (12 and 20 days) and two organic loading rates (2.0 and 2.5 kg TS/m^3 day). Metagenomics and metabolomics approaches were then used to characterize the impact of operational shocks in regard to temperature and fats, oil and grease addition, as determined through monitoring of biogas production, the microbial profile and their metabolism. Results showed that AD performance was not greatly affected by temperature shocks. Post-data analysis of the metagenomics and metabolomics data indicated that methanogens and methane oxidizing bacteria were low in relative abundance, and that the ratio of oxidizing bacteria (methane, sulphide and sulphate), with respect to sulphite reducing bacteria had a noticeable influence on biogas production. Furthermore, increased biogas production correlated with an increase in short chain fatty acids, a product of the addition of 20 % fat, oil and grease.

As demonstrated by the works of Gomez-Alvarez et al. (2012), Mohan et al. (2014), Beale et al. (2016b), the application of omics-based techniques to characterize the microbial community and their function provides insight to the resilience of crucial microbial populations within water sources, pipes and treatment process. In addition, it is anticipated that through multi-omics approaches, new insights into microbial populations in terms of diversity, resilience and activity when exposed to shocks and stresses (such as illegal discharges within sewer networks and operational shock during treatment) will be obtained. Furthermore, multi-omics may provide insight into driving higher biogas and methane production during treatment and collection systems for reuse.

3.4 Biofilms and Biofilms Known to Influence Corrosion

The safety and quality of potable water is often assumed and taken for granted by consumers in most developed countries. However, our understanding of potable water biofilms is limited, partly as they are difficult to access and traditional microbiology approaches fail to provide sufficient information on their composition and activity (Douterelo et al. 2014). To date, numerous researchers have applied omics-based techniques to characterize and assess aquatic biofilms, in a range of environments. For example, Shaw et al. (2014), Chao et al. (2013) assessed the community structure of potable water biofilms using metagenomics when exposed to different water treatment strategies. Metagenomics was also used to investigate intertidal marine biofilm communities (Zhang et al. 2013) and black fungal biofilms growth in domestic water fixtures (Heinrichs et al. 2013). Metabolomics has also been used to investigate biofilms found on the surface of copper pipes (Beale et al. 2010; Beale et al. 2012) and within potable water distribution networks (Beale et al. 2013). However, the application of multi-omics approaches has been limited. Leary et al. (2014) used a metagenomic and metaproteomic approach to analyze the composition and function of marine biofilms; others have investigated microbial communities in extreme environments using metagenomics and proteomics (Singer 2012; Schneider and Riedel 2010). Nevertheless, it has been identified that there is a need for a multi-omics approaches to assess biofilms. This has resulted in the recently proposed term "Biofomics" and web-based interface for biofilm data (Lourenço et al. 2012). Biofomics provides a framework, database depository and selection of statistical tools for researchers to follow, merge and examine data from different approaches such as metagenomics, transcriptomics, proteomics and key metabolites (metabolomics) (Nozhevnikova et al. 2015). The biofilm data results from experiments involving several types of bacterial genera such as *Salmonella* spp., *Escherichia coli* and *Candida* spp., and is supported by the minimum information about a biofilm experiment (MIABiE) initiative (Lourenço et al. 2012). To date, the majority of biofomics work has been related to the biomedical and clinical fields.

An issue of particular concern to infrastructure asset managers is biofilms that cause biocorrosion/biodeterioration (also known as MIC). Due to their potential financial and economic impact to infrastructure, there has been a considerable amount of research published on the role of microorganisms in promoting corrosion. The majority of this work has been undertaken to address the problem of MIC in offshore oil and gas pipelines, and concrete structures with some preliminary research on microbial/metal surface interactions in water pipes. As such, a number of extensive reviews have been compiled on MIC mechanisms over the past 20 years (Videla and Herrera 2009; Edyvean and Videla 1991; Videla 2003; Beech and Gaylarde 1999; Flemming 1994; Beech et al. 2014).

Corrosion is the result of a series of chemical, physical and (micro) biological processes leading to the deterioration of materials. The mechanisms of MIC and MIC inhibition are not completely understood, because they cannot be linked to a single biochemical reaction or specific microbial species or group (Kip and van Veen 2015). As such, MIC biofilm communities can be studied at both their compositional and functional levels through the use of multi-omics. A number of traditional techniques, such as clone libraries and genetic fingerprinting, along with more recent metagenomics and transcriptomics, are being used to characterize and understand MIC biofilms (Chakraborty et al. 2014).

Relatively few metabolomic-MIC studies have been reported. The role of corrosion products on MIC of carbon steel has been investigated by gas chromatography-mass spectrometry (GC-MS) (Liu et al. 2000), where S^{2-} and organic acids were found to destroy the protective layer and promote hydrogen permeation. Furthermore, GC-MS-based metabolomics has also been explored as a potential tool in monitoring MIC in copper pipes in water distribution systems (Beale et al. 2010, 2012). It was found that the biofilm metabolites of bacteria causing copper pipe MIC comprised a combination of organic acids, amino acids and lipids. These are common in biological metabolic processes, specifically those relating to soluble monomers and sulphite reducing substrates. In addition, the range of carboxylic acids released from the isolates (*Bosea*, *Methylobacterium*, *Paenibacillus*, *Sphingomonas* and *Variovorax*) suggests that the corrosion potential of these biofilms varies, which would account for localized pitting corrosion commonly observed in metallic pipes (Beale et al. 2014a). Kouremenos et al. (2014) investigated the metabolic profile of *Pseudomonas putida* in potable water exposed to high and low doses of soluble and insoluble iron using LC-TOFMS and identified metabolites that included nucleobases, nucleosides, dipeptides, amino acids, fatty acids, sugars and phospholipids as a response to exposure. While not directly related to MIC, the approach by Kouremenos et al. (2014) and the preliminary work of Beale et al. (2010, 2012) demonstrate the feasibility of GC and LC-based metabolomics techniques to assess microbial populations exposed to metals or isolated from biofilms known to cause MIC in potable water networks. In a study by Booth et al. (2011), the difference in response to metal stress between sessile and planktonic bacterial populations was characterized. The planktonic population had an oxidative stress response to copper ion exposure, whereas the same species within a biofilm environment responded by shifts in exo-polysaccharide-related metabolism. This finding suggests that microbial responses pertinent to corrosion are different between sessile and planktonic populations, and through more research using metabolomic-based techniques, linkages between the metabolite activity of both sessile and planktonic populations and their release of extracellular metabolites from a biofilm can be achieved. While biofilms, in general, and their influence on corrosion, in particular, have been subject to extensive research, there has been limited experimental work performed on the in situ characterization of the organic compounds within biofilms (Beech et al. 2014; Graeber et al. 2014).

With DNA sequencing costs decreasing, omics-based techniques are enabling an increasing number of laboratories to taxonomically and functionally classify a wide range of the organisms that are present in biofilms, and the extension of proteomics and metabolomics techniques enables the assessment of biofilm microbiological communities more broadly (Douterelo et al. 2014). The study of functional genes involved in metabolic pathways is essential when attempting to link microbial diversity with specific ecological functions. In the context of biofilms, better knowledge of the role microorganisms play in MIC and MIC processes such as biofilm formation and corrosion is required and through the application of multi-omics research, such knowledge will be realized.

3.5 Medical and Clinical

Metabolomics studies have been used to investigate the gut microbial population structure associated with a wide range of human diseases (Alam et al. 2014; Garrett 2015; Gevers et al. 2014; Goldman et al. 2014; Ley et al. 2006; Smith et al. 2013). Expression profiling studies have contributed significantly to the understanding of underlying molecular mechanisms of several diseases. In the context of cancer research, this has resulted in the improved classification of tumour subtypes (Karnovsky et al. 2012). However, the lack of early diagnostic markers still remains a problem (Diamandis 2010). Proteomics and metabolomics have the potential to provide additional biological insight for solving this problem (Enjalbert et al. 2011).

The multi-omics approach has been applied to identify markers related to retinoblastoma which is caused by the RB1 gene inactivation. The miRNA pathway analysis, coupled with miRNA microarray indicated a presence of 18 novel miRNAs responsible for the onset of this type of retinal cancer (Guha et al. 2015). In addition to cancer studies, multi-omics approach has been applied to other studies such as asthma-COPD overlap syndrome (Trentacoste et al.), autism disorders, sickle cell anemia and kidney disorders among many (Megan et al. 2016; Zeidan-Chulia et al. 2014; Goodman et al. 2016; Cisek et al. 2015). A very recent research conducting ACOS pattern among asthma patients recorded a trend of increased Immunoglobulin E (IgE) antibody. A differential gene expression of Toll-like receptor 10 (TLR10), an asthma related gene was observed. Also, a further study of single nucleotide polymorphisms (SNPs) indicated a role of another gene IL21R in ACOS. Based on pathway analysis, the pattern was also observed to affect the activities related to cytochrome P450 system (Megan et al. 2016).

A recent review by Higdon et al. (2015) indicated numerous domains (genes, RNA and proteins) related to autism spectrum disorders (ASD) such as fragile-X mental retardation protein (FMRP) and chromatin modifying genes and postsynaptic and embryonic development proteins. Expression analysis models such as linear models for microarray analysis (LIMMA) and significance analysis of microarrays (SAM) have been used to identify differential expressions based on several databases and networks such as KEGG, HMDB, BioCyc, Panther among many others. The former modelling approach (LIMMA) was used to identify Alzheimer's related genes and signal transduction pathways such as NOTCH and Wnt in mitochondrial expressional system of ASD patients (Zeidan-Chulia et al. 2014).

Overall, it has been observed that multi-omics research has been at fairly advanced levels in cancer research, and has made inroads into other clinical researches. It is expected that a rapid growth in those studies will be observed in near future, with a further expansion in other related fields, thereby leading to 'personalized medicine' development.

3.6 Biofuels

In recent years, one of the prominent fields of multi-omics applications is algal origin biofuels. Although, the studies involving these applications have been generally referred to as 'metabolic engineering', they involve more than two approaches (mostly, genomics or transcriptomics and metabolomics). A recent study reported by Trentacoste et al. (2013) was conducted to improve the lipid accumulation in microalga *Thalassiosira pseudonana*. Transcriptomic analysis indicated the role of hydrolase Thaps3_264297. A knocking out of this gene by lipase enzyme overexpression by RNAi and antisense approach resulted in a 3-5 fold increased output of lipids.

Another approach of multi-omics in improving lipid accumulation is by nitrogen starving of algal cells. The responses based on triacylglycerol (TAG) synthesis by *Chlamydomonas reinhardtii* showed a switching on of gluconeogenic phase (≤ 30 min), followed by a transition to glycolytic phase (≥ 4 h). Additionally, a transduction to Carbon–Nitrogen responsive pathways occurred (due to alterations in cw15 proteome), causing increased carbon assimilation and nitrogen metabolism. This led to an increased TAG synthesis in two stages of 'before TAG synthesis (BTS)' and 'after TAG synthesis (ATS)' (Park et al. 2015). Similar approaches have been previously reported for alga mediated high density biofuels (biodiesel fuel) (Beer et al. 2009; Hossain et al. 2008), indicating an estimated 13 % in energy surplus (Dassey et al. 2014).

Similar results to that of algal research mentioned above have also been observed in other organisms regarding biofuel production capabilities of microbial cells. A recent example is that of metabolome and proteome analysis of the yeast *Yarrowia lipolytica*. Nitrogen depletion in this yeast induced an alteration in 133 phosphorylation sites, thereby enriching phosphorylating proteins, causing up-regulation of lipid synthesis (Pomraning et al. 2016).

4 Conclusion

Enhanced knowledge through the application of 'omics'-based techniques provides a holistic opportunity to measure biological systems and the impact of these systems under stress. Furthermore, through the application of multi-omics, researchers are able to obtain a greater depth of knowledge that otherwise would not be achieved through a single 'omics'-based study. Combining metagenomic and transcriptomic data with proteomics and metabolomic datasets will provide researchers and clinicians details on the microbial population present and their metabolic functions. In addition, multi-omics research provides a path for self-validating findings through a combined parallel 'omics' approach, and will ultimately result in the accelerated understanding of these complex populations and processes, driving better diagnostic techniques and drug therapies, environmental remediation strategies and the development of innovative synthetic/engineered biological products.

It is anticipated that the rise of multi-omics research will enable biological systems to be well characterized, enabling model-based assessments to be generated and future research undertaken with only a subset of the data needed for clinical and experimental research, resulting in cost-effective more affordable research (through data mining of available literature and datasets and reducing the amount of experimental data needed). Although it is currently considered to be a nascent field, various potential applications and data integration tools are either currently underway or are expected to appear on the scientific horizon in the near future. It is expected that the exponential development in overall 'omics' knowledge coupled with rapidly developing technology in this field, will assist in the exploration of various multi-omics applications and related scientific discoveries in the fields of environmental, industrial and clinical biology.

Reference

Abram F (2015) Systems-based approaches to unravel multi-species microbial community functioning. Comput Struct Biotechnol J 13:24–32

Alam MZ, Alam Q, Kamal MA, Abuzenadah AM, Haque A (2014) A possible link of gut microbiota alteration in type 2 diabetes and alzheimer's disease pathogenicity: an update. Cns, Neurol Disord-Drug Targets 13:383–390

Albertsen M, Hansen LBS, Saunders AM, Nielsen PH, Nielsen KL (2012) A metagenome of a full-scale microbial community carrying out enhanced biological phosphorus removal. ISME J 6:1094–1106

Barh D, Zambare V, Azevedo V (2013) OMICS: applications in biomedical, agricultural, and environmental sciences. CRC Press, Boca Raton

Beale DJ, Karpe AV, Jadhav S, Muster TH, Palombo EA (2016a) Omics-based approaches and their use in the assessment of microbial-influenced corrosion of metals. Corros Rev 34:1–15

Beale DJ, Barratt R, Marlow DR, Dunn MS, Palombo EA, Morrison PD, Key C (2013) Application of metabolomics to understanding biofilms in water distribution systems: a pilot study. Biofouling 29:283–294

Beale DJ, Dunn MS, Marney D (2010) Application of GC–MS metabolic profiling to 'blue-green water' from microbial influenced corrosion in copper pipes. Corros Sci 52:3140–3145

Beale DJ, Dunn MS, Morrison PD, Porter NA, Marlow DR (2012) Characterisation of bulk water samples from copper pipes undergoing microbially influenced corrosion by diagnostic metabolomic profiling. Corros Sci 55:272–279

Beale DJ, Karpe AV, Mcleod JD, Gondalia SV, Muster TH, Othman MZ, Palombo EA, Joshi D (2016b) An 'omics' approach towards the characterisation of laboratory scale anaerobic digesters treating municipal sewage sludge. Water Res 88:346–357

Beale DJ, Morrison PD, Key C, Palombo EA (2014a) Metabolic profiling of biofilm bacteria known to cause microbial influenced corrosion. Water Sci Technol 69:1–8

Beale DJ, Morrison PD, Palombo EA (2014b) Detection of Listeria in milk using non-targeted metabolic profiling of *Listeria monocytogenes*: a proof-of-concept application. Food Control 42:343–346

Beech IB, Gaylarde CC (1999) Recent advances in the study of biocorrosion—an overview. Rev De Microbiol 30:177–190

Beech IB, Sztyler M, Gaylarde CC, Smith WL, Sunner J (2014) 2—Biofilms and biocorrosion. In: Liengen T, Féron D, Basséguy R, Beech IB (eds) Understanding biocorrosion. Woodhead Publishing, Cambridge

Beer LL, Boyd ES, Peters JW, Posewitz MC (2009) Engineering algae for biohydrogen and biofuel production. Curr Opin Biotechnol 20:264–271

Blanchet L, Smolinska A (2016) Data fusion in metabolomics and proteomics for biomarker Discovery. In: Jung K (ed) Statistical analysis in proteomics. Springer, New York

Booth SC, Workentine ML, WEN J, Shaykhutdinov R, Vogel HJ, Ceri H, Turner RJ, Weljie AM (2011) Differences in metabolism between the biofilm and planktonic response to metal stress. J Proteome Res 10:3190–3199

Chakraborty A, Dasgupta CK, Bhadury P (2014) 4—application of molecular techniques for the assessment of microbial communities in contaminated sites. In: Das S (ed) Microbial Biodegradation and bioremediation. Elsevier, Amsterdam

Chao Y, Ma L, Yang Y, JU F, Zhang XX, Wu WM, Zhang T (2013) Metagenomic analysis reveals significant changes of microbial compositions and protective functions during drinking water treatment. Scientific Reports, vol 3

Chariton AA, Ho KT, Proestou D, Bik H, simpson SL, Portis LM, Cantwell MG, Baguley JG, Burgess RM, Pelletier MM, Perron M, Gunsch C, Matthews RA (2014) A molecular-based approach for examining responses of eukaryotes in microcosms to contaminant-spiked estuarine sediments. Environ Toxicol Chem 33:359–369

Cisek K, Krochmal M, Klein J, Mischak H (2015) The application of multi-omics and systems biology to identify therapeutic targets in chronic kidney disease. Nephrol Dial Transplant gfv364

Cohen A, Bont L, Engelhard D, Moore E, Fernandez D, Kreisberg-Greenblatt R, Oved K, Eden E, Hays JP (2015) A multifaceted 'omics' approach for addressing the challenge of antimicrobial resistance. Future Microbiol 10:365–376

Dassey AJ, Hall SG, Theegala CS (2014) An analysis of energy consumption for algal biodiesel production: comparing the literature with current estimates. Algal Res 4:89–95

Diamandis EP (2010) Cancer biomarkers: can we turn recent failures into success? JNCI J Natl Cancer Inst 102:1462–1467

Douterelo I, Boxall JB, Deines P, Sekar R, Fish KE, Biggs CA (2014) Methodological approaches for studying the microbial ecology of drinking water distribution systems. Water Res 65:134–156

Edyvean RGJ, Videla HA (1991) Biological corrosion. Interdis Sci Rev 16:267–282

El-Serehy HA, AL-Rasheid KA, AL-Misned FA, AL-Talasat AAR, Gewik MM (2016) Microbial–meiofaunal interrelationships in coastal sediments of the Red Sea. Saudi J Biol Sci 23:327–334

Enjalbert B, Jourdan F, Portais JC (2011) Intuitive visualization and analysis of multi-omics data and application to *Escherichia coli* carbon metabolism. PLoS ONE 6:e21318

Erickson AR, Cantarel BL, Lamendella R, Darzi Y, Mongodin EF, Pan C, Shah M, Halfvarson J, Tysk C, Henrissat B, Raes J, Verberkmoes NC, Fraser CM, Hettich RL, Jansson JK (2012) Integrated metagenomics/metaproteomics reveals human host-microbiota signatures of crohn's disease. PLoS ONE 7:e49138

Field D, Garrity G, Gray T, Morrison N, Selengut J, Sterk P, Tatusova T, Thomson N, Allen MJ, Angiuoli SV, Ashburner M, Axelrod N, Baldauf S, Ballard S, Boore J, Cochrane G, COLE J, Dawyndt P, De Vos P, Depamphilis C, Edwards R, Faruque N, Feldman R, Gilbert J, Gilna P, Glockner FO, Goldstein P, Guralnick R, Haft D, Hancock D, Hermjakob H, Hertz-Fowler C, Hugenholtz P, Joint I, Kagan L, Kane M, Kennedy J, Kowalchuk G, Kottmann R, Kolker E, Kravitz S, Kyrpides N, Leebens-Mack J, Lewis SE, Li K, Lister AL, Lord P, Maltsev N, Markowitz V, Martiny J, Methe B, Mizrachi I, Moxon R, Nelson K, Parkhill J, Proctor L, White O, Sansone SA, Spiers A, Stevens R, Swift P, Taylor C, Tateno Y, Tett A, Turner S, Ussery D, Vaughan B, Ward N, Whetzel T, San Gil I, Wilson G, Wipat A (2008) The minimum information about a genome sequence (MIGS) specification. Nat Biotechnol 26: 541–547

Flemming HC (1994) Microbial deterioration of materials—fundamentals—economical and technical overview. Werkst Korros-Mater Corros 45:5–9

Fondi M, Liò P (2015) Multi -omics and metabolic modelling pipelines: challenges and tools for systems microbiology. Microbiol Res 171:52–64

Forsythe IJ, Wishart DS (2009) Exploring human metabolites using the human metabolome database. In: Current protocols in bioinformatics, Chap. 14, Unit 14.8

Garcia-Alcalde F, Garcia-Lopez F, Dopazo J, Conesa A (2011) Paintomics: a web based tool for the joint visualization of transcriptomics and metabolomics data. Bioinformatics 27:137–139

Garrett WS (2015) Cancer and the microbiota. Science 348:80–86

Gevers D, Kugathasan S, Denson LA, Vazquez-Baeza Y, Van Treuren W, Ren B, Schwager E, Knights D, Song SJ, Yassour M, Morgan XC, Kostic AD, Luo C, Gonzalez A, Mcdonald D, Haberman Y, Walters T, Baker S, Rosh J, Stephens M, Heyman M, Markowitz J, Baldassano R, Griffiths A, Sylvester F, Mack D, Kim S, Crandall W, Hyams J, Huttenhower C, Knight R, Xavier RJ (2014) The treatment-naive microbiome in new-onset Crohn's disease. Cell Host Microbe 15:382–392

Goldman SM, Kamel F, Ross GW, Jewell SA, Marras C, Hoppin JA, Umbach DM, Bhudhikanok GS, Meng C, Korell M, Comyns K, Hauser RA, Jankovic J, Factor SA, Bressman S, Lyons KE, Sandler DP, Langston JW, Tanner CM (2014) Peptidoglycan recognition protein genes and risk of parkinson's disease. Mov Disord 29:1171–1180

Gomez-Alvarez V, Revetta R, Domingo JW (2012) Metagenome analyses of corroded concrete wastewater pipe biofilms reveal a complex microbial system. BMC Microbiol 12:122

Goodman SR, Pace BS, Hansen KC, D'alessandro A, Xia Y, Daescu O, Glatt SJ (2016) Minireview: multiomic candidate biomarkers for clinical manifestations of sickle cell severity: early steps to precision medicine. Exp Biol Med 241:772–781

Graeber M, Boehm S, Kuever J (2014) 3—molecular methods for studying biocorrosion. In: Liengen T, Féron D, Basséguy R, Beech IB (eds) Understanding biocorrosion. Woodhead Publishing, Cambridge

Grimplet J, Cramer GR, Dickerson JA, Mathiason K, Van Hemert J, Fennell AY (2009) VitisNet: "Omics" integration through grapevine molecular networks. PLoS ONE 4:e8365

Guha N, Deepak SA, Lateef S, Padmanabhan A, Gundimeda S, Ghosh A, Mallipatna A, Suresh Babu V (2015) A multi-omics approach to identify biomarkers of clinically advanced retinoblastoma for diagnostics and therapeutic applications. FASEB J 29:417.7

Günther OP, Shin H, Ng RT, Mcmaster WR, Mcmanus BM, Keown PA, Tebbutt SJ, lê Cao KA (2014) Novel multivariate methods for integration of genomics and proteomics data: applications in a kidney transplant rejection study. OMICS: J Integr Biol 18:682–695

Handelsman J (2004) Metagenomics: application of genomics to uncultured microorganisms. Microbiol Mol Biol Rev 68:669–685

Heinrichs G, Hübner I, Schmidt CK, De hoog GS, Haase G (2013) Analysis of black fungal biofilms occurring at domestic water taps (I): compositional analysis using tag-encoded flx amplicon pyrosequencing. Mycopathologia 175:387–397

Higdon R, Earl RK, Stanberry L, Hudac CM, Montague E, Stewart E, Janko I, Choiniere J, Broomall W, Kolker N, Bernier RA, Kolker E (2015) The promise of multi-omics and clinical data integration to identify and target personalized healthcare approaches in autism spectrum disorders. OMICS 19:197–208

Hook SE, Osborn HL, Spadaro DA, Simpson SL (2014) Assessing mechanisms of toxicant response in the amphipod *Melita plumulosa* through transcriptomic profiling. Aquat Toxicol 146:247–257

Hossain A, Salleh A, Boyce AN, Chowdhury P, Naqiuddin M (2008) Biodiesel fuel production from algae as renewable energy. Am J Biochem Biotechnol 4:250–254

Hultman J, Waldrop MP, Mackelprang R, David MM, Mcfarland J, Blazewicz SJ, Harden J, Turetsky MR, Mcguire AD, Shah MB, Verberkmoes NC, Lee LH, Mavrommatis K, Jansson JK (2015) Multi-omics of permafrost, active layer and thermokarst bog soil microbiomes. Nature 521:208–212

Jadhav S, Gulati V, Fox EM, Karpe A, Beale DJ, Sevior D, Bhave M, Palombo EA (2015) Rapid identification and source-tracking of Listeria monocytogenes using MALDI-TOF mass spectrometry. Int J Food Microbiol 202:1–9

Jadhav S, Sevior D, Bhave M, Palombo EA (2014) Detection of Listeria monocytogenes from selective enrichment broth using MALDI–TOF Mass Spectrometry. J Proteomics 97:100–106

Jansson J, Willing B, Lucio M, Fekete A, Dicksved J, Halfvarson J, Tysk C, Schmitt-Kopplin P (2009) Metabolomics reveals metabolic biomarkers of crohn's disease. PLoS ONE 4:e6386

Jones OAH, Sdepanian S, Lofts S, Svendsen C, Spurgeon DJ, Maguire ML, Griffin JL (2014) Metabolomic analysis of soil communities can be used for pollution assessment. Environ Toxicol Chem 33:61–64

Jozefczuk S, Klie S, Catchpole G, Szymanski J, Cuadros-Inostroza A, Steinhauser D, Selbig J, Willmitzer L (2010) Metabolomic and transcriptomic stress response of *Escherichia coli*. Mol Syst Biol 6:364

Junker BH, Klukas C, Schreiber F (2006) Vanted: a system for advanced data analysis and visualization in the context of biological networks. BMC Bioinf 7:1–13

Kaever A, Landesfeind M, Feussner K, Mosblech A, Heilmann I, Morgenstern B, Feussner I, Meinicke P (2015) MarVis-pathway: integrative and exploratory pathway analysis of non-targeted metabolomics data. Metabolomics 11:764–777

Kamburov A, Cavill R, Ebbels TM, Herwig R, Keun HC (2011) Integrated pathway-level analysis of transcriptomics and metabolomics data with IMPaLA. Bioinformatics 27:2917–2918

Karnovsky A, Weymouth T, Hull T, Tarcea VG, Scardoni G, Laudanna C, Sartor MA, Stringer KA, Jagadish HV, Burant C, Athey B, Omenn GS (2012) Metscape 2 bioinformatics tool for the analysis and visualization of metabolomics and gene expression data. Bioinformatics 28:373–380

Karsenti E, Acinas SG, Bork P, Bowler C, Vargas CD, Raes J, Sullivan M, Arendt D, Benzoni F, Claverie JM, Follows M, Gorsky G, Hingamp P, Iudicone D, Jaillon O, Kandels-Lewis S, Krzic U, Not F, Ogata H, Pesant S, Reynaud EG, Sardet C, Sieracki ME, Speich S, Velayoudon D, Weissenbach J, Wincker PW, Consortium TT (2011) A holistic approach to marine eco-systems biology. PLoS Biol 9:e1001177

Kato H, Takahashi S, Saito K (2011) Omics and integrated omics for the promotion of food and nutrition science. J Tradit Complement Med 1:25–30

Kimes NE, Callaghan AV, Aktas DF, Smith WL, Sunner J, Golding BT, Drozdowska M, Hazen TC, Suflita JM, Morris PJ (2013) Metagenomic analysis and metabolite profiling of deep-sea sediments from the gulf of mexico following the deepwater horizon oil spill. Front Microbiol 4:50

King ZA, Dräger A, Ebrahim A, Sonnenschein N, Lewis NE, Palsson BO (2015) Escher: a web application for building, sharing, and embedding data-rich visualizations of biological pathways. PLoS Comput Biol 11:e1004321

Kip N, Van Veen JA (2015) The dual role of microbes in corrosion. ISME J 9:542–551
Kohl M, Megger DA, Trippler M, Meckel H, Ahrens M, Bracht T, Weber F, Hoffmann AC, Baba HA, Sitek B, Schlaak JF, Meyer HE, Stephan C, Eisenacher M (2014) A practical data processing workflow for multi-OMICS projects. Biochim Biophys Acta 1844:52–62
Kouremenos KA, Beale DJ, Antti H, Palombo EA (2014) Liquid chromatography time of flight mass spectrometry based environmental metabolomics for the analysis of *Pseudomonas putida* bacteria in potable water. J Chromatogr B 966:179–186
Kovács E, Wirth R, Maróti G, Bagi Z, Nagy K, Minárovits J, Rákhely G, Kovács KL (2015) Augmented biogas production from protein-rich substrates and associated metagenomic changes. Bioresour Technol 178:254–261
Krämer A, Green J, Pollard J, Tugendreich S (2013) Causal analysis approaches in ingenuity pathway analysis. Bioinform 30:523–530
Kuo TC, Tian TF, Tseng YJ (2013) 3Omics: a web-based systems biology tool for analysis, integration and visualization of human transcriptomic, proteomic and metabolomic data. BMC Syst Biol 7:64
Kutmon M, Van Iersel MP, Bohler A, Kelder T, Nunes N, Pico AR, Evelo CT (2015) PathVisio 3: an extendable pathway analysis toolbox. PLoS Comput Biol 11:e1004085
Leary DH, Li RW, Hamdan LJ, Hervey WJ, Lebedev N, Wang Z, Deschamps JR, Kusterbeck AW, Vora GJ (2014) Integrated metagenomic and metaproteomic analyses of marine biofilm communities. Biofouling 30:1211–1223
Ley RE, Turnbaugh PJ, Klein S, Gordon JI (2006) Microbial ecology—human gut microbes associated with obesity. Nature 444:1022–1023
Lin K, Kools H, De Groot PJ, Gavai AK, Basnet RK, Cheng F, Wu J, Wang X, Lommen A, Hooiveld GJ, Bonnema G, Visser RG, Muller MR, Leunissen JA (2011) Madmax—MANAGEMENT and analysis database for multiple ~omics experiments. J Integr Bioinform 8:160
Liu H, Xu L, Zeng J (2000) Role of corrosion products in biofilms in microbiologically induced corrosion of carbon steel. Br Corros J 35:131–135
Llewellyn CA, Sommer U, Dupont CL, Allen AE, Viant MR (2015) Using community metabolomics as a new approach to discriminate marine microbial particulate organic matter in the western English Channel. Prog Oceanogr 137 Part B:421–433
Lourenço A, Ferreira A, Veiga N, Machado I, Pereira MO, Azevedo NF (2012) BiofOmics: a web platform for the systematic and standardized collection of high-throughput biofilm data. PLoS One 7:e39960
Ma X, Zheng X, Zhang M, Yang X, Chen L, Huang L, Cao S (2014) Electron beam irradiation of bamboo chips: degradation of cellulose and hemicelluloses. Cellulose 21:3865–3870
Machado D, Herrgård M (2014) Systematic evaluation of methods for integration of transcriptomic data into constraint-based models of metabolism. PLoS Comput Biol 10:e1003580
Megan EH, Michael HC, Merry-Lynn M, Peter JC, Ruth TS, Diego JM, Barry JM, James DC, Russell PB, Edwin KS, Craig PH (2016) A multi-Omics Approach Identifies Inflammatory Patterns In The asthma-COPD overlap syndrome. D91. Genetics and Genomics of COPD. American Thoracic Society
Mohan AM, Bibby KJ, Lipus D, Hammack RW, Gregory KB (2014) The functional potential of microbial communities in hydraulic fracturing source water and produced water from natural gas extraction characterized by metagenomic sequencing. PLoS ONE 9:e107682
Morrison N, Wood AJ, Hancock D, Shah S, Hakes L, Gray T, Tiwari B, Kille P, Cossins A, Hegarty M, Allen MJ, Wilson WH, Olive P, Last K, Kramer C, Bailhache T, Reeves J, Pallett D, Warne J, Nashar K, Parkinson H, Sansone SA, Rocca-Serra P, Stevens R, Snape J, Brass A, Field D (2006) Annotation of environmental OMICS data: application to the transcriptomics domain. OMICS 10:172–178
Neria-González I, Wang ET, Ramírez F, Romero JM, Hernández-Rodríguez C (2006) Characterization of bacterial community associated to biofilms of corroded oil pipelines from the southeast of Mexico. Anaerobe 12:122–133

Neuweger H, Persicke M, Albaum SP, Bekel T, Dondrup M, Huser AT, Winnebald J, Schneider J, Kalinowski J, Goesmann A (2009) Visualizing post genomics data-sets on customized pathway maps by ProMeTra-aeration-dependent gene expression and metabolism of *Corynebacterium glutamicum* as an example. BMC Syst Biol 3:82

Noecker C, Eng A, Srinivasan S, Theriot CM, Young VB, Jansson JK, Fredricks DN, Borenstein E (2016) Metabolic model-based integration of microbiome taxonomic and metabolomic profiles elucidates mechanistic links between ecological and metabolic variation. mSystems 1: e00013-15

Nozhevnikova AN, Botchkova EA, Plakunov VK (2015) Multi-species biofilms in ecology, medicine, and biotechnology. Microbiology 84:731–750

Orchard S, Kerrien S (2010) Molecular interactions and data standardisation. Methods Mol Biol 604:309–318

Osborn HL, Hook SE (2013) Using transcriptomic profiles in the diatom *Phaeodactylum tricornutum* to identify and prioritize stressors. Aquat Toxicol 138–139:12–25

Pap B, Györkei Á, Boboescu IZ, Nagy IK, Bíró T, Kondorosi É, Maróti G (2015) Temperature-dependent transformation of biogas-producing microbial communities points to the increased importance of hydrogenotrophic methanogenesis under thermophilic operation. Bioresour Technol 177:375–380

Park JJ, Wang H, Gargouri M, Deshpande RR, Skepper JN, Holguin FO, Juergens MT, Shachar-Hill Y, Hicks LM, Gang DR (2015) The response of *Chlamydomonas reinhardtii* to nitrogen deprivation: a systems biology analysis. Plant J 81:611–624

Pascault N, Loux V, Derozier S, Martin V, Debroas D, Maloufi S, Humbert J-F, Leloup J (2015) Technical challenges in metatranscriptomic studies applied to the bacterial communities of freshwater ecosystems. Genetica 143:157–167

Pertea M, Salzberg SL (2010) Between a chicken and a grape: estimating the number of human genes. Genome Biol 11:206

Pomraning KR, Kim Y-M, Nicora CD, Chu RK, Bredeweg EL, Purvine SO, Hu D, Metz TO, Baker SE (2016) Multi-omics analysis reveals regulators of the response to nitrogen limitation in *Yarrowia lipolytica*. BMC Genom 17:1–18

Röling WFM, Ferrer M, Golyshin PN (2010) Systems approaches to microbial communities and their functioning. Curr Opin Biotechnol 21:532–538

Saito K, Arai S, Kato H (2005) A nutrigenomics database–integrated repository for publications and associated microarray data in nutrigenomics research. Br J Nutr 94:493–495

Sansone SA, Fan T, Goodacre R, Griffin JL, Hardy NW, Kaddurah-Daouk R, Kristal BS, Lindon J, Mendes P, Morrison N, Nikolau B, Robertson D, Sumner LW, Taylor C, Van Der Werf M, Van Ommen B, Fiehn O (2007) The metabolomics standards initiative. Nat Biotechnol 25:846–848

Santo Domingo JW, Revetta RP, Iker B, Gomez-Alvarez V, Garcia J, Sullivan J, Weast J (2011) Molecular survey of concrete sewer biofilm microbial communities. Biofouling 27:993–1001

Schneider MV, Orchard S (2011) Omics technologies, data and bioinformatics principles. In: Mayer B (ed) Bioinformatics for omics data: methods and protocols. Humana Press, New York

Schneider T, Riedel K (2010) Environmental proteomics: analysis of structure and function of microbial communities. Proteomics 10:785–798

Shaw JLA, Monis P, Fabris R, Ho L, Braun K, Drikas M, Cooper A (2014) Assessing the impact of water treatment on bacterial biofilms in drinking water distribution systems using high-throughput DNA sequencing. Chemosphere 117:185–192

Singer SW (2012) Targeted isolation of proteins from natural microbial communities living in an extreme environment. Methods Mol Biol 881:63–72

Smith PM, Howitt MR, Panikov N, Michaud M, Gallini CA, Bohlooly Y, Glickman JN, Garrett WS (2013) The microbial metabolites, short-chain fatty acids, regulate colonic treg cell homeostasis. Science 341:569–573

Srinivasan S, Morgan MT, Fiedler TL, Djukovic D, Hoffman NG, Raftery D, Marrazzo JM, Fredricks DN (2015) Metabolic signatures of bacterial vaginosis. mBio 6:e00204–e00215

Stasiewicz MJ, Den Bakker HC, Wiedmann M (2015) Genomics tools in microbial food safety. Curr Opin Food Sci 4:105–110

Sumner LW, Amberg A, Barrett D, Beale MH, Beger R, Daykin CA, Fan TWM, Fiehn O, Goodacre R, Griffin JL, Hankemeier T, Hardy N, Harnly J, Higashi R, Kopka J, Lane AN, Lindon JC, Marriott P, Nicholls AW, Reily MD, Thaden JJ, Viant MR (2007) Proposed minimum reporting standards for chemical analysis. Metabolomics 3:211–221

Sunagar K, Morgenstern D, Reitzel AM, Moran Y (2016) Ecological venomics: how genomics, transcriptomics and proteomics can shed new light on the ecology and evolution of venom. J Proteomics 135:62–72

Takahashi S, Saito K, Jia H, Kato H (2014) An integrated multi-omics study revealed metabolic alterations underlying the effects of coffee consumption. PLoS One 9:e91134

Talbot G, Topp E, Palin MF, Massé DI (2008) Evaluation of molecular methods used for establishing the interactions and functions of microorganisms in anaerobic bioreactors. Water Res 42:513–537

Thakur NL, Jain R, Natalio F, Hamer B, Thakur AN, Müller WEG (2008) Marine molecular biology: an emerging field of biological sciences. Biotechnol Adv 26:233–245

Theriot CM, Koenigsknecht MJ, Carlson JRPE, Hatton GE, Nelson AM, Li B, Huffnagle GBZ, Li J, Young VB (2014) Antibiotic-induced shifts in the mouse gut microbiome and metabolome increase susceptibility to *Clostridium difficile* infection. Nat Commun 5

Thimm O, Blasing O, Gibon Y, Nagel A, Meyer S, Kruger P, Selbig J, Muller LA, Rhee SY, Stitt M (2004) Mapman: a user-driven tool to display genomics data sets onto diagrams of metabolic pathways and other biological processes. Plant J 37:914–939

Tokimatsu T, Sakurai N, Suzuki H, Ohta H, Nishitani K, Koyama T, Umezawa T, Misawa N, Saito K, Shibata D (2005a) KaPPA-view: a web-based analysis tool for integration of transcript and metabolite data on plant metabolic pathway maps. Plant Physiol 138:1289–1300

Tokimatsu T, Sakurai N, Suzuki H, Ohta H, Nishitani K, Koyama T, Umezawa T, Misawa N, Saito K, Shibatanenell D (2005b) KaPPA-view. a web-based analysis tool for integration of transcript and metabolite data on plant metabolic pathway maps. Plant Physiol 138:1289–1300

Trentacoste EM, Shrestha RP, Smith SR, glé C, Hartmann AC, Hildebrand M, Gerwick WH (2013) Metabolic engineering of lipid catabolism increases microalgal lipid accumulation without compromising growth. Proc Natl Acad Sci USA 110:19748–19753

Turnbaugh PJ, Gordon JI (2008) An invitation to the marriage of metagenomics and metabolomics. Cell 134:708–713

Usadel B, Nagel A, Thimm O, Redestig H, Blaesing OE, Palacios-Rojas N, Selbig J, Hannemann J, Piques MC, Steinhauser D, Scheible WR, Gibon Y, Morcuende R, Weicht D, Meyer S, Stitt M (2005) Extension of the visualization tool mapman to allow statistical analysis of arrays, display of corresponding genes, and comparison with known responses. Plant Physiol 138:1195–1204

Valdés A, Ibáñez C, Simó C, García-Cañas V (2013) Recent transcriptomics advances and emerging applications in food science. TrAC Trends Anal Chem 52:142–154

Vanwonterghem I, Jensen PD, Ho DP, Batstone DJ, Tyson GW (2014) Linking microbial community structure, interactions and function in anaerobic digesters using new molecular techniques. Curr Opin Biotechnol 27:55–64

Videla HA (2003) Biocorrosion and biofouling of metals and alloys of industrial usage. Present state of the art at the beginning of the new millennium. Rev Metal 34:256–264

Videla HA, Herrera LK (2009) Understanding microbial inhibition of corrosion. a comprehensive overview. Int Biodeterior Biodegradation 63:896–900

Vincke E, Boon N, Verstraete W (2001) Analysis of the microbial communities on corroded concrete sewer pipes—a case study. Appl Microbiol Biotechnol 57:776–785

Wagele B, Witting M, Schmitt-Kopplin P, Suhre K (2012) Mass TRIX reloaded: combined analysis and visualization of transcriptome and metabolome data. PLoS One 7:e39860

Wheelock AM, Wheelock CE (2013) Trials and tribulations of 'omics data analysis: assessing quality of SIMCA-based multivariate models using examples from pulmonary medicine. Mol Bio Syst 9:2589–2596

Wildburger NC, Lichti CF, Leduc RD, Schmidt M, Kroes RA, Moskal JR, Nilsson CL (2015) Quantitative proteomics and transcriptomics reveals metabolic differences in attracting and non-attracting human-in-mouse glioma stem cell xenografts and stromal cells. EuPA Open Proteomics 8:94–103

Xia J, Fjell CD, Mayer ML, Pena OM, Wishart DS, Hancock RE (2013) Inmex–a web-based tool for integrative meta-analysis of expression data. Nucleic Acids Res 41:W63–W70

Xia T, Hemert JV, Dickerson JA (2010) OmicsAnalyzer: a Cytoscape plug-in suite for modeling omics data. Bioinformatics 26:2995–2996

Xu Y, Dou D, Ran X, Liu C, Chen J (2015) Integrative analysis of proteomics and metabolomics of anaphylactoid reaction induced by Xuesaitong injection. J Chromatogr A 1416:103–111

Yanagawa K, kouduka M, Nakamura Y, Hachikubo A, Tomaru H, Suzuki Y (2014) Distinct microbial communities thriving in gas hydrate-associated sediments from the eastern Japan Sea. J Asian Earth Sci 90:243–249

Yizhak K, Benyamini T, Liebermeister W, Ruppin E, Shlomi T (2010) Integrating quantitative proteomics and metabolomics with a genome-scale metabolic network model. Bioinformatics 26:i255–i260

Zeidan-Chulia F, De Oliveira B-H, Salmina A, Casanova M, Gelain D, Noda M, Verkhratsky A, Moreira JC (2014) Altered expression of Alzheimer's disease-related genes in the cerebellum of autistic patients: a model for disrupted brain connectome and therapy. Cell Death Dis 5: e1250

Zhang W, Li F, Nie L (2010) Integrating multiple 'omics' analysis for microbial biology: application and methodologies. Microbiology 156:287–301

Zhang W, Wang Y, Lee OO, Tian R, Cao H, Gao Z, Li Y, Yu L, Xu Y, Qian PY (2013) Adaptation of intertidal biofilm communities is driven by metal ion and oxidative stresses. Scientific reports, vol 3

Index

Note: Page numbers followed by *f* and *t* indicate figures and tables respectively

A

ABC transmembrane proteins (ATP-binding cassette), 6
Abiotic stress adaptation, 167
Adipose tissue form (ATF), 92
Adulteration of food material, 231, 245–246
After TAG synthesis (ATS), 304
Akkermansia muciniphila, 24, 28, 58
Alkylphosphocoline, 95
Aminosugars, 93
Ammonia fibre explosion (AFEX), 264
Anaerobic digesters, 300
Anaerobic digestion, 209, 210
Anaerobic reactors, 210
Analytical instruments in metabolomics, advancements of, 234–239
Anthropogenic effects on soil microbes, 174–175
  climatic change effects, 181–183
  engineered nanomaterials (ENM), 175–177
  heavy metal contamination, 177–179
  organic contaminants, 179–181
Antibiotics, 27, 67, 162, 167, 173, 180–181, 241, 243
Antimonials, pentavalent and trivalent, 94
Arabidopsis–Pseudomonas rhizosphere model, 164
Arabidopsis thaliana, 93, 164–165, 167, 169
Arbuscular mycorrhizal fungi (AMF), 161, 163, 170–171, 178, 180, 181–182
Arginine deprivation response (ADR), 91
Aspergillus nidulans, 274
Aspergillus oryzae, 265
Aspergillus spp., 265, 267
Atmospheric pressure ionisation (API), 236
Autism spectrum disorders (ASD), 14, 28, 303

B

Bacillus subtilis, 134, 174, 181
Bacteria, 149–150
  bacterial soil inoculants, 166–167
  and metal toxicity, 204
    biofilm versus planktonic response to copper, 208
    investigations of biofilms and morphological variants, 207–208
    methodological considerations and conclusions, 209
    tellurite resistance, 205–206, 205*f*
Bacteroidetes, 14
Bayesian probabilistic approach, 214
Before TAG synthesis (BTS), 304
β-oxidation inhibitors, 97
Bifidobacteria, 16
Bifidobacterium, 24, 28
Biodefense and Emerging Infections Research Resources Repository (BEI Resources), 61
Biofertilizers, 166
Biofilms, 162, 178, 301–303
  and morphological variants, 207–28
  versus planktonic response to copper, 208
Biofomics, 294*t*, 301
Biofuels, 304
Biomarkers, 233, 240, 242–243, 245
Biomass waste management, microbial metabolomics in, 261, 263–265
  biomass degradation/conversion methods, 265–267
  future research, 280–281
  metabolic engineering approaches, 276
    to develop fungal bioprocessing abilities, 277–280
  microbial process metabolomics and its application to biomass, 268
    classification of metabolomic approaches, 268–269
    metabolic fingerprinting, 272–275
    metabolic flux (fluxomics), 275–276

D.J. Beale et al. (eds.), *Microbial Metabolomics*,
DOI 10.1007/978-3-319-46326-1

metabolic profiling: targeted and untargeted approaches, 269–272
Biopesticides, 180–181
Bodonidae, 83
Botrytis allii, 242
Burkholderia cepacia, 242
Butyrivibrio fibrisolvens, 270

C

^{13}C-acetate, 131, 159
Cadmium contamination, 178–179
^{13}C-amino acids, 99, 101
cAMP, 33
Capillary electrophoresis and mass spectrometry (CE-MS), 125, 236, 237
Carbohydrates, dietary, 21–22
Carbon fullerenes and nanotubes, 176
^{13}C-cellulose, 131, 132
Cellulomonas flavigena, 264
^{13}C-fatty acids, 99, 101
^{13}C-glucose, 101, 131, 276
$^{13}CH_4$, 131
Challenges, for microbial metabolomics, 1–2, 5–6
Chemical contaminants and xenobiotic molecules, 229
Chemical food safety, 240–241
Chicory, 20
$^{13}CH_3OH$, 132
^{13}C-labeling approaches, 103–104
^{13}C-labelled carbon sources, 203
^{13}C-labelled glucose, 183
Claudin-1, 31
^{13}C-lignin, 131
Climatic change effects, on soil microbes, 181–183
Clostridium cellulolyticum, 267
Clostridium difficile VPI 10463, 59
Clostridium proteoclasticum, 270
Clostridium thermocellum, 265, 267, 271
Clostridium thermohydrosulfuricum, 265
^{13}C-naphthalene, 132
Codes of good practices (CGP), 227
Codex Alimentarius Commission (CAC), 226
Colorectal cancer (CRC), 16–17, 29, 49, 50
Community metabolomics, 199, 201, 204
 bacteria and metal toxicity, 204
  biofilm versus planktonic response to copper, 208
  investigations of biofilms and morphological variants, 207–208
  methodological considerations and conclusions, 209
  tellurite resistance, 205–206, 205*f*
 potential industrial uses of, 209
  anaerobic waste conversion applications, 209–211
  ecosystems biology, integrated approach for, 213–214
  state of the science, 211–213
 soil science, 201–204
 upcoming challenges, 215–216
  data storage and sharing, 217–218
  knowledge transfer to applied research outcomes, 218–219
  meta-data, importance of, 217
  need for better reference samples, 216
Consolidated bioprocessing (CBP), 266, 267, 277
Corrosion, 301–303
Corynebacterium glutamicum, 279
Coxiella brunetti, 87
Crenarchaeota, 212
^{13}C-ribose, 131
CRISPR (clustered regularly interspaced short palindromic repeat)-Cas9 system, 93
Critical control points (CCPs), 227
Crohn's disease (CD), 31
Cryptococcus neoformans, 87
Cultivation-based multiplex phenotyping, 59
Culture-dependent analysis of gut microbiota, 57–61
Culturomics, 59–60
Cutaneous infections (CL), 83–84
Cutting-edge applications, 7
 assessing metabolic state of microbial community, 7
 future perspective, 8
 host–microbe interactions, 7
 new diagnosis of microbial infection through metabolomics, 7–8
Cutting-edge technologies, 1
Cytoscape with MODAM and Cytoscape with OmicsAnalyzer, 295*t*

D

Data integration, 138
Data storage and sharing, 217–218
Defensins, 15
Desorption electrospray analysis (DESI), 241
Deuterium oxide, 103
Deutsche Sammlung von Mikroorganismen und Zellkulturen (DSMZ) GmbH, 61
Dietary fibre, 19–23, 34
 and gut health, 16–19
 and short chain fatty acids, 26–29
 interaction with gut microbiota, 23–26
Dietary fibre hypothesis, 17

Diffuse cutaneous infections (DCL), 83, 84
Direct analysis in real time (DART), 241
Direct infusion (DI), 236
Direct infusion-ESI-MS (DI-ESI-MS), 280
Direct insertion probe/electron ionisation/ion trap detection (DIP/EI/ITD) mass spectrometry, 274
Dissolved organic matter (DOM), 122, 125
DNA sequencing, 56
D_2O, 103
Drug resistance, 88, 94

E
EcoGenomics, 215
Ecological intensification, 147
Eco-metabolomics, 157
Ecosystems biology, 213–214
  integrated approach for, 213–214
Electrospray ionization (ESI), 125
Embden-Meyerhof-Parnas pathway, 279
Engineered nanomaterials (ENM), 175–177
Enteroids/colonoids, 63
Enteropathogenic microbes, 28
Environmental contaminants, 293–298
Environmental metabolomics, 216
Era pre-metabolome, 3*t*
Escher, 294*t*
Escherichia coli, 2, 28, 230
  Metabolome Database (ECMDB), 4
  O157:H7, 15
  transconjugants, 68
Essential oil (EO), 166
Euglena mutabilis, 128–129
European Prospective Investigation into Cancer and Nutrition (EPIC), 16
Exometabolomics, 119
  for analysis of
    soil organic matter, 120–126
    whole microbial communities, 126–128
  characterizing metabolism of individual members, 128–131
  future outlook, 136–138
  metabolite imaging, 133–136
  stable isotope probing (SIP) techniques, 131–133
Exopolysaccharides, 162
Expression analysis models, 304
Extracellular enzymes (EEs), 173–174
Extracellular polymeric substance (EPS), 210
Exudates, 162
  rhizoengineering, 163–164

F
Faecalibacterium prausnitzii, 66, 69–71
Faecal microbial transplantation (FMT), 28
Fast food, 225
Fatty acid methyl esters (FAMEs), 177
Fatty acid β-oxidation, 89
Fermentable Oligo- Di- and Monosaccharides And Polyols (FODMAP), 24
Firmicutes phylum, 14
FISH-microautoradiography, 134
Flavonoids, 161
Fluorescence in situ hybridisation (FISH), 212
Fluxomics, 276–277
Food adulteration, 231, 245–246
Food and nutrition, 298
Food-borne illness, 230
Food-borne pathogens, 231
Food metabolome, workflow for analysis of, 235
Food metabolomics, 233, 233–234
Foodomics, 233
Food safety and current practices, 226–229
Food safety and quality research, 225, 229–230
  application of metabolomics in, 240
    chemical food safety, 240–241
    food pathogens and food spoilage microorganisms, 242–243
    food quality and traceability, 246–247
    microbial food toxins, 243
    study of GM crops and food materials, 244–245
  current state, 230
    pathogenic microbes and their stress responses, 231–232
    study of food composition and its importance, 231
  metabolomics and, 233
    food metabolomics, 233–234
    recent advancements of analytical instruments in metabolomics, 234–235
Food safety regulatory model, 227*f*
Fourier transform infrared spectroscopy (FTIR), 235, 237, 272
Fourier transform ion cyclotron resonance (FTICR), 236
Fourier transform-ion cyclotron resonance-mass spectrometry (FT-ICR-MS), 125
Fragile-X mental retardation protein (FMRP), 304

Fructooligosaccharides (FOS), 20, 23–24
Fructose 1,6-bisphosphatase (FBPase), 92
Fructose 6-phosphate, 93
Fungal bioprocessing abilities, metabolic engineering to develop, 277–280
Fungal soil inoculants, 167
Fungi, 265, 267, 268–269, 274, 279, 280–281
Fusarium circinatuum , 274
Future perspective, 8

G

Galactooligosaccharides (GOS), 23–24
Gamma amino butyric acid (GABA), 15
Gas chromatography and mass spectrometry (GC-MS), 4, 123, 126, 169, 208, 231, 236–237, 242, 245, 272
Genetically modified (GM) food products, 225, 229
Genetic knockouts, using metabolite profiling, 92–93
Geobacillus thermoglucosidasius, 267
Geobacter sulfurreducens, 212
GIM3E, 294*t*
Glucagon-like peptide-2 (GLP-2), 32
Gluconacetobacter xylinus, 273
Gluconeogenesis, 89
Glucosamine (GlcN), 93
Glucosamine 6-phosphate deaminase (GND), 93
Glutamic acid, 172
Glycolysis, 89
Glycosomal succinate fermentation (GSF) pathway, 97
GM crops and food materials, study of, 244–245
G-protein-coupled receptors (GPCRs), 30
Green revolution, 229
Growth rate, 96, 104
Gut-associated disorders, 14
Gut intestinal barrier, 31
Gut microbiota, 14
  bioactive landscape of (*see* Gut microbiota, bioactive landscape of)
  dietary fibres interaction with, 23–26
Gut microbiota, bioactive landscape of, 49
  culture-dependent analysis, 57–61
  human holobiont, 51–53
  NF-κB suppressive gut microbes
    bioprospecting for, 69–71
    genetic-based strategies to identify, 66–69
    in vitro approaches to identify, 62–64
    metabolomic-based strategies to identify, 64–66
  structure-function capacity of human gut microbiota, 53–57

H

^{1}H-nuclear magnetic resonance spectroscopy (^{1}H-NMR), 150*t*, 173, 174–175, 178, 274
  -based community metabolic profiling, 202
  metabolites found in soil-related samples using, 150*t*–156*t*
^2H_2O, 103–106, 275
"Habitat simulating" media, use of, 58
Hazard Analysis Critical Control Point (HACCP), 227, 228*f*
Heavy metal contamination, 177–179
Heavy water, 103, 104
Hemicelluloses, 19, 20
High-throughput screening approaches, 63
Histone deacetylases (HDACs), 30
Holobiont concept, 51–53
Host cells, 85–86, 86, 96
Host–microbe interactions, 7
Human colonic microbiota, 13
Human holobiont, 51–53
Human Microbiome Project (HMP), 51
Humic substances, 120
Humus, 168–169
Hydrothermal treatment (HT), 264
Hydroxymethyl furfural (HMF), 272

I

IMPaLA (integrated molecular pathway level analysis), 294*t*
Inflammatory bowel disease (IBD), 27, 28, 31, 49, 50
Ingenuity pathway analysis, 294*t*
INMEX, 295*t*
Intestinal barrier, 31
Inulin, 20–21, 23
IOMA, 295*t*
Ion-mobility spectrometry (IMS), 236
Isobaric tags for relative and absolute quantification/liquid chromatography-mass spectrometry/mass spectrometry (iTRAQ/LC-MS/MS), 91
Isotope labelling, 183
Isotope-ratio MS (IRMS), 245

J

Juvenile spondyloarthritis, 70

K

KaPPA-view, 295*t*
Kinetoplastida, 83
Klebsiella mobilis CIAM 880, 177

Knowledge transfer to applied research outcomes, 218–219
Koji process, 265
Kyoto Encyclopedia of Genes and Genomes (KEGG), 88

L
Lactones, 163, 173
Labelling techniques, 174–175
Laetiporus spp., 265
Laser ablation electrospray ionization (LAESI), 8
LdAAP3, 91
LeishCyc, 88
Leishmania, 83
  central carbon metabolism in, 98*f*
  characterization of genetic knockouts using metabolite profiling, 92–93
  characterizing metabolism of, 88
  genome studies and metabolism mapping, 88–89
  insights into metabolism using metabolomics, 91–92
  key differences between developmental stages, 97
    flux analyses, 99–102
    footprinting approaches, 97–98
    intracellular metabolite levels, 98–99
  and leishmaniasis, 83–84
  lifecycle, 85–87
    stages, 105*f*
  studying a drug's mode of action and the development of resistance in, 94–96
  studying metabolism in vivo, 102–106
  transcriptomic and proteomic studies, 89–91
  treatment of leishmaniasis, 84–85
Leishmania aethiopica, 83
Leishmania amastigotes, 87, 104
Leishmania amazonensis, 83, 93
Leishmania braziliensis, 83
Leishmania chagasi, 83
Leishmania donovani, 83
Leishmania guyanensis, 84
Leishmania infantum, 83, 89, 94
Leishmania major, 83, 106
Leishmania mexicana, 83, 87, 91, 99, 101, 104
Leishmania peruviana, 83
Leishmania spp., 87, 92
Lentinula edodes, 267
Lignin, 21, 261, 263
Lignin peroxidase (LiP), 266
Linear models for microarray analysis (LIMMA), 304
Liquid chromatography and mass spectrometry (LC-MS), 94, 125, 126, 166, 209, 236–237
Liquid chromatography with electrospray ionisation-mass spectrometry, 274
Listeria monocytenes, 230
Listeria monocytogenes, 87, 232, 243
Lolium perenne, 273
Lubricus rubellus, 203–204
Lubricus terrestris, 203–204
Lutzomyia, 85

M
Macroinvertebrate communities, 218
Macrophages, 32, 87, 93
MADMAX (management and analysis database for multiple ~ omics experiments), 295*t*
MAM (microbial anti-inflammatory molecule) protein, 70
Manganese peroxidase, 266
MapMan (software tool), 295*t*
Marine subsurface environments, 293
MarVis-Pathway (marker visualization pathway), 295*t*
Mass Spectrometry (MS), 2, 8, 123, 134, 136, 235, 236
MassTrix, 295*t*
Matrix-assisted laser desorption/ionisation-time of flight mass spectrometry (MALDI-TOF-MS), 60, 61, 243
Matrix-assisted laser desorption/ionization (MALDI), 8
Matrix effect (ME), 233
Metabolic engineering approaches, 276, 278
  to develop fungal bioprocessing abilities, 277–280
Metabolic fingerprinting, 272–275
Metabolic flux (fluxomics), 275–276
Metabolic footprinting, 98, 122
Metabolic profiling, targeted and untargeted approaches, 269–272
Metabolic state of microbial community, assessing, 7
Metabolite imaging, 133–136
Metabolite profiling, characterization of genetic knockouts using, 92–93
Metabolites found in soil-related samples using 1H-NMR, 150*t*–156*t*
Metabolome, 2–4, 3*t*
Metabolomics, 1, 7–8, 297, 300, 303
  classification of, 268–269
  experimental setup, 200*f*
  insights into *Leishmania* metabolism using, 91–92

Metabolomics (*cont.*)
revealing key differences between *Leishmania* developmental stages, 97
flux analyses, 99–102
footprinting approaches, 97–98
intracellular metabolites levels, 98–99
to identify NF-κB suppressive gut microbes, 62–64
untargeted, 93
Meta-data, importance of, 217
Metagenomes, 125–126, 300, 301
Metagenomic DNA, 215
Meta-metabolomics, 201
Metaparental mating, 66–68
Methionine sulfoxime (MSO), 101
Methyl chloroformate (MCF) derivatization, 4
MetScape 2 (software tool), 295*t*
MFS (Major Facilitator Superfamily), 6
MIAME (Minimal Information about a Microarray Experiment) standard, 217
MIC biofilms, 302, 303
Microbes, metabolites and health, 13
dietary fibre, 19–23
and gut health, 16–19
and short chain fatty acids, 26–29
interaction with gut microbiota, 23–26
future prospects, 34–35
short chain fatty acids, inflammation and allergy, 29–33
Microbial biomass processing, 269
Microbial food toxins, 243
Microbial fuel cells (MFCs), 209, 210, 211
and anaerobic waste conversion applications, 209–211
Microbial interactions through space, 133–136
Microbial systems, 1
Microcoleus vaginatus, 129
Mid-infrared (MIR) spectroscopy, 127
Mid-infrared spectroscopy (MIR), 174, 203
MIENS (Minimal Information about an Environmental Sequence), 217
MIGS (Minimal Information about a Genome Sequence), 217
Miltefosine, 95
MIMS (Minimal Information about a Metagenome Sequence), 217
Mini-guts, 63
Minimum information about a Biofilm experiment (MIABiE) initiative, 301
MiRNA pathway analysis, 303
Mitogen activated protein kinase 2 (MAPK2)-dependent signaling pathway, 91
MixOmics (R package), 295*t*
Model-based integration methods, 291–293
Monocarboxylate transporter-1 (MCT-1), 30
Most probable number (MPN) approach, 59
Mucocutaneous leishmaniasis (MCL), 83
Multi-omic data analysis challenge, 291–293
Multi-omics-based approaches, 289
applications, 293
biofilms and corrosion, 301–303
biofuels, 304
environmental contaminants, 293–298
food and nutrition, 298
medical and clinical, 303–304
water and wastewater systems, 299–301
challenges for, 291–293
tools, 294*t*–296*t*
Mycobacterium tuberculosis, 93

N

N-acetylglucosamine acetyltransferase (GNAT), 93
Nanostructure-initiator mass spectrometry (NIMS), 134
Nascent field, 157–159
Nitrogen, in rhizosphere, 170–171
Nitrogen starving of algal cells, 304
NMR metabolomics, 174
N-nitroso compounds (NOC), 29
Non-digestible oligosaccharides, 20
Non-starch polysaccharides (NSP), 16, 19, 20, 22, 26
"Non-targeted" community metabolomics, 204
Nuclear factor-kappa B (NF-κB) suppressive gut microbes, 50, 53
bioprospecting for, 69–71
genetic-based strategies to identify, 66–69
in vitro approaches to identify, 62–64
metabolomic-based strategies to identify, 64–66
Nuclear magnetic resonance (NMR), 123, 126, 127, 231, 235–236
Nutrients in rhizosphere, 170–172

O

Obesity, 49, 50
role of the microbiota in, 14
Oil organic matter (SOM) metabolites, 148
Oligofructose, 21
Omics-based techniques, 289
Omics information, 213
Orbitrap, 236
Organic contaminants, 179–181
Organosolvent treatments, 264
Orthogonal partial least square-discriminant analysis (OPLS-DA), 206

P
PaintOmics (software tool), 296*t*, 298
Parasitic disease, 84
Partial least squares discriminant analysis (PLS-DA), 207
Particulate organic matter (POM), 204
Pathogenic microbes and their stress responses, 231–232
PathVisio 3 (software tool), 296*t*
pEHR5 vector system, 67
Penicillium chrysogenum, 265, 280
Penicillium commune, 272
Pentavalent and trivalent antimonials, 94
Pentose phosphate pathway (PPP), 89
Persistent organic pollutants (POP), 180
Pesticide reduction, 180
Pesticides, 229, 240, 241
Phanerochaete chrysosporium, 266
Phanerochaete spp., 265
Phenolic compounds, 161
Phlebia spp., 265
Phlebotomus , 85
Phosphoenolpyruvate (PEP), 97
Phospholipid fatty acids (PLFAs), 131, 176
Phosphorus, in rhizosphere, 170
p-hydroxybenzoic acid (PHBA), 280
Phytotoxin, 163
Plant biomass, 261
Plant growth-promoting rhizobacteria (PGPR), 166, 171
Plant metabolomics, 2
Plasmopara viticola, 271
Plate wash PCR, 59
Pleurotus sapidus, 276
Polybrominated diphenyl ethers (PBDEs), 241
Polychlorinated biphenyls (PCBs), 164, 180, 241
Polycyclic aromatic hydrocarbons (PAHs), 180, 241
Polymerase chain reaction (PCR)-amplifying 16S rRNA genes, 211
Polymorphonuclear leucocytes (PMNs), 85, 97
Post-data analysis versus integrated analysis approaches, 292*f*
Postia spp., 265
Principal component analysis (PCA), 127, 207, 273
Promastigote, motile, 85
Promastigote to amastigote differentiation, 91
ProMeTra, 296*t*
Protected designation of origin (PDO), 246
Protected geographic indication (PGI), 246
Proteomics, 89–91, 303
Pseudomonas fluorescens, 207, 208
Pseudomonas pseudoalcaligenes, 157
Pseudomonas pseudoalcaligenes KF707, 205
Purine salvage, 95
Pyrimidine synthesis, 89
Pyruvate, 97

Q
Quadrupole (Q), 236
Quadrupole ion-trap (QIT), 236
Quality control, 226. *See also* Food safety and quality research
Quantitative microbial metabolomics, 4
Quercus suber, 274
Quorum sensing (QS), 162, 163

R
Rare biosphere, 213
Ready-to-eat food, 225
Replica extraction transfer, 134
Resistant starch (RS), 17–18, 20, 23–24
REX-NIMS approach, 136
Rhizoengineering, 163–164
Rhizosphere, 149, 159
  map of, 159–161
  metabolomics, 161–162
  mRNA pyrosequencing, 165
  nutrients in, 170–172
Ribosomal Database Project, 214
Rugose small-colony variants (RSCVs), 207
Ruminococcus bromii, 25, 27

S
Saccharomyces cerevisiae, 2, 4, 93, 267, 271, 272, 279
Salinity, 182
Salmonella enterica, 232
Salmonella spp., 230
Salmonella typhimurium, 242
Sampling (quenching) methods, 5
Saprotrophic microbes, 161
Schewanella oneidensis MR-1, 134
Secondary ion mass spectrometry (SIMS), 8, 134
Separation techniques, 234, 237–238, 241
Shigella spp., 87
Short chain fatty acid (SCFA), 17, 18, 24, 27, 29–33, 273
  and NF-κB activation, 64–65
Short chain fatty acids, inflammation and allergy, 29–33
Significance analysis of microarrays (SAM), 304
SIMCA (software tool), 296*t*
Single-cell metabolomics, 8

Single nucleotide polymorphisms (SNPs), 303
16S rRNA gene, 53–54, 59–61, 211–212
Small-colony variants (SCVs), 207
Sodium fluoroacetate (NaFAc), 101
Soil, complexity of, 147–157
Soil community, 128–131
Soil exometabolomics studies, 122, 127
Soil inoculants, 165–168
Soil matrix, 148
Soil microbial metabolomics, 147
  anthropogenic effects on soil microbes, 174–175
    climatic change effects, 181–182
    engineered nanomaterials (ENM), 175–177
    heavy metal contamination, 171–173
    organic contaminants, 179–181
  complexity of soil, 147–157
  extracellular enzymes, 173–174
  exudates, 162
    rhizoengineering, 163–164
  humus, 168–169
  LC–MS analyses, 158*t*
  metabolite coverage over the lifespan of plants, 165
  nascent field, 157–159
  rhizosphere, 159
    map of, 159–161
    metabolomics, 161–162
    nutrients in, 170–172
  soil inoculants, 165–168
  terroir, 169
  tools for, 183–184
Soil microorganisms, 119, 133
  metabolomics (*see* Soil microbial metabolomics)
Soil organic matter (SOM), 120, 122
  exometabolomics for analysis of, 120–126
Soil science, 201–204
Solid phase extraction (SPE), 234
Solid phase microextraction (SPME), 234
Solid state fermentation (SSF), 265–266
Stable isotope labeling approaches, 99, 103
Stable isotope probing (SIP) techniques, 131–133, 134
Steam explosion (SE) experiment, 264
Streptococcus bovis, 270
Streptomyces coelicolor, 134
Submerged fermentation/shake flask fermentation (SmF), 265
SWCNT (single-walled carbon nanotubes), 176
Symbiotic fermentation, 267
Synechococcus sp. PCC 7002, 128
Systems biology, 214, 278

T
T5 cells, 206
Targeted metabolomics methods, 271
Tellurite resistance, 205–206, 205*f*
Tellurium oxide, 205
Terroir, 169
Thermoanaerobacter ethanolicus, 265
Thermophilic bacteria, 265, 267
Thin layer chromatography (TLC), 272
3Omics, 294*t*
Time of flight (ToF), 236
Time-of-flight secondary ion mass spectrometry (TOF-SIMS), 136
Toll-like receptor 10 (TLR10), 303
Toll-like receptors (TLRs), 62, 63
Total dietary fibre (TDF), 17
Toxoplasma gondii, 87
Trametes spp., 265
Transmembrane protein family PulD, 6
Tree litter, 172
Triacylglycerol (TAG) synthesis, 304
Trichoderma, 167, 168
Trichoderma hamatum, 274
Trichoderma *reesei*, 276
Trichoderma spp., 265, 267
"Trojan Horse" strategy, 86
"Trojan rabbit" model, 86
Trypanosoma, 83
Trypanosoma brucei, 83, 92
Trypanosoma cruzi, 83, 87
Trypanosomatidae, 83
Type IV secretion systems (T4SS), 6

U
Ulcerative colitis (UC), 27
Ultra-high-performance liquid chromatography coupled to electrospray ionisation-high resolution mass spectrometry (UHPLC-ESI-HRMS), 274
Ultrasensitive analytical methods, 8
Uncinula necator, 271
Upflow anaerobic sludge blanket (UASB) reactor, 210

V
VANTED (visualization and analysis of networks with related experimental data), 296*t*
Visceral leishmaniasis (VL), 83

VitisNet (software tool), 296*t*, 298
Volatile fatty acids (VFAs), 273
Volatile organic compounds (VOC), 242

W
Water and wastewater systems, 299–301
Water extractable organic matter (WEOM), 122
Water soluble fibres, 21
Web-based tools, 4
Whole microbial communities, exometabolomics for analysis of, 126–128
Wrinkly spreaders (WS), 207

X
xcp secretion pathway, 6

Y
Yarrowia lipolytica, 304
Yeast Metabolome Database (YMDB), 4

Z
Zwitterionic HILIC chromatography, 125
Zymomonas mobilis, 275

FSC
www.fsc.org
MIX
Papier | Fördert
gute Waldnutzung
FSC® C083411

Kolibri 360 GmbH
Ferdinand-Jühlke-Straße 7
99095 Erfurt, Deutschland
produktsicherheit@kolibri360.de